VOLUME SEVENTY TWO

ADVANCES IN MARINE BIOLOGY

Humpback Dolphins (*Sousa* spp.): Current Status and Conservation, Part 1

ADVANCES IN MARINE BIOLOGY

VOLUME SEVENTY TWO

ADVANCES IN
MARINE BIOLOGY

Humpback Dolphins (*Sousa* spp.): Current Status and Conservation, Part 1

Edited by

THOMAS A. JEFFERSON
Clymene Enterprises, Lakeside, California, USA

BARBARA E. CURRY
Physiological Ecology and Bioenergetics Laboratory
Conservation Biology Program
University of Central Florida
Orlando, FL 32816
USA

ELSEVIER

AMSTERDAM • BOSTON • HEIDELBERG • LONDON
NEW YORK • OXFORD • PARIS • SAN DIEGO
SAN FRANCISCO • SINGAPORE • SYDNEY • TOKYO
Academic Press is an imprint of Elsevier

Academic Press is an imprint of Elsevier
125 London Wall, London, EC2Y 5AS, UK
The Boulevard, Langford Lane, Kidlington, Oxford OX5 1GB, UK
525 B Street, Suite 1800, San Diego, CA 92101-4495, USA
225 Wyman Street, Waltham, MA 02451, USA

First edition 2015

ISBN: 978-0-12-803258-9
ISSN: 0065-2881

CONTRIBUTORS TO VOLUME 72

Boris Andrianantenaina
Institut Halieutique et des Sciences Marines, Universite de Toliara, Toliara, Madagascar

Norbert Andrianarivelo
Institut Halieutique et des Sciences Marines, Universite de Toliara, Toliara, Madagascar

Robert Baldwin
Five Oceans Environmental Services LLC, Sultanate of Oman

Isha Bopardikar
Konkan Cetacean Research Team, Mumbai, Maharashtra, India

Gill T. Braulik
Wildlife Conservation Society—Tanzania Program, Zanzibar, Tanzania

Salvatore Cerchio
New England Aquarium, Central Wharf, Boston, and Woods Hole Oceanographic Institution, Woods Hole, Massachusetts, USA

Victor G. Cockcroft
Department of Zoology, Nelson Mandela Metropolitan University, Port Elizabeth, South Africa

Tim Collins
Wildlife Conservation Society, Marine Program, Global Conservation, Bronx, New York, USA

Barbara E. Curry
Physiological Ecology and Bioenergetics Laboratory, University of Central Florida, Orlando, Florida, USA

Ken Findlay
Mammal Research Institute Whale Unit, University of Pretoria, Pretoria, South Africa

William P. Froneman
Department of Zoology and Entomology, Rhodes University, Grahamstown, South Africa

Thomas A. Jefferson
Clymene Enterprises, Lakeside, California, USA

Ketki Jog
Konkan Cetacean Research Team, Mumbai, Maharashtra, India

Muhammad Shoaib Kiani
Institute of Marine Science, University of Karachi, Karachi, Pakistan

David Lundquist
New Zealand Department of Conservation, Wellington, New Zealand

Rahul Muralidharan
Ashoka Trust for Research in Ecology and the Environment, Bangalore, Karnataka, India

Divya Panicker
Affiliated to Wildlife Programme at National Centre for Biological Sciences, Bangalore, Karnataka, India

Sarah Piwetz
Department of Marine Biology, Texas A&M University at Galveston, Galveston, Texas, USA

Stephanie Plön
Coastal and Marine Research Institute, Nelson Mandela Metropolitan University, Port Elizabeth, South Africa

Mihir Sule
Konkan Cetacean Research Team, Mumbai, Maharashtra, India

Dipani Sutaria
College of Marine and Environmental Sciences, James Cook University, Townsville, Queensland, Australia

Koen Van Waerebeek
Peruvian Centre for Cetacean Research (CEPEC), Lima, Peru

Caroline R. Weir
Ketos Ecology, Kingsbridge, Devon, United Kingdom

Bernd Würsig
Department of Marine Biology, and Department of Wildlife and Fisheries Sciences, Texas A&M University at Galveston, Galveston, Texas, USA

CONTENTS

Contributors to Volume 72 v
Series Contents for Last Fifteen Years xi
Preface xxiii
List of Abbreviations xxvii

1. Humpback Dolphins: A Brief Introduction to the Genus *Sousa* **1**
Thomas A. Jefferson and Barbara E. Curry
1. Introduction 2
2. Taxonomy 3
3. Geographic Range 6
4. Abundance and Trends 7
5. Habitat and Ecology 8
6. Threats 10
7. Conservation Actions and Status 11
References 13

2. Humpback Dolphin (Genus *Sousa*) Behavioural Responses to Human Activities **17**
Sarah Piwetz, David Lundquist, and Bernd Würsig
1. Introduction 18
2. Understanding Animal Behaviour 19
3. Anthropogenic Activities Affecting Marine Mammal Behaviour 22
4. Anthropogenic Effects on Humpback Dolphins 26
5. Conclusions and Future Research 34
Acknowledgements 36
References 37

3. Re-assessment of the Conservation Status of the Atlantic Humpback Dolphin, *Sousa teuszii* (Kükenthal, 1892), Using the IUCN Red List Criteria **47**
Tim Collins
1. Introduction 48
2. Taxonomic Notes 50
3. Geographic Range 51
4. Population 53
5. Habitat and Ecology 57

6. Use and Trade 64
7. Threats 64
8. Conservation Actions 68
9. IUCN Red List Justification 69
Acknowledgements 71
References 71

4. A Review of the Geographical Distribution and Habitat of the Atlantic Humpback Dolphin (*Sousa teuszii*) 79

Caroline R. Weir and Tim Collins

1. Introduction 80
2. Methods 81
3. Distribution 85
4. Group Size 98
5. Habitat 101
6. Conclusions and Future Work 105
Acknowledgements 106
Appendix A 106
References 116

5. Assessment of the Conservation Status of the Indian Ocean Humpback Dolphin (*Sousa plumbea*) Using the IUCN Red List Criteria 119

Gill T. Braulik, Ken Findlay, Salvatore Cerchio, and Robert Baldwin

1. Introduction 120
2. Taxonomy 121
3. Geographic Range 122
4. Population 125
5. Habitat and Ecology 128
6. Use and Trade 131
7. Threats 132
8. Conservation Actions 135
9. IUCN Red List Justification 136
Acknowledgments 137
References 137

6. The Natural History and Conservation of Indian Ocean Humpback Dolphins (*Sousa plumbea*) in South African Waters 143

Stephanie Plön, Victor G. Cockcroft, and William P. Froneman

1. Background: Previous Studies in South African Waters 144
2. Age, Growth and Reproduction 146

3. Diet 151
4. Habitat Utilisation 152
5. Threats 153
6. Health 155
7. Conclusions 156
References 157

7. Ecology and Conservation Status of Indian Ocean Humpback Dolphins (*Sousa plumbea*) in Madagascar 163

Salvatore Cerchio, Norbert Andrianarivelo, and Boris Andrianantenaina

1. Introduction 164
2. Dolphin Surveys on the West Coast of Madagascar 167
3. Assessing By-catch and Hunting Through Interview Surveys 179
4. General Discussion 187
Acknowledgements 196
References 196

8. A Review of the Status of the Indian Ocean Humpback Dolphin (*Sousa plumbea*) in Pakistan 201

Muhammad Shoaib Kiani and Koen Van Waerebeek

1. Introduction 202
2. Distribution in Pakistan 203
3. Stock Structure 208
4. Abundance 211
5. Habitat 212
6. Biology and Behaviour 213
7. Parasites and Pathology 217
8. Anthropogenic Threats and Conservation 218
9. Conclusions and Recommendations 223
Acknowledgements 224
References 224

9. Humpback Dolphins (Genus *Sousa*) in India: An Overview of Status and Conservation Issues 229

Dipani Sutaria, Divya Panicker, Ketki Jog, Mihir Sule, Rahul Muralidharan, and Isha Bopardikar

1. Introduction 230
2. Taxonomy 233
3. Distribution Records 235
4. Population Status 237

5. Human Interactions and Threats 242
6. Conservation and Management 252
References 253

Subject Index *257*
Taxonomic Index *265*

SERIES CONTENTS FOR LAST FIFTEEN YEARS*

Volume 38, 2000.
Blaxter, J. H. S. The enhancement of marine fish stocks. pp. 1–54.
Bergström, B. I. The biology of Pandalus. pp. 55–245.

Volume 39, 2001.
Peterson, C. H. The "Exxon Valdez" oil spill in Alaska: acute indirect and chronic effects on the ecosystem. pp. 1–103.
Johnson, W. S., Stevens, M. and Watling, L. Reproduction and development of marine peracaridans. pp. 105–260.
Rodhouse, P. G., Elvidge, C. D. and Trathan, P. N. Remote sensing of the global light-fishing fleet: an analysis of interactions with oceanography, other fisheries and predators. pp. 261–303.

Volume 40, 2001.
Hemmingsen, W. and MacKenzie, K. The parasite fauna of the Atlantic cod, Gadus morhua L. pp. 1–80.
Kathiresan, K. and Bingham, B. L. Biology of mangroves and mangrove ecosystems. pp. 81–251.
Zaccone, G., Kapoor, B. G., Fasulo, S. and Ainis, L. Structural, histochemical and functional aspects of the epidermis of fishes. pp. 253–348.

Volume 41, 2001.
Whitfield, M. Interactions between phytoplankton and trace metals in the ocean. pp. 1–128.
Hamel, J.-F., Conand, C., Pawson, D. L. and Mercier, A. The sea cucumber Holothuria scabra (Holothuroidea: Echinodermata): its biology and exploitation as beche-de-Mer. pp. 129–223.

Volume 42, 2002.
Zardus, J. D. Protobranch bivalves. pp. 1–65.
Mikkelsen, P. M. Shelled opisthobranchs. pp. 67–136.
Reynolds, P. D. The Scaphopoda. pp. 137–236.
Harasewych, M. G. Pleurotomarioidean gastropods. pp. 237–294.

*The full list of contents for volumes 1–37 can be found in volume 38

Volume 43, 2002.
Rohde, K. Ecology and biogeography of marine parasites. pp. 1–86.
Ramirez Llodra, E. Fecundity and life-history strategies in marine invertebrates. pp. 87–170.
Brierley, A. S. and Thomas, D. N. Ecology of southern ocean pack ice. pp. 171–276.
Hedley, J. D. and Mumby, P. J. Biological and remote sensing perspectives of pigmentation in coral reef organisms. pp. 277–317.

Volume 44, 2003.
Hirst, A. G., Roff, J. C. and Lampitt, R. S. A synthesis of growth rates in epipelagic invertebrate zooplankton. pp. 3–142.
Boletzky, S. von. Biology of early life stages in cephalopod molluscs. pp. 143–203.
Pittman, S. J. and McAlpine, C. A. Movements of marine fish and decapod crustaceans: process, theory and application. pp. 205–294.
Cutts, C. J. Culture of harpacticoid copepods: potential as live feed for rearing marine fish. pp. 295–315.

Volume 45, 2003.
Cumulative Taxonomic and Subject Index.

Volume 46, 2003.
Gooday, A. J. Benthic foraminifera (Protista) as tools in deep-water palaeoceanography: environmental influences on faunal characteristics. pp. 1–90.
Subramoniam, T. and Gunamalai, V. Breeding biology of the intertidal sand crab, Emerita (Decapoda: Anomura). pp. 91–182.
Coles, S. L. and Brown, B. E. Coral bleaching—capacity for acclimatization and adaptation. pp. 183–223.
Dalsgaard J., St. John M., Kattner G., Müller-Navarra D. and Hagen W. Fatty acid trophic markers in the pelagic marine environment. pp. 225–340.

Volume 47, 2004.
Southward, A. J., Langmead, O., Hardman-Mountford, N. J., Aiken, J., Boalch, G. T., Dando, P. R., Genner, M. J., Joint, I., Kendall, M. A., Halliday, N. C., Harris, R. P., Leaper, R., Mieszkowska, N., Pingree, R. D., Richardson, A. J., Sims, D.W., Smith, T., Walne, A. W. and Hawkins, S. J. Long-term oceanographic and ecological research in the western English Channel. pp. 1–105.

Queiroga, H. and Blanton, J. Interactions between behaviour and physical forcing in the control of horizontal transport of decapod crustacean larvae. pp. 107–214.

Braithwaite, R. A. and McEvoy, L. A. Marine biofouling on fish farms and its remediation. pp. 215–252.

Frangoulis, C., Christou, E. D. and Hecq, J. H. Comparison of marine copepod outfluxes: nature, rate, fate and role in the carbon and nitrogen cycles. pp. 253–309.

Volume 48, 2005.

Canfield, D. E., Kristensen, E. and Thamdrup, B. Aquatic Geomicrobiology. pp. 1–599.

Volume 49, 2005.

Bell, J. D., Rothlisberg, P. C., Munro, J. L., Loneragan, N. R., Nash, W. J., Ward, R. D. and Andrew, N. L. Restocking and stock enhancement of marine invertebrate fisheries. pp. 1–358.

Volume 50, 2006.

Lewis, J. B. Biology and ecology of the hydrocoral Millepora on coral reefs. pp. 1–55.

Harborne, A. R., Mumby, P. J., Micheli, F., Perry, C. T., Dahlgren, C. P., Holmes, K. E., and Brumbaugh, D. R. The functional value of Caribbean coral reef, seagrass and mangrove habitats to ecosystem processes. pp. 57–189.

Collins, M. A. and Rodhouse, P. G. K. Southern ocean cephalopods. pp. 191–265.

Tarasov, V. G. Effects of shallow-water hydrothermal venting on biological communities of coastal marine ecosystems of the western Pacific. pp. 267–410.

Volume 51, 2006.

Elena Guijarro Garcia. The fishery for Iceland scallop (Chlamys islandica) in the Northeast Atlantic. pp. 1–55.

Jeffrey, M. Leis. Are larvae of demersal fishes plankton or nekton? pp. 57–141.

John C. Montgomery, Andrew Jeffs, Stephen D. Simpson, Mark Meekan and Chris Tindle. Sound as an orientation cue for the pelagic larvae of reef fishes and decapod crustaceans. pp. 143–196.

Carolin E. Arndt and Kerrie M. Swadling. Crustacea in Arctic and Antarctic sea ice: Distribution, diet and life history strategies. pp. 197–315.

Volume 52, 2007.
Leys, S. P., Mackie, G. O. and Reiswig, H. M. The Biology of Glass Sponges. pp. 1–145.
Garcia E. G. The Northern Shrimp (Pandalus borealis) Offshore Fishery in the Northeast Atlantic. pp. 147–266.
Fraser K. P. P. and Rogers A. D. Protein Metabolism in Marine Animals: The Underlying Mechanism of Growth. pp. 267–362.

Volume 53, 2008.
Dustin J. Marshall and Michael J. Keough. The Evolutionary Ecology of Offspring Size in Marine Invertebrates. pp. 1–60.
Kerry A. Naish, Joseph E. Taylor III, Phillip S. Levin, Thomas P. Quinn, James R. Winton, Daniel Huppert, and Ray Hilborn. An Evaluation of the Effects of Conservation and Fishery Enhancement Hatcheries on Wild Populations of Salmon. pp. 61–194.
Shannon Gowans, Bernd Würsig, and Leszek Karczmarski. The Social Structure and Strategies of Delphinids: Predictions Based on an Ecological Framework. pp. 195–294.

Volume 54, 2008.
Bridget S. Green. Maternal Effects in Fish Populations. pp. 1–105.
Victoria J. Wearmouth and David W. Sims. Sexual Segregation in Marine Fish, Reptiles, Birds and Mammals: Behaviour Patterns, Mechanisms and Conservation Implications. pp. 107–170.
David W. Sims. Sieving a Living: A Review of the Biology, Ecology and Conservation Status of the Plankton-Feeding Basking Shark *Cetorhinus Maximus*. pp. 171–220.
Charles H. Peterson, Kenneth W. Able, Christin Frieswyk DeJong, Michael F. Piehler, Charles A. Simenstad, and Joy B. Zedler. Practical Proxies for Tidal Marsh Ecosystem Services: Application to Injury and Restoration. pp. 221–266.

Volume 55, 2008.
Annie Mercier and Jean-Francois Hamel. Introduction. pp. 1–6.
Annie Mercier and Jean-Francois Hamel. Gametogenesis. pp. 7–72.
Annie Mercier and Jean-Francois Hamel. Spawning. pp. 73–168.
Annie Mercier and Jean-Francois Hamel. Discussion. pp. 169–194.

Volume 56, 2009.

Philip C. Reid, Astrid C. Fischer, Emily Lewis-Brown, Michael P. Meredith, Mike Sparrow, Andreas J. Andersson, Avan Antia, Nicholas R. Bates, Ulrich Bathmann, Gregory Beaugrand, Holger Brix, Stephen Dye, Martin Edwards, Tore Furevik, Reidun Gangst, Hjalmar Hatun, Russell R. Hopcroft, Mike Kendall, Sabine Kasten, Ralph Keeling, Corinne Le Quere, Fred T. Mackenzie, Gill Malin, Cecilie Mauritzen, Jon Olafsson, Charlie Paull, Eric Rignot, Koji Shimada, Meike Vogt, Craig Wallace, Zhaomin Wang and Richard Washington. Impacts of the Oceans on Climate Change. pp. 1–150.

Elvira S. Poloczanska, Colin J. Limpus and Graeme C. Hays. Vulnerability of Marine Turtles to Climate Change. pp. 151–212.

Nova Mieszkowska, Martin J. Genner, Stephen J. Hawkins and David W. Sims. Effects of Climate Change and Commercial Fishing on Atlantic Cod *Gadus morhua*. pp. 213–274.

Iain C. Field, Mark G. Meekan, Rik C. Buckworth and Corey J. A. Bradshaw. Susceptibility of Sharks, Rays and Chimaeras to Global Extinction. pp. 275–364.

Milagros Penela-Arenaz, Juan Bellas and Elsa Vazquez. Effects of the *Prestige* Oil Spill on the Biota of NW Spain: 5 Years of Learning. pp. 365–396.

Volume 57, 2010.

Geraint A. Tarling, Natalie S. Ensor, Torsten Fregin, William P. Good-all-Copestake and Peter Fretwell. An Introduction to the Biology of Northern Krill (*Meganyctiphanes norvegica* Sars). pp. 1–40.

Tomaso Patarnello, Chiara Papetti and Lorenzo Zane. Genetics of Northern Krill (*Meganyctiphanes norvegica* Sars). pp. 41–58.

Geraint A. Tarling. Population Dynamics of Northern Krill (*Meganyctiphanes norvegica* Sars). pp. 59–90.

John I. Spicer and Reinhard Saborowski. Physiology and Metabolism of Northern Krill (*Meganyctiphanes norvegica* Sars). pp. 91–126.

Katrin Schmidt. Food and Feeding in Northern Krill (*Meganyctiphanes norvegica* Sars). pp. 127–172.

Friedrich Buchholz and Cornelia Buchholz. Growth and Moulting in Northern Krill (*Meganyctiphanes norvegica* Sars). pp. 173–198.

Janine Cuzin-Roudy. Reproduction in Northern Krill. pp. 199–230.

Edward Gaten, Konrad Wiese and Magnus L. Johnson. Laboratory-Based Observations of Behaviour in Northern Krill (*Meganyctiphanes norvegica* Sars). pp. 231–254.

Stein Kaartvedt. Diel Vertical Migration Behaviour of the Northern Krill (*Meganyctiphanes norvegica* Sars). pp. 255–276.
Yvan Simard and Michel Harvey. Predation on Northern Krill (*Meganyctiphanes norvegica* Sars). pp. 277–306.

Volume 58, 2010.
A. G. Glover, A. J. Gooday, D. M. Bailey, D. S. M. Billett, P. Chevaldonné, A. Colaço, J. Copley, D. Cuvelier, D. Desbruyères, V. Kalogeropoulou, M. Klages, N. Lampadariou, C. Lejeusne, N. C. Mestre, G. L. J. Paterson, T. Perez, H. Ruhl, J. Sarrazin, T. Soltwedel, E. H. Soto, S. Thatje, A. Tselepides, S. Van Gaever, and A. Vanreusel. Temporal Change in Deep-Sea Benthic Ecosystems: A Review of the Evidence From Recent Time-Series Studies. pp. 1–96.
Hilario Murua. The Biology and Fisheries of European Hake, *Merluccius merluccius*, in the North-East Atlantic. pp. 97–154.
Jacopo Aguzzi and Joan B. Company. Chronobiology of Deep-Water Decapod Crustaceans on Continental Margins. pp. 155–226.
Martin A. Collins, Paul Brickle, Judith Brown, and Mark Belchier. The Patagonian Toothfish: Biology, Ecology and Fishery. pp. 227–300.

Volume 59, 2011.
Charles W. Walker, Rebecca J. Van Beneden, Annette F. Muttray, S. Anne Böttger, Melissa L. Kelley, Abraham E. Tucker, and W. Kelley Thomas. p53 Superfamily Proteins in Marine Bivalve Cancer and Stress Biology. pp 1–36.
Martin Wahl, Veijo Jormalainen, Britas Klemens Eriksson, James A. Coyer, Markus Molis, Hendrik Schubert, Megan Dethier, Anneli Ehlers, Rolf Karez, Inken Kruse, Mark Lenz, Gareth Pearson, Sven Rohde, Sofia A. Wikström, and Jeanine L. Olsen. Stress Ecology in *Fucus*: Abiotic, Biotic and Genetic Interactions. pp. 37–106.
Steven R. Dudgeon and Janet E. Kübler. Hydrozoans and the Shape of Things to Come. pp. 107–144.
Miles Lamare, David Burritt, and Kathryn Lister. Ultraviolet Radiation and Echinoderms: Past, Present and Future Perspectives. pp. 145–187.

Volume 60, 2011.
Tatiana A. Rynearson and Brian Palenik. Learning to Read the Oceans: Genomics of Marine Phytoplankton. pp. 1–40.
Les Watling, Scott C. France, Eric Pante and Anne Simpson. Biology of Deep-Water Octocorals. pp. 41–122.

Cristián J. Monaco and Brian Helmuth. Tipping Points, Thresholds and the Keystone Role of Physiology in Marine Climate Change Research. pp. 123–160.
David A. Ritz, Alistair J. Hobday, John C. Montgomery and Ashley J.W. Ward. Social Aggregation in the Pelagic Zone with Special Reference to Fish and Invertebrates. pp. 161–228.

Volume 61, 2012.
Gert Wörheide, Martin Dohrmann, Dirk Erpenbeck, Claire Larroux, Manuel Maldonado, Oliver Voigt, Carole Borchiellini and Denis Lavrov. Deep Phylogeny and Evolution of Sponges (Phylum Porifera). pp. 1–78.
Paco Cárdenas, Thierry Pérez and Nicole Boury-Esnault. Sponge Systematics Facing New Challenges. pp. 79–210.
Klaus Rützler. The Role of Sponges in the Mesoamerican Barrier-Reef Ecosystem, Belize. pp. 211–272.
Janie Wulff. Ecological Interactions and the Distribution, Abundance, and Diversity of Sponges. pp. 273–344.
Maria J. Uriz and Xavier Turon. Sponge Ecology in the Molecular Era. pp. 345–410.

Volume 62, 2012.
Sally P. Leys and April Hill. The Physiology and Molecular Biology of Sponge Tissues. pp. 1–56.
Robert W. Thacker and Christopher J. Freeman. Sponge–Microbe Symbioses: Recent Advances and New Directions. pp. 57–112.
Manuel Maldonado, Marta Ribes and Fleur C. van Duyl. Nutrient Fluxes Through Sponges: Biology, Budgets, and Ecological Implications. pp. 113–182.
Grégory Genta-Jouve and Olivier P. Thomas. Sponge Chemical Diversity: From Biosynthetic Pathways to Ecological Roles. pp. 183–230.
Xiaohong Wang, Heinz C. Schröder, Matthias Wiens, Ute Schloßmacher and Werner E. G. Müller. Biosilica: Molecular Biology, Biochemistry and Function in Demosponges as well as its Applied Aspects for Tissue Engineering. pp. 231–272.
Klaske J. Schippers, Detmer Sipkema, Ronald Osinga, Hauke Smidt, Shirley A. Pomponi, Dirk E. Martens and René H. Wijffels. Cultivation of Sponges, Sponge Cells and Symbionts: Achievements and Future Prospects. pp. 273–338.

Volume 63, 2012.

Michael Stat, Andrew C. Baker, David G. Bourne, Adrienne M. S. Correa, Zac Forsman, Megan J. Huggett, Xavier Pochon, Derek Skillings, Robert J. Toonen, Madeleine J. H. van Oppen, and Ruth D. Gates. Molecular Delineation of Species in the Coral Holobiont. pp. 1–66.

Daniel Wagner, Daniel G. Luck, and Robert J. Toonen. The Biology and Ecology of Black Corals (Cnidaria: Anthozoa: Hexacorallia: Antipatharia). pp. 67–132.

Cathy H. Lucas, William M. Graham, and Chad Widmer. Jellyfish Life Histories: Role of Polyps in Forming and Maintaining Scyphomedusa Populations. pp. 133–196.

T. Aran Mooney, Maya Yamato, and Brian K. Branstetter. Hearing in Cetaceans: From Natural History to Experimental Biology. pp. 197–246.

Volume 64, 2013.

Dale Tshudy. Systematics and Position of *Nephrops* Among the Lobsters. pp. 1–26.

Mark P. Johnson, Colm Lordan, and Anne Marie Power. Habitat and Ecology of *Nephrops norvegicus*. pp. 27–64.

Emi Katoh, Valerio Sbragaglia, Jacopo Aguzzi, and Thomas Breithaupt. Sensory Biology and Behaviour of *Nephrops norvegicus*. pp. 65–106.

Edward Gaten, Steve Moss, and Magnus L. Johnson. The Reniform Reflecting Superposition Compound Eyes of *Nephrops norvegicus*: Optics, Susceptibility to Light-Induced Damage, Electrophysiology and a Ray Tracing Model. pp. 107–148.

Susanne P. Eriksson, Bodil Hernroth, and Susanne P. Baden. Stress Biology and Immunology in *Nephrops norvegicus*. pp. 149–200.

Adam Powell and Susanne P. Eriksson. Reproduction: Life Cycle, Larvae and Larviculture. pp. 201–246.

Anette Ungfors, Ewen Bell, Magnus L. Johnson, Daniel Cowing, Nicola C. Dobson, Ralf Bublitz, and Jane Sandell. *Nephrops* Fisheries in European Waters. pp. 247–314.

Volume 65, 2013.

Isobel S.M. Bloor, Martin J. Attrill, and Emma L. Jackson. A Review of the Factors Influencing Spawning, Early Life Stage Survival and Recruitment Variability in the Common Cuttlefish (*Sepia officinalis*). pp. 1–66.

Dianna K. Padilla and Monique M. Savedo. A Systematic Review of Phenotypic Plasticity in Marine Invertebrate and Plant Systems. pp. 67–120.

Leif K. Rasmuson. The Biology, Ecology and Fishery of the Dungeness crab, *Cancer magister*. pp. 121–174.

Volume 66, 2013.

Lisa-ann Gershwin, Anthony J. Richardson, Kenneth D. Winkel, Peter J. Fenner, John Lippmann, Russell Hore, Griselda Avila-Soria, David Brewer, Rudy J. Kloser, Andy Steven, and Scott Condie. Biology and Ecology of Irukandji Jellyfish (Cnidaria: Cubozoa). pp. 1–86.

April M. H. Blakeslee, Amy E. Fowler, and Carolyn L. Keogh. Marine Invasions and Parasite Escape: Updates and New Perspectives. pp. 87–170.

Michael P. Russell. Echinoderm Responses to Variation in Salinity. pp. 171–212.

Daniela M. Ceccarelli, A. David McKinnon, Serge Andréfouët, Valerie Allain, Jock Young, Daniel C. Gledhill, Adrian Flynn, Nicholas J. Bax, Robin Beaman, Philippe Borsa, Richard Brinkman, Rodrigo H. Bustamante, Robert Campbell, Mike Cappo, Sophie Cravatte, Stéphanie D'Agata, Catherine M. Dichmont, Piers K. Dunstan, Cécile Dupouy, Graham Edgar, Richard Farman, Miles Furnas, Claire Garrigue, Trevor Hutton, Michel Kulbicki, Yves Letourneur, Dhugal Lindsay, Christophe Menkes, David Mouillot, Valeriano Parravicini, Claude Payri, Bernard Pelletier, Bertrand Richer de Forges, Ken Ridgway, Martine Rodier, Sarah Samadi, David Schoeman, Tim Skewes, Steven Swearer, Laurent Vigliola, Laurent Wantiez, Alan Williams, Ashley Williams, and Anthony J. Richardson. The Coral Sea: Physical Environment, Ecosystem Status and Biodiversity Assets. pp. 213–290.

Volume 67, 2014.

Erica A.G. Vidal, Roger Villanueva, José P. Andrade, Ian G. Gleadall, José Iglesias, Noussithé Koueta, Carlos Rosas, Susumu Segawa, Bret Grasse, Rita M. Franco-Santos, Caroline B. Albertin, Claudia Caamal-Monsreal, Maria E. Chimal, Eric Edsinger-Gonzales, Pedro Gallardo, Charles Le Pabic, Cristina Pascual, Katina Roumbedakis, and James Wood. Cephalopod Culture: Current Status of Main Biological Models and Research Priorities. pp. 1–98.

Paul G.K. Rodhouse, Graham J. Pierce, Owen C. Nichols, Warwick H.H. Sauer, Alexander I. Arkhipkin, Vladimir V. Laptikhovsky, Marek R. Lipiński, Jorge E. Ramos, Michaël Gras, Hideaki Kidokoro, Kazuhiro Sadayasu, João Pereira, Evgenia Lefkaditou, Cristina Pita, Maria Gasalla, Manuel Haimovici, Mitsuo Sakai, and Nicola Downey. Environmental

Effects on Cephalopod Population Dynamics: Implications for Management of Fisheries. pp. 99–234.
Henk-Jan T. Hoving, José A.A. Perez, Kathrin Bolstad, Heather Braid, Aaron B. Evans, Dirk Fuchs, Heather Judkins, Jesse T. Kelly, José E.A.R. Marian, Ryuta Nakajima, Uwe Piatkowski, Amanda Reid, Michael Vecchione, and José C.C. Xavier. The Study of Deep-Sea Cephalopods. pp. 235–362.
Jean-Paul Robin, Michael Roberts, Lou Zeidberg, Isobel Bloor, Almendra Rodriguez, Felipe Briceño, Nicola Downey, Maite Mascaró, Mike Navarro, Angel Guerra, Jennifer Hofmeister, Diogo D. Barcellos, Silvia A.P. Lourenço, Clyde F.E. Roper, Natalie A. Moltschaniwskyj, Corey P. Green, and Jennifer Mather. Transitions During Cephalopod Life History: The Role of Habitat, Environment, Functional Morphology and Behaviour. pp. 363–440.

Volume 68, 2014.
Paul K.S. Shin, Siu Gin Cheung, Tsui Yun Tsang, and Ho Yin Wai. Ecology of Artificial Reefs in the Subtropics. pp. 1–64.

Hrafnkell Eiríksson. Reproductive Biology of Female Norway Lobster, *Nephrops norvegicus* (Linnaeus, 1758) Leach, in Icelandic Waters During the Period 1960–2010: Comparative Overview of Distribution Areas in the Northeast Atlantic and the Mediterranean. pp. 65–210.

Volume 69, 2014.
Ray Hilborn. Introduction to Marine Managed Areas. pp. 1–14.
Philip N. Trathan, Martin A. Collins, Susie M. Grant, Mark Belchier, David K.A. Barnes, Judith Brown, and Iain J. Staniland. The South Georgia and the South Sandwich Islands MPA: Protecting A Biodiverse Oceanic Island Chain Situated in the Flow of the Antarctic Circumpolar Current. pp. 15–78.
Richard P. Dunne, Nicholas V.C. Polunin, Peter H. Sand, and Magnus L. Johnson. The Creation of the Chagos Marine Protected Area: A Fisheries Perspective. pp. 79–128.
Michelle T. Schärer-Umpierre, Daniel Mateos-Molina, Richard Appeldoorn, Ivonne Bejarano, Edwin A. Hernández-Delgado, Richard S. Nemeth, Michael I. Nemeth, Manuel Valdés-Pizzini, and Tyler B. Smith. Marine Managed Areas and Associated Fisheries in the US Caribbean. pp. 129–152.
Alan M. Friedlander, Kostantinos A. Stamoulis, John N. Kittinger, Jeffrey C. Drazen, and Brian N. Tissot. Understanding the Scale of

Marine Protection in Hawai'i: From Community-Based Management to the Remote Northwestern Hawaiian Islands. pp. 153–204.
Louis W. Botsford, J. Wilson White, Mark H. Carr, and Jennifer E. Caselle. Marine Protected Area Networks in California, USA. pp. 205–252.
Bob Kearney and Graham Farebrother. Inadequate Evaluation and Management of Threats in Australia's Marine Parks, Including the Great Barrier Reef, Misdirect Marine Conservation. pp. 253–288.
Randi Rotjan, Regen Jamieson, Ben Carr, Les Kaufman, Sangeeta Mangubhai, David Obura, Ray Pierce, Betarim Rimon, Bud Ris, Stuart Sandin, Peter Shelley, U. Rashid Sumaila, Sue Taei, Heather Tausig, Tukabu Teroroko, Simon Thorrold, Brooke Wikgren, Teuea Toatu, and Greg Stone. Establishment, Management, and Maintenance of the Phoenix Islands Protected Area. pp. 289–324.
Alex J. Caveen, Clare Fitzsimmons, Margherita Pieraccini, Euan Dunn, Christopher J. Sweeting, Magnus L. Johnson, Helen Bloomfield, Estelle V. Jones, Paula Lightfoot, Tim S. Gray, Selina M. Stead, and Nicholas V. C. Polunin. Diverging Strategies to Planning an Ecologically Coherent Network of MPAs in the North Sea: The Roles of Advocacy, Evidence and Pragmatism in the Face of Uncertaintya. pp. 325–370.
Carlo Pipitone, Fabio Badalamenti, Tomás Vega Fernández, and Giovanni D'Anna. Spatial Management of Fisheries in the Mediterranean Sea: Problematic Issues and a Few Success Stories. pp. 371–402.

Volume 70, 2015.
Alex D. Rogers, Christopher Yesson, and Pippa Gravestock. A Biophysical and Economic Profile of South Georgia and the South Sandwich Islands as Potential Large-Scale Antarctic Protected Areas. pp. 1–286.

Volume 71, 2015.
Ricardo Calado and Miguel Costa Leal. Trophic Ecology of Benthic Marine Invertebrates with Bi-Phasic Life Cycles: What Are We Still Missing? pp. 1–70.
Jesse M.A. van der Grient and Alex D. Rogers. Body Size Versus Depth: Regional and Taxonomical Variation in Deep-Sea Meio- and Macrofaunal Organisms. pp. 71–108.
Lorena Basso, Maite Vázquez-Luis, José R. García-March, Salud Deudero, Elvira Alvarez, Nardo Vicente, Carlos M. Duarte, and Iris E. Hendriks. The Pen Shell, *Pinna nobilis*: A Review of Population Status and Recommended Research Priorities in the Mediterranean Sea. pp. 109–160.

PREFACE

Until recently, there has been relatively little information available on the biology and conservation of the humpback dolphins (genus *Sousa*), a group of coastal small cetaceans found in the eastern Atlantic, Indian, and western Pacific oceans. This is particularly true in comparison to our knowledge of many other well-known delphinid species. For example, the scientific literature is rich in details about the biology and ecology of the bottlenose dolphins, *Tursiops* spp., common dolphins, *Delphinus* spp., and members of the genus *Stenella*, including information on abundance, distribution, behavior, life history parameters, taxonomy, and phylogenetics. Humpback dolphins occupy the coastal waters of some fairly remote and/or logistically difficult to access areas (e.g. parts of coastal India, Africa, Northern and Western Australia, and the coasts of many Southeast Asian countries). In many instances, very little research has been conducted, and only limited biological material or scientific knowledge has been obtained from these areas. Nonetheless, in recent years dedicated researchers in these regions have labored, sometimes under difficult circumstances, to document new information on humpback dolphin populations and their conservation status. Humpback dolphins have become iconic 'flagship' species in some areas, perhaps especially so in parts of Asia, and now figure prominently in the wildlife conservation efforts of many developing nations.

Although the deteriorating habitats and declining conservation prospects of humpback dolphins of the world have been a major concern of the IUCN Cetacean Specialist Group for many years, it was the recent (2014) revision of the taxonomy of the genus that prompted a thorough review of their current conservation status. The division of the genus from two species into four species, including the description of a new species from Australia and New Guinea (the Australian humpback dolphin *Sousa sahulensis*), made it clear that the old IUCN Red List assessments were now badly out of date, and a fresh examination was urgently needed. Plans for a workshop to do just that were laid almost immediately.

The Workshop to Assess/Re-assess International Union for Conservation of Nature (IUCN) Red Listings for Indo-Pacific Genera of Coastal Marine Small Cetaceans, which was held on 20–21 May 2015, in San Diego, California, USA, provided a forum to discuss and refine the IUCN status assessments for the four species of humpback dolphins. Experts from around the world were assembled for the task (which also covered some species and populations of the genera *Orcaella* and *Neophocaena*). Primary funding for this workshop was provided by the Ocean Park Conservation Foundation, Hong Kong, and logistics and on-site coordination were provided by the International Whaling Commission Secretariat. Many people contributed, and Randall R. Reeves and Robert L. Brownell, Jr. were instrumental in organizing the workshop.

Participants in that workshop on coastal marine small cetaceans contributed their expertise to preliminary assessments and re-assessments of IUCN Red List conservation status for the four currently recognized species of humpback dolphins, and these form the basis of the present volumes. The amount of new information that came together as a result of the workshop was so great that we could not fit it into a single volume, and therefore, this publication now consists of two volumes. Volume 72 of *Advances in Marine Biology*, Part 1 of *Humpback Dolphins* (*Sousa Spp.*): *Current Status and Conservation*, covers general introductions to the genus and the species that occur in the western Atlantic and Indian Oceans. Volume 73, Part 2, covers the more eastern species that have their distributions mainly in the Pacific Ocean.

The new IUCN Red List status assessments presented herein are preliminary and have not yet gone through the full approval process. So, we emphasize that at the date of publication of these volumes, these assessments are not yet been officially accepted and published on the Red List (http://www.iucnredlist.org). Until they are, the official IUCN Red List status for humpback dolphins remains unchanged.

We appreciate the efforts of the many experts who contributed their knowledge and time to various chapters in these volumes, including M. Amano, R. Baldwin, I. Beasley, F. Bonacorso, C. Boyd, G. Braulik, R.L. Brownell, Jr., B. Chen, S.J. Chivers, T. Collins, K. Danil, S.M. Dawson, L. Dolar, J. Durban, D. Fertl, S.L. Huang, L. Karczmarski, J. Kiszka, S. Mesnick, J. Moore, R. Nanayakkara, C. Palmer, G.J. Parra, W.F. Perrin, T.J. Ragen, R.R. Reeves, H.C. Rosenbaum, E. Slooten, B.D. Smith, B.L. Taylor, K. Van Waerebeek, D. Wang, T. Whitty, A. Wright, and B. Würsig.

We hope and expect that this two-volume publication will form a cornerstone for humpback dolphin conservation and management plans throughout the entire range of the genus for years to come. In particular, we urge managers to take heed of the assessments and recommendations contained herein, and to provide the protection that these animals deserve.

The optimists in us look forward to a day when we can again review humpback dolphin conservation status and 'downlist' all of the species and populations as no longer in any danger of extinction or extirpation. We hope that day will come, and that it is not too far off.

Thomas A. Jefferson
Barbara E. Curry

APPENDIX: ABBREVIATIONS AND ACRONYMS USED THROUGHOUT THIS VOLUME

Abbreviation	Explanation
AFCD	Hong Kong Agriculture, Fisheries and Conservation Department
AMSA	Australian Maritime Safety Authority
ArcGIS	Geographic Information System software
BIA	Biologically Important Area
CEBEL	Cetacean Ecology, Behaviour and Evolution Lab
CI	Confidence interval
CITES	Convention on International Trade in Endangered Species
CMS	Convention in Migratory Species
CV	Coefficient of Variation
CW	Coastal Walkabout
DA	Dampier Archipelago
DDT	Dichlorodiphenyltrichloroethane
DMP	Western Australian Department of Mines and Petroleum
DNA	Deoxyribonucleic acid
DoE	Australian Government Department of the Environment
DoECC	United Kingdom Department of Energy and Climate Change
DPaW	Western Australian Department of Parks and Wildlife
DPMC	Australian Government Department of Prime Minister and Cabinet
DRE	Dafengjiang River and Nanliujiang River estuary, China
EEZ	Exclusive Economic Zone
EIA	Environmental Impact Assessment
EPA	Environmental Protection Authority
ETS	Eastern Taiwan Strait
GBR	Great Barrier Reef
GIS	Geographic Information Service
GPS	Global Positioning System
HCH	Hexachlorocyclohexane
HK/PRE	Hong Kong/Pearl River Estuary
HSF	high speed ferry
IFAW	International Fund for Animal Welfare
IUCN	International Union for the Conservation of Nature
IWC	International Whaling Commission
K	Carrying capacity
MMA	Marine Management Area
MP	Marine Park
MPA	Marine Protected Area

Continued

Abbreviation	Explanation
MSC	Marine Stewardship Council
mtDNA	Mitochondrial DNA
MUCRU	Murdoch University Cetacean Research Unit
n	Sample size
N	Population size
NGO	Non-governmental organization
NMFS	US National Marine Fisheries Service
NOAA	US National Oceanic and Atmospheric Administration
NT	Northern Territory
nuDNA	Nuclear DNA
NWC	North West Cape
NWCDRP	North West Cape Dolphin Research Project
PAM	Passive acoustic monitoring
PCB	Polychlorinated biphenyl
PIMRG	Pilbara Indigenous Marine Reference Group
POP	Persistent organic pollutant
PVA	Population Viability Analysis
QLD	Queensland
SC	Shatian/Caotan, China
SD	Standard deviation
SE	Standard error
SEA	Strategic Environmental Assessment
SMM	Society for Marine Mammalogy
SST	Sea Surface Temperature
SWIO	Southwestern Indian Ocean
TEK	Traditional ecological knowledge
TO	Traditional Owner
UAV	Unmanned aerial vehicle
UK	United Kingdom
US	United States
WA	Western Australia
WAMSI	Western Australian Marine Science Institution
WAPC	Western Australian Planning Commission (Department of Planning)

CHAPTER ONE

Humpback Dolphins: A Brief Introduction to the Genus *Sousa*

Thomas A. Jefferson*[,1], Barbara E. Curry[†]
*Clymene Enterprises, Lakeside, California, USA
[†]Physiological Ecology and Bioenergetics Laboratory, University of Central Florida, Orlando, Florida, USA
[1]Corresponding author: e-mail address: sclymene@aol.com

Contents

1. Introduction 2
2. Taxonomy 3
3. Geographic Range 6
4. Abundance and Trends 7
5. Habitat and Ecology 8
6. Threats 10
7. Conservation Actions and Status 11
References 13

Abstract

The delphinid genus *Sousa* has recently undergone a major revision, and currently contains four species, the Atlantic humpback (*Sousa teuszii*), Indian Ocean humpback (*Sousa plumbea*), Indo-Pacific humpback (*Sousa chinensis*), and Australian humpback (*Sousa sahulensis*) dolphins. Recent molecular evidence suggests that humpback dolphins in the Bay of Bengal may comprise a fifth species. These moderate-sized dolphin species are found in shallow (<30 m), coastal waters of the eastern Atlantic, Indian, and western Pacific oceans. Abundance and trends have only been studied in a few areas, mostly in eastern Africa, China, and northern Australia. No global, empirically derived abundance estimates exist for any of the four species, but none appear to number more than about 20,000 individuals. Humpback dolphins feed mostly on small fishes, and sometimes shrimps; occur for the most part in small groups (mostly 12 or less); have limited nearshore movements; and in most parts of their range exhibit a fission/fusion type of social organization. Major threats that affect all the species are entanglement in fishing gear, and habitat degradation/destruction from various forms of coastal development. Impacts from vessel traffic (including behavioural disturbance and displacement, as well as mortality and morbidity from collisions with vessels) appear to be significant in most areas. Several other threats are apparently significant only in particular parts of the range of some species (e.g. high levels of organochlorine contaminants affecting Indo-Pacific humpback dolphins in Hong Kong). Direct hunting only occurs in limited areas and primarily on a small scale. Conservation actions so far have been limited, with most populations receiving little study and almost no management attention. Much more

Advances in Marine Biology, Volume 72
ISSN 0065-2881
http://dx.doi.org/10.1016/bs.amb.2015.04.001

work is needed on humpback dolphin population status, threats, and how the major threats can be reduced or eliminated. Extinction risks for the four species and some populations are preliminarily re-assessed using the IUCN Red List criteria in the current volume. The results suggest that all four species in the genus are threatened at some level (suggested Red List status ranges from Vulnerable for *S. chinensis* and *S. sahulensis* to Critically Endangered for *S. teuszii*).

1. INTRODUCTION

Humpback dolphins (*Sousa* spp.) are not particularly well known to most people in western countries, though coastal residents in Africa and the Indo-Pacific region are likely more aware of them. In these areas, humpback dolphins are often the most conspicuous and most common species of coastal small cetacean. Despite this fact, and perhaps partly owing to the developing state of many nations in these areas, humpback dolphins are generally not well studied and there is far less known of their biology and natural history than there is, for instance, for the coastal bottlenose dolphins (*Tursiops* spp.).

Unfortunately, the nearshore habitat of humpback dolphins and their presence in some of the most intensively developed, and conversely, most impoverished (protein-poor) parts of the world, means that these animals face a diverse set of threats. Many populations appear to be in immediate danger of extirpation. At least three of the species (*S. teuszii*, *S. plumbea*, and the newly described *S. sahulensis*) have ranges that are restricted enough and have threats severe enough, that they may potentially be threatened with extinction within the next few decades.

This volume contains preliminary assessments of the International Union for the Conservation of Nature (IUCN) Red List conservation status of the four currently recognized species of humpback dolphins; current listings are summarized in Table 1. It is important to recognize that, at the date of publication of this volume (late 2015), these new assessments had not yet been formally accepted and published in the Red List, http://www.iucnredlist.org. Until they are, the official IUCN Red List status of humpback dolphins remains unchanged, i.e. Atlantic humpback dolphin (*S. teuszii*) as Vulnerable and Indo-Pacific humpback dolphin (*S. chinensis*) as Near Threatened. The Taiwan Strait 'subpopulation' of *S. chinensis* is the only unit below the species level that has been assessed and red-listed. This subpopulation was assessed in 2008 and listed as Critically Endangered (Reeves et al.,

Table 1 Summary of the Known Differences Among the Four Species of *Sousa*

Characteristic	*S. teuszii*	*S. plumbea*	*S. chinensis*	*S. sahulensis*
Range	Eastern Atlantic from W. Sahara to Angola	Western Indian Ocean from S. Africa to Myanmar	Eastern Indian Ocean and western Pacific from east India to China and SE Asia	Western Pacific from southern Australia to New Guinea
IUCN Red List status				
Current	Vulnerable	Not listed	Near Threatened (includes all three Indo-Pacific species)	Not listed
External morphology				
Dorsal hump	Prominent dorsal hump	Prominent dorsal hump	No dorsal hump	No dorsal hump
Sexual dimorphism	Sexually dimorphic	Sexually dimorphic	Little or no dimorphism	Slight dimorphism
Coloration	Uniform grey, with lighter belly	Uniform brownish grey with lighter belly	Mostly white as adults	Dark grey back and lighter belly, curved dorsal 'cape'

2008a). More recently, these dolphins were described as a new subspecies, *S. c. taiwanensis* (Wang et al., 2016). The data currently available on humpback dolphins in other portions of their range, such as the Hong Kong/Pearl River Estuary are considered insufficient to determine whether they fit the Red List definition of subpopulation.

In this chapter, we provide a general introduction to dolphins of the genus *Sousa*. More detailed regional information about biology and conservation status for each of the species will be presented throughout this volume.

2. TAXONOMY

The taxonomy of the humpback dolphins has been in a state of disarray for several centuries, including much controversy about even such basic issues as how many species comprise the genus. The first species of humpback dolphins were described in the late 1700s and 1800s, in the genus *Delphinus*, which was used at the time as a repository for virtually any new marine species of dolphin. However, it later became clear that these animals were distinctive enough to warrant their own genus, and first *Sotalia* (along with the South American dolphins that were thought to be closely related), and then later *Sousa*, were erected to separate these animals from other marine dolphin species (see review in Jefferson and Rosenbaum, 2014).

Over the centuries, many different putative species of humpback dolphins were described and placed into the new genus. Nine nominal species were proposed through the end of the twentieth century (Jefferson and Rosenbaum, 2014). Reviews conducted in the 1900s generally concluded that most of these species were not valid, and eventually only two (*S. teuszii* and *S. chinensis*) or sometimes three (the previous two, plus *S. plumbea*) species were recognized. This state of affairs remained the *status quo* for many decades, in fact until 2013.

Earlier indications, from morphological and molecular studies, that *S. plumbea* might in fact be valid, and that a new species of humpback dolphin from Australian waters was justified, were finally documented with multiple lines of evidence and confirmed in late 2013 (Mendez et al., 2013). The following year, a comprehensive review of the taxonomy of the genus, including a detailed historical review of all the relevant nomenclature, was published (Jefferson and Rosenbaum, 2014). That review described or re-described four species in the genus: *S. teuszii* (the Atlantic humpback dolphin of West Africa), *S. plumbea* (the Indian Ocean humpback dolphin), *S. chinensis* (the Indo-Pacific humpback dolphin, from the Bay of Bengal and throughout Southeast Asia), and the new species, *S. sahulensis* (the Australian humpback dolphin, which is now also known to occur in at least southern New Guinea (see Beasley et al., 2016)).

The four-species arrangement was quickly accepted and adopted by the Society for Marine Mammology's Ad Hoc Committee on Taxonomy (now considered by most to be the global authority on marine mammal taxonomy—Society for Marine Mammalogy (SMM, 2014)). The four species are illustrated in Figure 1 and the differences among them are summarized in Table 1.

For centuries, one of the most perplexing issues in *Sousa* taxonomy has been the status of humpback dolphins in the Bay of Bengal (eastern India, Bangladesh, and Myanmar). This region contains dolphins that in some ways show characteristics associated with both *S. chinensis* and *S. plumbea* (e.g. some evidence of a dorsal hump, wide-based dorsal fins, extensive spotting on the body, and large unpigmented areas on the sides and back; Muralidharan, 2013; Smith et al., 2015; Sutaria and Jefferson, 2004; and

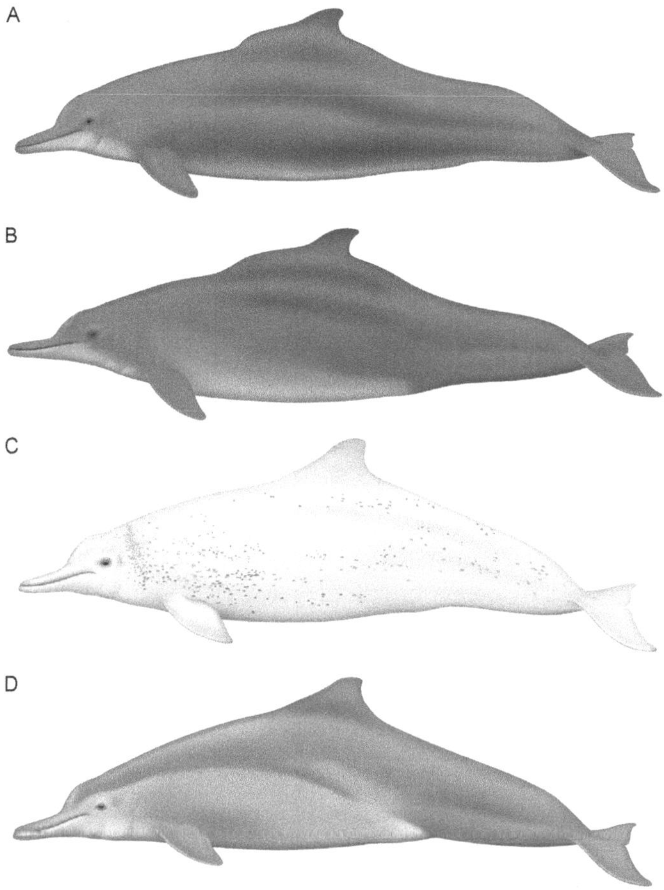

Figure 1 The four currently recognized species of humpback dolphins, genus *Sousa*: *S. teuszii* (A), *S. plumbea* (B), *S. chinensis* (C), and *S. sahulensis* (D). Illustration by U. Gorter.

see Sutaria et al., 2015). Similarly, due to seemingly intermediate characteristics, the affinity of the nominal species *S. lentiginosa* (Owen, 1866), from Vishakhapatnam in eastern India, has been difficult to ascertain (see Jefferson and Rosenbaum, 2014). Very few specimens of humpback dolphins from the Bay of Bengal have been available for taxonomic studies, and samples for molecular work have been almost non-existent. The information available at the time, led Jefferson and Rosenbaum (2014) to suggest that the Bay of Bengal may be an area of overlap in distribution of *S. chinensis* and *S. plumbea*, or that there may be some type of intermediate form there.

The recent collection and analysis of a reasonably large number ($n = 15$) of genetic samples from humpback dolphins off Bangladesh has shed some light on this issue (Amaral et al., 2015). Most of the samples were highly divergent from other humpback dolphins previously sampled, and did not appear to be closely associated with either *S. chinensis* or *S. plumbea* (Amaral et al., 2015). Although more work, including the use of additional genetic markers needs to be completed to confirm these preliminary results, the new information has led to suggestions that there may be a distinct species of humpback dolphin in the Bay of Bengal. Due to some similarities in morphology and coloration, it seems possible that this may correspond to *S. lentiginosa* (Owen, 1866), but comparisons of Owen's (1866) type specimen with the newly obtained genetic material remain necessary. There is even some preliminary evidence of a length difference between eastern Indian and western Indian/Sri Lankan humpback dolphins (the latter presumably corresponding to *S. plumbea*) (Jefferson, unpublished data). If indeed found to be a distinct species, and regardless of the eventual nomenclature (note that in addition to *S. lentiginosa*, the type of *S. plumbea* G. Cuvier, 1829, is also from the Bay of Bengal), this would add a fifth species to the genus *Sousa*, making it one of the most speciose of delphinid genera. In fact, if the paraphyletic genera *Lagenorhynchus* and *Stenella* are eventually split (see Perrin et al., 2013), then it would likely contain more species than any other in the family Delphinidae.

3. GEOGRAPHIC RANGE

Humpback dolphins are found in shallow (mostly <30 m), coastal waters of the Indo-Pacific region, and the eastern Atlantic (Parra and Ross, 2009; Ross et al., 1994; Figure 2). The Atlantic humpback dolphin occurs in the West African region from Morrocco/Western Sahara south to Angola, apparently as several isolated populations or meta-populations (see Collins, 2015; Weir and Collins, 2015). The Indian Ocean humpback dolphin ranges from South Africa, around the rim of the Indian Ocean, north and east to at least the southern tip of India (and possibly to Myanmar), although the range limits in the eastern part of the distribution are not currently well known (see Braulik et al., 2015).

The Indo-Pacific humpback dolphin is found in waters of Southeast Asia, and is currently known from at least Myanmar (and possibly eastern India) eastward to the island of Borneo, and northward to central China (Jefferson and Smith, 2016). Finally, the newly described Australian humpback dolphin is found in the warmer waters of central and northern Australia, extending north to at the least the southern parts of the island

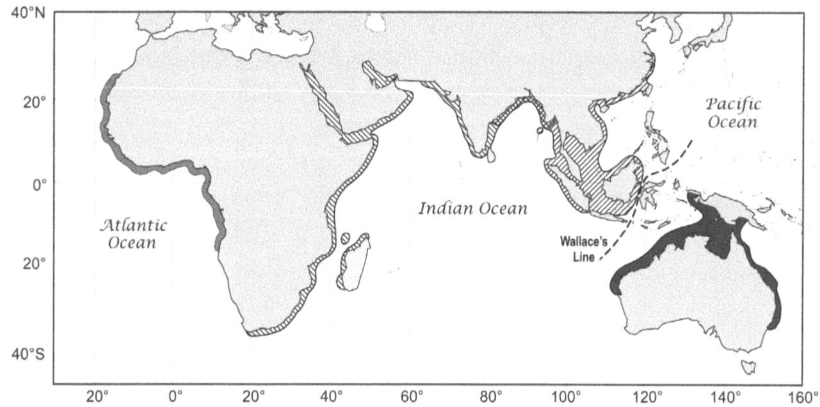

Figure 2 Approximate range of humpback dolphins, genus *Sousa*: *S. teuszii* (light grey shading), *S. plumbea* (135° hatching), *S. chinensis* (45° hatching), and *S. sahulensis* (dark grey shading). *From Jefferson and Rosenbaum (2014). Note that some revisions to these ranges are proposed in chapters in this volume.*

of New Guinea (Beasley et al., 2016; Parra and Cagnazzi, 2016). Whether this latter species also occurs along the northern coasts of the island of New Guinea is currently unknown, but doubtful.

4. ABUNDANCE AND TRENDS

There are no empirically based global estimates of abundance for any of the four species of humpback dolphins. Moreover, many of the available estimates were made more than 10 years ago. Population size, or abundance, has only been numerically estimated via systematic studies for a few very specific areas of the genus' range. However, it is possible to estimate population sizes within very broad parameters for the four species by taking existing estimates and interpreting them in the context of each species' known range.

For the Atlantic humpback dolphin, there have been virtually no comprehensive population assessments, and all that is known of their population biology comes from rough 'approximations' or small-scale photo-identification catalogue sizes. Rough guesses of about 100 animals in the Saloum River delta, Senegal, of 'several hundred' in Guinea Bissau, and a group of about 10 photo-identified in Angolan waters are all that exist (Van Waerebeek et al., 2004; Weir, 2009). Total population size is thought to be no more than about 3,000, and a population decline for the species is inferred (Collins, 2015; Reeves et al., 2012).

The Indian Ocean humpback dolphin has been studied extensively only in southern/eastern Africa (i.e. South Africa, Mozambique, and Zanzibar), although there are other, more-recent, or cursory studies currently ongoing in Mayotte, Madagascar, the Persian Gulf area, Pakistan, and Sri Lanka. A few estimates of abundance are currently available (ca. 450 individuals in Algoa Bay, South Africa (Karczmarski et al., 1999); 207 in Richard's Bay, South Africa (Atkins and Atkins, 2002); 105 in Maputo Bay, Mozambique (Guissamulo and Cockcroft, 2004); ca. 60 in the Bazaruto Archipelago, Mozambique (Guissamulo and Cockcroft, 1997); and 62 in Zanzibar (Stensland et al., 2006)). Considering these in light of the species', overall range suggests that the total population is not likely to be much greater than 10,000 individuals, and a population decline can be inferred (Braulik et al., 2015; Reeves et al., 2008b).

Population biology of the Indo-Pacific humpback dolphin has been studied extensively in Chinese waters, starting about 20 years ago (e.g. Chen et al., 2008, 2010; Jefferson, 2007; Wang et al., 2012), but there are only a few studies that provide abundance estimates for areas outside of China/Taiwan (one in Borneo (Minton et al., 2016) and two in Thailand (Jaroensutasinee et al., 2010; Jutapruet et al., 2015)). The available abundance/population estimates add up to about 5500 animals (Jefferson and Smith, 2016), and taking account of the unstudied portions of the specie's range, it appears that the total species population is unlikely to number much greater than 16,000 individuals. Where studied, most populations are confirmed or suspected to be declining, and thus a declining population should be inferred for *S. chinensis* as a whole (Jefferson and Smith, 2016).

The newly described Australian humpback dolphin has not been well studied, but in the past decade or so there has been work on its population biology in Queensland (abundance estimates of 44 in Cleveland Bay (Parra et al., 2006) and 141 in Moreton Bay (Corkeron et al., 1997)), and more recently in Western Australia (see Parra and Cagnazzi, 2016). Although these represent only a very small portion of the species' overall range, it appears unlikely that the species would number over 10,000 individuals, and it is probably much smaller. A population decline can reasonably be inferred, although this is uncertain (Parra and Cagnazzi, 2016).

5. HABITAT AND ECOLOGY

The habitat of each of the four species is very similar. They all occur in tropical to warm temperate regions, and show a strong preference for shallow (<20–30 m deep) nearshore waters. Although most of the four species also seem to prefer brackish waters associated with estuarine systems, in some areas they have been found to occur in mangrove areas, around rocky and coral reefs, over sandy bottoms, and along rocky shorelines. They do not inhabit deep, oceanic areas, and their movements appear to be limited by water depth (ca. 40 m appears to be the limit) more than distance from shore (e.g. Hanf et al., 2016; Weir and Collins, 2015).

Group size and structure for humpback dolphins is similar among the four species. All occur mostly in small groups, from pairs to groups of up to about a dozen. Larger aggregations do develop on occasion. In Bangladesh, some groups may number as many as 205 individuals (Jefferson and Smith, 2016), and in Hong Kong and adjacent waters, when dolphins aggregate behind active trawlers to feed on prey stirred up by the nets, they may gather into groups as large as 44 individuals (Hung, 2008). Group structure has not been studied in most areas, but where it has, most humpback dolphin populations appear to have a fission/fusion type social structure, with only short-term associations beyond the mother/calf bond (see Jefferson and Karczmarski, 2001). However, in a few areas (e.g. Mozambique and off the west coast of Taiwan), these dolphins exhibit more stability in group structure (Guissamulo and Cockcroft, 2004; Wang et al., 2016).

There have only been a few long-term, detailed studies of the behaviour of humpback dolphins (e.g. South Africa (Karczmarski, 1999; Karczmarski and Cockcroft, 1999; Karczmarski et al., 1999, 2000), Hong Kong (Hung, 2014; Würsig et al., 2016)), and much of what is known about their behavioural ecology comes from relatively superficial, often-opportunistic observations. Certain general rules seem to hold throughout the range of the genus, however. Humpback dolphins rarely ride the bow or stern waves of vessels, and exhibit moderate levels of aerial activity (see Jefferson and Karczmarski, 2001). Breaching, acrobatic leaps, spyhopping, and other active surface behaviours are not uncommon, and are seen especially frequently during the times of year when reproductive activity is at its peak. Responses to human activities (in particular to boats) are variable, with populations in most areas having been described as shy and difficult to approach, but in other areas (e.g. Hong Kong), dolphins may have become accustomed to human activities and may tolerate human presence with relatively little sign of obvious disturbance (see Jefferson and Karczmarski, 2001; Piwetz et al., 2015; Würsig et al., 2016).

Life history and reproductive parameters have only been studied in detail for two regions: for Indian Ocean humpback dolphins in South African waters (Cockcroft, 1989; Nolte, 2013; Plön et al., 2015), and for Indo-Pacific humpback dolphins in the Pearl River Estuary of China (Jefferson et al., 2012). Because most life history parameters for these two populations at different ends of the genus's range were found to be similar, there is most likely similarity throughout the entire range, although this is not confirmed. Humpback dolphins appear to live to be in their 40s, and become sexually mature at about 9–13 years of age (a few years earlier for females than males) (however, see Plön et al., 2015, for a discussion of a potential different interpretation of South African life history parameters). Gestation is about 10–11 months, and a single calf is born, and then nursed for 2–5 years. Calving likely occurs year-round in most parts of the range, but where studied, there appears to be a peak in calving in spring/summer months (see Jefferson et al., 2012; Plön et al., 2015).

Feeding habits of humpback dolphins are known mostly from stomach contents studies on just a few populations, but the general rule appears to be that humpback dolphins are opportunistic feeders, taking a wide variety of prey. They seem to prefer small schooling and benthic/demersal fishes, although shrimps are sometimes taken in some areas. Squids and other cephalopods do not seem to be a major part of their diet anywhere (Jefferson and Karczmarski, 2001).

The natural predators of humpback dolphins throughout their range are likely to be killer whales (*Orcinus orca*) and large sharks, although killer whales tend be rare in the very nearshore shallow waters of the tropics/subtropics that humpback dolphins primarily inhabit. Shark bites are commonly seen on Indian Ocean humpback dolphins in South African waters (Cockcroft et al., 1989), but are very rare on Indo-Pacific humpback dolphins in waters of the Pearl River Estuary or elsewhere in China (see Jefferson et al., 2006). Diseases and associated pathology have not been well studied in humpback dolphins, and as these dolphins are rarely held captive, little is known, other than the documentation of a few diseases, and that these animals are hosts to a number of parasites (both internal and external) (Jefferson and Karczmarski, 2001; also see Plön et al., 2015).

6. THREATS

The coastal habitats of humpback dolphins, especially in developing nations, involve a large number of potential threats (see Piwetz et al., 2015). Two threats that appear to be major factors for virtually all humpback dolphin populations, whether they occur in highly developed or in less-impacted areas, are entanglement in fishing gear and impacts of vessel traffic (the latter encompasses both behavioural disturbance from vessel movements and noise, and the more serious issue of mortality and morbidity from collisions with vessels—see Karczmarski et al. (2016). A third threat that appears to affect most, if not all, populations of humpback dolphins is habitat degradation/destruction from coastal development and other anthropogenic activities (e.g. Karczmarski et al., 2016; Sutaria et al., 2015).

Serious threats that may only be a major factor for some populations of the genus are the deleterious effects of environmental contaminants (especially organochlorines; e.g. Hong Kong (Jefferson et al., 2006; and see Karczmarski et al., 2016)), and prey depletion from overfishing (e.g. Wang et al., 2016). Direct hunting is only known to occur in a few places in the range of the genus, but where it does, the catches always appear to be unregulated, and thus this threat has the potential to be a major one for the affected populations (e.g. West Africa (see Collins, 2015); Pakistan (see Kiani and Van Waerebeek, 2015); Madagascar (see Cerchio et al., 2015)). A few other threats are sometimes mentioned, but at present, are probably not major factors for most populations of humpback dolphins. These include habitat alteration from climate change, and harassment from dolphin-watching operations.

7. CONSERVATION ACTIONS AND STATUS

Although dolphins in many human societies are thought of as 'charismatic megafauna' and therefore enjoy popular status among the general public, the unfortunate reality is that, even with laws protecting them in place, most humpback dolphin populations appear to be negatively affected by many human activities. Where legislation exists, often laws are not enforced, and in many cases there is simply inadequate information on the biology of these animals to conserve and manage them effectively, whatever the attitudes of the government officials or local citizens may be (e.g. Karczmarski et al., 2016). The proposed new IUCN Red List assessments are presented in Table 2.

In poor, developing countries with inadequate legislation (e.g. Madagascar—see Cerchio et al., 2015), and in relatively wealthy, developed countries that may have laws in place for wildlife protection (e.g. Hong Kong—Jefferson et al., 2009; Karczmarski et al., 2016), humpback dolphins are facing serious threats, and populations are apparently declining. An

Table 2 Summary of the New IUCN Red List Status Assessments Proposed for the Four Species of *Sousa*

Species	Proposed Status	Criteria	Reference
Atlantic humpback dolphin *S. teuszii*	Critically Endangered	A3cd	Collins (2015)
Indian Ocean humpback dolphin *S. plumbea*	Endangered	A4cd	Braulik et al. (2015)
Indo-Pacific humpback dolphin *S. chinensis*	Vulnerable	A4cd	Jefferson and Smith (2016)
Australian humpback dolphin *S. sahulensis*	Vulnerable	C2a(i)	Parra and Cagnazzi (2016)

These are not yet official.

example of the former situation is found throughout West Africa, where Atlantic humpback dolphins are poorly studied, and where there has been almost no attempt at management of stocks or populations. There, these dolphins are sometimes killed directly by humans as a food source (see Collins, 2015). Populations in West Africa are also impacted by pollution, shipping, oil and gas operations, fishing activities, and coastal development, all in the absence of effective environmental impact assessment (Collins, 2015; Weir et al., 2011). Examples of the serious threats facing humpback dolphins that are found in the territories of relatively wealthy nations, are evident along Chinese coastlines (e.g. Hong Kong, Guangdong Province, Xiamen, and Taiwan), where Indo-Pacific humpback dolphins apparently occur as a collection of distinct populations (though, as previously mentioned, some of these may have formerly shared connections, and may now be separated only as a result of human impacts), some very small (<100 individuals) and some quite large (>2000 animals)).

In China and Taiwan, well-developed environmental impact assessment legislation exists, wildlife protective laws are in place, marine parks and reserves have been established (often with the express goal of protecting dolphin habitat), dolphin populations are relatively well studied (with basic information on population size, and sometimes even trends, available in most areas), and there is a segment of society that is willing to advocate for the welfare of these animals. Despite all of this, humpback dolphins are declining in Xiamen, Taiwan, and in Hong Kong and the Pearl River Estuary region (Jefferson and Smith, 2016; Karczmarski et al., 2016). The federal and provincial governments have failed to provide adequate protection for the animals to avoid population declines. In Hong Kong, dolphin numbers have been declining for over a decade (although the overall abundance, of humpback dolphins occurring throughout the Pearl River Estuary, remains large, and is probably still viable; see Jefferson and Smith, 2016), and in Taiwan, dolphin numbers continue to dwindle (currently less than 75 dolphins; Wang et al., 2016).

Because so little is known about humpback populations in some areas (e.g. see Collins, 2015 for West Africa; Cerchio et al., 2015 for Madagascar; and Minton et al., 2016 for Borneo) and research work has been scant, some populations of humpback dolphins may have already been extirpated, without us even being aware of it. Only by thoroughly evaluating and recognizing the conservation status of the various species, subspecies, and populations of humpback dolphins will we have any chance of preserving the biological diversity that remains today. There is an urgent need to take stock of what we know about these dolphins, acknowledge what still remains to be learned, and identify the most important risk factors affecting them. Beyond this, we also need to assess how these threats impact the various taxa, and aggressively pursue solutions to eliminate, or at least reduce, these threats to sustainable levels. The approaches used will need to be tailored to each area and taxon to be effective. It is our hope that the chapters that follow in this volume represent significant steps towards achieving this important goal.

REFERENCES

Amaral, A.R., Smith, B.D., Mansur, R.M., Brownell, R.L., Rosenbaum, H.C., 2015. Genetic identity of Indo-Pacific humpback dolphins (*Sousa* spp.) in the northern Bay of Bengal, Bangladesh. International Whaling Commission Scientific Committee Report, 13 pp (unpublished).

Atkins, S., Atkins, B.L., 2002. Abundance and site fidelity of Indo-Pacific hump-backed dolphins (*Sousa chinensis*) at Richard's Bay, South Africa. International Whaling Commission Scientific Committee Report, 11 pp (unpublished).

Beasley, I., Jedensjö, M., Wijaya, G.M., Anamiato, J., Kahn, B., Kreb, D., 2016. Observations on Australian humpback dolphins (*Sousa sahulensis*) in waters of the Pacific Islands and New Guinea. Adv. Mar. Biol. 73, (in press).

Braulik, G.T., Findlay, K., Cerchio, S., Baldwin, R., 2015. Assessment of the conservation status of the Indian Ocean humpback dolphin (*Sousa plumbea*) using the IUCN Red List Criteria. Adv. Mar. Biol. 72, 119–141.

Cerchio, S., Andrianarivelo, N., Andrianantenaina, B., 2015. Ecology and conservation status of Indian Ocean humpback dolphins (*Sousa plumbea*) in Madagascar. Adv. Mar. Biol. 72, 163–199.

Chen, B., Zheng, D., Zhai, F., Xu, X., Wang, Q., Yang, G., 2008. Abundance, distribution and conservation of Chinese white dolphins (*Sousa chinensis*) in Xiamen, China. Mammal. Biol. (Zeit. für Saug.) 73, 156–164.

Chen, T., Hung, S.K., Qiu, Y., Jia, X., Jefferson, T.A., 2010. Distribution, abundance, and individual movements of Indo-Pacific humpback dolphins (*Sousa chinensis*) in the Pearl River Estuary, China. Mammalia 74, 117–125.

Cockcroft, V.G., 1989. Biology of Indopacific humpback dolphins (*Sousa plumbea*) off Natal, South Africa. Paper presented at the Eight Biennial Conference on the Biology of Marine Mammals, Pacific Grove, California, USA (unpublished).

Cockcroft, V.G., Cliff, G., Ross, G.J.B., 1989. Shark predation on Indian Ocean bottlenose dolphins *Tursiops truncatus* off Natal, South Africa. S. Afr. J. Zool. 24, 305–310.

Collins, T., 2015. Re-assessment of the conservation status of the Atlantic humpback dolphin *Sousa teuszii* (Kükenthal, 1892), using the IUCN Red List Criteria. Adv. Mar. Biol. 72, 47–77.

Corkeron, P., Morissette, N.M., Porter, L., Marsh, H., 1997. Distribution and status and of hump-backed dolphins, *Sousa chinensis*, in Australian waters. Asian Mar. Biol. 14, 49–59.

Guissamulo, A., Cockcroft, V.G., 1997. Dolphin and dugong occurrence and distribution and fisheries internations in Maputo and Bazaruto Bays, Mozambique. International Whaling Commission Scientific Committee Report, 11 pp (unpublished) (SC/49/SM24).

Guissamulo, A., Cockcroft, V.G., 2004. Ecology and population estimates of Indo-Pacific humpback dolphins (*Sousa chinensis*) in Maputo Bay, Mozambique. Aquat. Mamm. 30, 94–102.

Hanf, D.M., Hunt, T., Parra, G.J., 2016. Humpback dolphins of Western Australia: a review of current knowledge and recommendations for future management. Adv. Mar. Biol. 73, (in press).

Hung, S.K., 2008. Habitat use of Indo-Pacific humpback dolphins (*Sousa chinensis*) in Hong Kong. Ph.D. thesis, University of Hong Kong.

Hung, S.K., 2014. Monitoring of marine mammals in Hong Kong waters (2013–14): Final report to the Agriculture, Fisheries and Conservation Department, Hong Kong SAR Government. (unpublished).

Jaroensutasinee, M., Jutapruet, S., Jaroensutasinee, K., 2010. Population size of Indo-Pacific humpback dolphins (*Sousa chinensis*) at Khanom, Thailand. Walailak J. Sci. Tech. 7, 115–126.

Jefferson, T.A., 2007. Monitoring of Chinese white dolphins (*Sousa chinensis*) in Hong Kong waters: biopsy sampling and population data analysis. Unpublished contract report to the Agriculture, Fisheries and Conservation Department, Hong Kong SAR Government.

Jefferson, T.A., Karczmarski, K., 2001. *Sousa chinensis*. Mammal. Spec. 655, 1–9.

Jefferson, T.A., Rosenbaum, H.R., 2014. Taxonomic revision of the humpback dolphins (*Sousa* spp.), and description of a new species from Australia. Mar. Mamm. Sci 30, 1494–1541.

Jefferson, T.A., Smith, B.D., 2016. Re-assessment of the conservation status of the Indo-Pacific humpback dolphin (*Sousa chinensis*) using the IUCN Red List Criteria. Adv. Mar. Biol. 73, (in press).

Jefferson, T.A., Hung, S.K., Lam, P.K.S., 2006. Strandings, mortality and morbidity of Indo-Pacific humpback dolphins in Hong Kong, with emphasis on the role of environmental contaminants. J. Cetacean Res. Manag. 8, 181–193.

Jefferson, T.A., Hung, S.K., Würsig, B., 2009. Protecting small cetaceans from coastal development: impact assessment and mitigation experience in Hong Kong. Mar. Policy 33, 305–311.

Jefferson, T.A., Hung, S.K., Robertson, K.M., Archer, F.I., 2012. Life history of the Indo-Pacific humpback dolphin in the Pearl River Estuary, southern China. Mar. Mamm. Sci. 28, 84–104.

Jutapruet, S., Huang, S.-L., Li, S., Lin, S., Kittiwattanawong, M., Pradit, K., 2015. Population size and habitat characteristics of the Indo-Pacific Humpback dolphin (*Sousa chinensis*) off Donsak, Surat Thani, Thailand. Aquat. Mamm. 41, 129–142.
Karczmarski, L., 1999. Group dynamics of humpback dolphins (*Sousa chinensis*) in the Algoa Bay region, South Africa. J. Zool., London 249, 283–293.
Karczmarski, L., Cockcroft, V.G., 1999. Daylight behaviour of humpback dolphins, *Sousa chinensis* in Algoa Bay, South Africa. Zeit. für Saug. 64, 19–29.
Karczmarski, L., Cockcroft, V.G., McLachlan, A., 1999. Group size and seasonal pattern of occurrence of humpback dolphins *Sousa chinensis* in Algoa Bay, South Africa. S. Afr. J. Mar. Sci. 21, 89–97.
Karczmarski, L., Cockcroft, V.G., McLachlan, A., 2000. Habitat use and preferences of Indo-Pacific humpback dolphins *Sousa chinensis* in Algoa Bay, South Africa. Mar. Mamm. Sci. 16, 65–79.
Karczmarski, L., Huang, S., Or, C.K.M., Gui, D., Chan, S.C.Y., Lin, W., Porter, L., Wong, W., Zheng, R., Ho, Y., Chui, S.Y.S., Tiongson, A.J.C., Mo, Y., Chang, W., Kwok, J.H.W., Tang, R.W.K., Lee, A.T.L., Yiu, S., Keith, M., Gailey, G., Wu, Y., 2016. Humpback dolphins in Hong Kong and the Pearl River Delta: status, threats and conservation challenges. Adv. Mar. Biol. 73, (in press).
Kiani, M.S., Van Waerebeek, K., 2015. A review of the status of the Indian Ocean humpback dolphin *Sousa plumbea* in Pakistan. Adv. Mar. Biol. 72, 201–227.
Mendez, M., Jefferson, T.A., Kolokotronis, S.O., Krützen, M., Parra, G.J., Collins, T., Minton, G., Baldwin, R., Berggren, P., Sarnblad, A., Amir, O.A., Peddemors, V.M., Karczmarski, L., Guissamula, A., Smith, B., Sutaria, D., Amato, G., Rosenbaum, H.C., 2013. Integrating multiple lines of evidence to better understand the evolutionary divergence of humpback dolphins along their entire distribution range: a new dolphin species in Australian waters? Mol. Ecol. 22, 5936–5948.
Minton, G., Poh, A.N.Z., Peter, C., Porter, L., Kreb, D., 2016. Indo-Pacific humpback dolphins in Borneo: A review of current knowledge with emphasis on Sarawak. Adv. Mar. Biol. 73, (in press).
Muralidharan, R., 2013. Sightings and behavioral observations of Indo-Pacific humpback dolphins *Sousa chinensis* (Osbeck, 1765) along Chennai coast, Bay of Bengal. J. Threat. Taxa 5, 5002–5006.
Nolte, Z., 2013. The Natural History of the Humpback Dolphin, *Sousa chinensis*, in KwaZulu-Natal, South Africa: Age, Growth and Reproduction. M.Sc. thesis, Rhodes University, South Africa, 158 pp.
Owen, R., 1866. On some Indian Cetacea collected by Walter Elliot, Esq. Trans. Zool. Soc., London 6, 17–47.
Parra, G.J., Cagnazzi, D., 2016. Conservation status of the Australian humpback dolphin (*Sousa sahulensis*) using the IUCN Red List Criteria. Adv. Mar. Biol. 73, (in press).
Parra, G.J., Ross, G.J.B., 2009. Humpback dolphins *Sousa chinensis* and *S. teuszii*. In: Perrin, W.F., Würsig, B., Thewissen, J.G.M. (Eds.), Encyclopedia of Marine Mammals, second ed. Academic Press, London, pp. 576–582.
Parra, G., Corkeron, P.J., Marsh, H., 2006. Population sizes, site fidelity and residence patterns of Austrlian snubfin and Indo-Pacific humpback dolphins: implications for conservation. Biol. Conserv. 129, 167–180.
Perrin, W.F., Rosel, P., Cipriano, F., 2013. How to contend with paraphyly in the taxonomy of the delphinine cetaceans? Mar. Mamm. Sci. 29, 567–588.
Piwetz, S., Lundquist, D., Würsig, B., 2015. Humpback dolphin (*genus Sousa*) behavioural responses to human activities. Adv. Mar. Biol. 72, 17–45.
Plön, S., Cockcroft, V.G., Froneman, W.P., 2015. The natural history and conservation of Indian Ocean humpback dolphins (*Sousa plumbea*) in South African waters. Adv. Mar. Biol. 72, 143–162.

Reeves, R.R., Dalebout, M.L., Jefferson, T.A., Karczmarski, L., Laidre, K., O'Corry-Crowe, G., Rojas-Bracho, L., Secchi, E.R., Slooten, E., Smith, B.D., Wang, J.Y., Zhou, K., 2008a. *Sousa chinensis* Eastern Tawian Strait subpopulation. The IUCN Red List of Threatened Species. Version 2014.3. Downloaded on 12 December 2014, www.iucnredlist.org.

Reeves, R.R., Dalebout, M.L., Jefferson, T.A., Karczmarski, L., Laidre, K., O'Corry-Crowe, G., Rojas-Bracho, L., Secchi, E.R., Slooten, E., Smith, B.D., Wang, J.Y., Zhou, K., 2008b. *Sousa chinensis*. The IUCN Red List of Threatened Species. Version 2014.3. Downloaded on 12 December 2014, www.iucnredlist.org.

Reeves, R.R., Collins, T., Jefferson, T.A., Karczmarski, L., Laidre, K., O'Corry-Crowe, G., Rojas-Bracho, L., Secchi, E.R., Slooten, E., Smith, B.D., Wang, J.Y., Zhou, K., 2012. *Sousa teuszii*. The IUCN Red List of Threatened Species. Version 2014.3. Downloaded on 12 December 2014, www.iucnredlist.org.

Ross, G.J.B., Heinsohn, G.E., Cockcroft, V.G., 1994. Humpack dolphins *Sousa chinensis* (Osbeck, 1765), *Sousa plumbea* (G. Cuvier, 1829) and *Sousa teuszii* (Kukenthal, 1892). In: Ridgway, S.H., Harrison, R. (Eds.), Handbook of Marine Mammals, Volume 5: The First Book of Dolphins. Academic Press, London, pp. 23–42.

Smith, B.D., Mansur, R.M., Strindberg, S., Redfern, J., Moore, T., 2015. Population demographics, habitat selection, and a spatial and photographic analysis of bycatch risk of Indo-Pacific humpback dolphins *Sousa chinensis* and bottlenose dolphins *Tursiops aduncus* in the northern Bay of Bengal, Bangladesh. International Whaling Commission Scientific Committee Report, 30 pp (unpublished).

Society for Marine Mammalogy, 2014. List of marine mammals species and subspecies. Downloaded on 12 December 2014, http://www.marinemammalscience.org/species-information/list-of-marine-mammal-species-subspecies/.

Stensland, E., Carlen, I., Sarnblad, A., Bignert, A., Berggren, P., 2006. Population size, distribution, and behavior of Indo-Pacific bottlenose (*Tursiops aduncus*) and humpback (*Sousa chinensis*) dolphins off the south coast of Zanzibar. Mar. Mamm. Sci. 22, 667–682.

Sutaria, D., Jefferson, T.A., 2004. Records of Indo-Pacific humpback dolphins (*Sousa chinensis*, Osbeck, 1765) along the coasts of India and Sri Lanka: an overview. Aquat. Mamm. 30, 125–136.

Sutaria, D., Panicker, D., Jog, K., Sule, M., Muralidharan, R., Bopardikar, I., 2015. Humpback dolphins (*genus Sousa*) in India: an overview of status and conservation issues. Adv. Mar. Biol. 72, 229–256.

Van Waerebeek, K., Barnett, L., Camara, A., Cham, A., Diallo, M., Djiba, A., Jallow, A., Ndiaye, E., Ould-Bilal, A.O.S., Bamy, I.L., 2004. Distribution, status, and biology of the Atlantic humpback dolphin, *Sousa teuszii* (Kukenthal, 1892). Aquat. Mamm. 30, 56–83.

Wang, J.Y., Yang, S.C., Fruet, P.F., Daura-Jorge, F.G., Secchi, E.R., 2012. Mark-recapture analysis of the Critically Endangered Eastern Taiwan Strait population of Indo-Pacific humpback dolphins (*Sousa chinensis*): implications for conservation. Bull. Mar. Sci. 88, 885–902.

Wang, J.Y., Yang, S.C., Hung, S.K., 2015. Diagnosability and description of a new subspecies of Indo-Pacific humpback dolphin, *Sousa chinensis* (Osbeck, 1765), from the Taiwan Strait. Zool. Stud. 54, 1–15.

Wang, J.Y., Riehl, K.N., Klein, M.N., Javdan, S., Hoffman, J.M., Dungan, S.Z., Dares, L.E., Araújo-Wang, C., 2016. Biology and conservation status of the Taiwanese humpback dolphin, *Sousa chinensis taiwanensis*. Adv. Mar. Biol. 73, (in press).

Weir, C.R., 2009. Distribution, behaviour and photo-identification of Atlantic humpback dolphins *Sousa teuszii* off Flamingos, Angola. Afr. J. Mar. Sci. 31, 319–331.

Weir, C.R., Collins, T., 2015. A review of the geographical distribution and habitat of the Atlantic humpback dolphin (*Sousa teuszii*). Adv. Mar. Biol. 72, 79–117.

Weir, C.R., Van Waerebeek, K., Jefferson, T.A., Collins, T., 2011. West Africa's Atlantic humpback dolphin (*Sousa teuszii*): endemic, enigmatic, and soon endangered? Afr. Zool. 46, 1–17.

Würsig, B., Parsons, E.C.M., Piwetz, S., Porter, L., 2016. The behavioural ecology of Indo-Pacific humpback dolphins in Hong Kong. Adv. Mar. Biol. 73, (in press).

CHAPTER TWO

Humpback Dolphin (Genus *Sousa*) Behavioural Responses to Human Activities

Sarah Piwetz*[,1], David Lundquist[†], Bernd Würsig*[,‡]

*Department of Marine Biology, Texas A&M University at Galveston, Galveston, Texas, USA
[†]New Zealand Department of Conservation, Wellington, New Zealand
[‡]Department of Wildlife and Fisheries Sciences, Texas A&M University at Galveston, Galveston, Texas, USA
[1]Corresponding author: e-mail address: sarahpiwetz@hotmail.com

Contents

1. Introduction 18
2. Understanding Animal Behaviour 19
3. Anthropogenic Activities Affecting Marine Mammal Behaviour 22
   3.1 Marine Vessel Traffic 23
   3.2 Dolphin-Based Tourism 23
   3.3 Cetacean–Fishery Interactions 24
   3.4 Noise Pollution 24
   3.5 Habitat Loss and Degradation 25
   3.6 Cumulative Effects 26
4. Anthropogenic Effects on Humpback Dolphins 26
   4.1 Marine Vessel Traffic 26
   4.2 Dolphin-Based Tourism 28
   4.3 Cetacean–Fishery Interactions 29
   4.4 Noise Pollution 30
   4.5 Habitat Loss and Degradation 32
   4.6 Cumulative Effects 33
5. Conclusions and Future Research 34
Acknowledgements 36
References 37

Abstract

Humpback dolphins (genus *Sousa*) use shallow, near-shore waters throughout their range. This coastal distribution makes them vulnerable to recreational and commercial disturbances, especially near heavily populated and industrialized areas. Most research focusing on *Sousa* and human activities has emphasized direct impacts and threats, involving injury and death, with relatively little focus on indirect effects on dolphins, such as changes in behaviour that may lead to deleterious effects. Understanding

Advances in Marine Biology, Volume 72
ISSN 0065-2881
http://dx.doi.org/10.1016/bs.amb.2015.08.007

behaviour is important in resolving human–wildlife conflict and is an important component of conservation. This chapter gives an overview of animal behavioural responses to human activity with examples from diverse taxa; reviews the scientific literature on behavioural responses of humpback dolphins to human activity throughout their range, including marine vessel traffic, dolphin tourism, cetacean–fishery interactions, noise pollution, and habitat alteration; and highlights information and data gaps for future humpback dolphin research to better inform behaviour-based management decisions that contribute to conservation efforts.

1. INTRODUCTION

Studying animal behaviour is critical to understanding human–wildlife conflict and to improving wildlife conservation and management (Mann and Würsig, 2014). Many conservation-based studies involving marine mammals and human activity focus on interactions that lead to direct injury or mortality, such as fisheries bycatch, entanglement, and vessel strike (Moore et al., 2009; Van Waerebeek et al., 2007; Weir and Pierce, 2013). Anthropogenic activity also affects marine mammals in ways that are less obvious, such as altering natural behaviour (Lima and Dill, 1990; Srinivasan et al., 2010). These effects on animal behaviour can be detected using behavioural indicators of disturbance (Berger-Tal et al., 2011), and such indicators may be particularly useful in natural settings involving highly mobile animals, where controlled experimental designs are not feasible (Beale, 2007). Identifying these responses, and incorporating knowledge into adaptive management schemes, can help resolve wildlife conservation issues (Cooke et al., 2014). The conservation behaviour framework by Berger-Tal et al. (2011) proposes that behaviour-oriented conservation studies should focus on: (1) human impacts on animal behaviour (potential stressor stimuli), (2) behavioural indicators (potential stress responses), and (3) behaviour-based management (Berger-Tal et al., 2011). Until recently, research involving humpback dolphins (genus *Sousa*) and human activities has focused on direct physical impacts and threats, with relatively little focus on behaviour. This chapter focuses on the first two themes of the conservation behaviour framework, potential human-generated stressor stimuli and potential animal stress responses, in reviewing the growing scientific literature on humpback dolphins. Before we do so, however, we begin with an overview of animal behavioural responses to human-generated activity with examples from diverse taxa. We then review behavioural responses of humpback dolphins to human activity including marine vessel traffic, dolphin tourism, cetacean–fishery interactions, noise pollution, and habitat alteration. Finally, we highlight information and data gaps for future humpback dolphin research to better inform behaviour-based management decisions that contribute to conservation efforts.

2. UNDERSTANDING ANIMAL BEHAVIOUR

When encountering an external stimulus, animals may avoid, ignore, or be attracted to it. Fleeing, for instance, is a form of avoidance and is the most common short-term behaviour observed by prey animals in response to predators (Humphries and Driver, 1970). Attraction, the opposite of avoidance, is defined by the movement of an animal toward a particular stimulus that may result in acquiring food, shelter, or other positive reinforcements (Whittaker and Knight, 1998). Responses likely reflect a ratio of costs (e.g. expending energy or abandoning resources) and benefits (e.g. reducing capture probability) (Frid and Dill, 2002). Over time, repeated short-term behavioural responses may result in changes in activity state, movement patterns, and habitat selection (Fernández-Juricic and Tellería, 2000; Frid and Dill, 2002; Sirot, 2010; Srinivasan et al., 2010), and may result in time and energy spent engaging in one biologically significant activity at the expense of another (Dugatkin, 2009; Frid and Dill, 2002). For example, defensive tactics such as increased vigilance may alter foraging patterns, including timing, rate, and location of feeding, which lead to reduced overall energy consumption (Brown, 1999; Christiansen and Lusseau, 2014; Dugatkin, 2009; Preisler et al., 2006). Several delphinid species have been reported to flee in response to natural predators, including spinner, *Stenella longirostris*, and common bottlenose, *Tursiops truncatus*, dolphins from sharks (Connor and Heithaus, 1996), and dusky dolphins, *Lagenorhynchus obscurus*, from killer whales, *Orcinus orca* (Markowitz et al., 2004; Würsig and Würsig, 1980; Würsig et al., 1997). The same pattern of response has been described for birds that are approached by stimuli perceived as threatening, such as humans on foot or in vehicles (e.g. all-terrain vehicles and boats) (Rodgers and Smith, 1997).

In behavioural studies, it is important to recognize that animals may respond to stimuli even when no short-term behavioural responses are observed. The stress response to an external stimulus begins with the perception of risk (Moberg, 2000). The acute response to an external stressor occurs via the autonomic nervous system, associated with the flight or fight response, by secretion of epinephrine (flight) and/or norepinephrine (fight) into the bloodstream. This response affects body physiology including changes in heart rate, respiratory rate, and blood pressure (Ewbank, 1985; Fair and Becker, 2000; Moberg, 2000). Response may also include a secretion of hormones from the hypothalamic–pituitary–adrenal axis, which may have longer-lasting effects on the body by affecting immune competence, reproduction, and metabolism (Moberg, 2000). These physiological and neurochemical responses are not easily detectable, particularly in behavioural studies of animals in natural settings (Stankowich, 2008).

Furthermore, animals that respond most noticeably to external stimuli may actually be less affected than individuals that display little to no behavioural changes (Beale and Monaghan, 2004). Weaker individuals may lack the capacity to respond in the same way or at the same intensity as healthy animals. Also, if displaced by a particular disturbance, there may be a greater proximate cost to leaving a prey-rich habitat for animals in poor condition. For example, ruddy turnstones, *Arenaria interpres*, in poor nutritional condition, showed reduced scanning, fled when threatening stimuli were closer, and decreased flight distance from threatening stimuli, when compared to those in better condition (Beale and Monaghan, 2004).

The context in which threatening stimuli appear may also influence responses. For example, ungulates, terrestrial mammals closely related to dolphins, respond to stimuli in a less threatened manner when they occur in a predictable context (e.g. humans hiking on designated trails, rather than random movement off trails) (Stankowich, 2008). Furthermore, ungulates tend to flee at greater distances when stimuli approach rapidly or directly versus slowly or indirectly (Stankowich, 2008). A comparable example in the marine environment might be a slow-moving cargo ship along a predictable shipping lane versus a fast-moving private motorboat with a less restricted route.

Often the goal of studying disturbance of wildlife is to identify impacts that have health consequences (Beale, 2007; Nisbet, 2000). However, long-term impacts are inherently more difficult to study logistically than short-term changes in animal behaviour (particularly for species with long-life expectancy and long inter-birth intervals), and long-term data sets are limited in the relatively young field of marine mammalogy. Examples of notable exceptions are the Sarasota Dolphin Research Program, which supports the longest-running study of a dolphin population, spanning nearly 45 years (see Wells, 2014), and the 30 years of research in Shark Bay, Western Australia (see Connor and Krützen, 2015). In some cases, studying behavioural changes allows long-term impacts on animals to be inferred, but as mentioned above, other effects can be problematic without causing noticeable changes in behaviour (Lusseau and Bejder, 2007). Studies may use variation in animal response magnitude, existing population biology data, and multiple response variables to make such inferences (Lusseau and Bejder, 2007). Furthermore, short-term effects of stimuli should not be discounted, since they contribute to the understanding of animal behaviour and lay the foundation on which long-term studies are built.

It is possible for animals to display a single short-term response to a stimulus without experiencing detrimental long-term impacts (Stankowich, 2008). However, repeated exposure to a particular stimulus over time may not be benign (Beale, 2007). These short-term behavioural changes can cause shifts in habitat use, temporary displacement, reduced energy consumption, and interference with communication. Without proper management, short-term changes may lead to long-term consequences not only to individuals but also to entire populations (e.g. long-term habitat displacement or decline in population abundance; Bejder et al., 2006a).

Behavioural habituation is "the relative persistent waning of a response as a result of repeated stimulation, which is not followed by any kind of reinforcement" (Thorpe, 1963). In other words, habituation is characterized by behavioural responses that decrease over time. Habituation is an essential survival skill for a number of wildlife species living in crowded urban environments (Burger and Gochfeld, 1991). However, one of the most common misinterpretations in animal studies is to describe animals that do not visibly respond to a stimulus as "habituated". Although the outward response decreases, animals may still respond in more discreet ways that are not easily detectable (as noted above) and these effects may become problematic with repeated exposure. For example, Humboldt penguins, *Spheniscus humboldti*, showed a significant increase in heart rate in the presence of humans, and a subsequent decrease in breeding success, with no observable behavioural reactions (Ellenberg et al., 2006).

Sensitization, as opposed to habituation, involves increased behavioural reactions to stimuli after exposure. At the individual level, behavioural responses increase over time when a repeated stimulus has consequences (Bejder et al., 2009). For example, hunted populations of artiodactyls, the closest related group of living land mammals to dolphins, were described as increasing responses (i.e. fleeing) to humans over time, whereas non-hunted populations did not (Stankowich, 2008).

Relative to habituation and sensitization, when individuals become habituated to a stimulus, their tolerance levels increase and when they become sensitized to a stimulus, their tolerance levels decrease and they react more often (Bejder et al., 2009). Selective tolerance also has been described in which a stimulus may elicit strikingly different behaviours based on species and context (e.g. season, age class, and group composition). For example, individual New Zealand sea lions, *Phocarctos hookeri*, are tolerant to human approach when hauled out on land, but in the water, respond intolerantly (display aggression) to approach by human kayakers (Higham and Shelton, 2011). Similarly, in one population of common bottlenose dolphins, groups of males show a higher avoidance response to vessel activity than do groups of females with offspring (Higham and Shelton, 2011). In terms of naturally occurring factors, animal behaviour varies based on life history traits, age, sex, group size and composition, nutritional condition, reproductive status, and intrinsic tolerance (Beale and Monaghan, 2004; Bejder et al., 2009; Blumstein, 2006; Constantine, 2001; Higham and Shelton, 2011; Lusseau et al., 2003; Stankowich, 2008).

3. ANTHROPOGENIC ACTIVITIES AFFECTING MARINE MAMMAL BEHAVIOUR

From an evolutionary perspective, anthropogenic activities in the marine environment are a recent occurrence and cetaceans may instinctually respond in the same manner to human presence that they would to predators (Bejder et al., 2009; Frid and Dill, 2002; Lima and Dill, 1990). Examples of human-generated stimuli that might be perceived as threatening include unfamiliar movements or sounds not previously encountered at the individual level or in the phylogenetic history (Dill, 1974; Frid and Dill, 2002). Short-term effects of human-generated activity on behaviour have been demonstrated across a broad suite of odontocete, or toothed whale, taxa and, in some cases, resemble responses to potential predators. Odontocetes have exhibited alterations in respiration rate in response to acoustic alarms (Kastelein et al., 2006), and alterations in linearity (i.e. swimming direction) and swimming speed in response to marine vessel traffic and dolphin-based tourism (Dans et al., 2008; Lundquist et al., 2012; Lusseau, 2006; Nowacek et al., 2001; Stensland and Berggren, 2007; Williams et al., 2002). Reduced or disrupted foraging, resting, and/or socializing bouts have also been documented relative to anthropogenic activity (Christiansen et al., 2010; Constantine et al., 2004; Dans et al., 2008; Lusseau et al., 2009; Stockin et al., 2008; Visser et al., 2011; Williams et al., 2006).

Here, we describe several potential human-generated stressor stimuli in the marine environment and marine mammal behavioural responses.

3.1 Marine Vessel Traffic

A variety of responses by marine mammals have been described relative to marine vessel traffic. Some dolphins respond to vessel activity by fleeing, including diving for longer durations or swimming away, or avoiding areas with vessel traffic all together (Higham and Shelton, 2011). Common bottlenose dolphins off Clearwater, Florida (Gulf of Mexico), avoided important foraging areas when motorboat traffic was high (Allen and Read, 2000). If adjacent vessel-free areas are of equal quality and require minimal energetic costs for animal movement, displacement from a habitat might not be of concern (Nowacek et al., 2007). However, avoidance behaviour in response to external stimuli (Beale, 2007) may prevent animals from exploiting a resource-rich environment or from utilizing their entire home range (Whittaker and Knight, 1998). Continued vessel activity may cause long-term displacement of some animals to a particular area, leading to a decrease in abundance in the local population (Lusseau, 2004). Although a decline in abundance may not have a negative outcome for large and genetically diverse populations, it may have biologically significant consequences for small or resident cetacean populations (Bejder et al., 2006b).

3.2 Dolphin-Based Tourism

Commercial marine mammal tourism has often been considered a benign and sustainable activity, especially as an alternative to directed hunting. However, the quick growth of the industry on a global scale in recent decades (Christiansen and Lusseau, 2014; O'Connor et al., 2009) has raised questions about potential effects on target populations. Research shows that tourism can affect animal behaviour, with broader biological and ecological implications (Bejder et al., 2006a). For example, a large scale avoidance of tourism vessels in Shark Bay, Australia, coincided with a decline in abundance of bottlenose dolphins (*T. cf. aduncus*) in a previously used habitat (Bejder et al., 2006b). The decline was partially attributed to displacement of those dolphins most sensitive to the vessel activity. Boat-based tourism was also implicated in long-term shifts in habitat use by common bottlenose dolphins off New Zealand (Lusseau, 2004). When boat frequency increased, dolphins switched from local avoidance to avoiding the area all together. However, this habitat shift may not have been due entirely to tourism (see end of this section).

In various locations around the world, tourism includes hand feeding dolphins from vessels and from shore. In such cases of wildlife food provisioning, animals may become dependent on easily obtained food sources that alter the quality of their natural diet and may become food-aggressive toward humans that do not offer food (Whittaker and Knight, 1998), creating negative interactions for both wildlife and humans. In Monkey Mia, Western Australia, calves of provisioned female bottlenose dolphins (*T. cf. aduncus*) received reduced care and increased mortality compared to those of non-provisioned females (Mann and Kemps, 2003; Mann et al., 2000). Calf survivorship increased with improved regulations, but calf behavioural development was affected and so the long-term impacts of provisioning, in this case, remain unclear (Foroughirad and Mann, 2013).

3.3 Cetacean–Fishery Interactions

Cetaceans are often attracted to fishing vessels where acquisition of concentrated prey sources may be facilitated (Fertl and Leatherwood, 1997). Many cetacean species exploit gillnets, commercial trawling nets, and fishing hooks that congregate food sources and make them easier to access than naturally occurring, patchily distributed prey (Fertl and Leatherwood, 1997; Gilman et al., 2007; Jefferson and Curry, 1994; Perrin et al., 1994). Attraction behaviour may seemingly benefit animals in the short term, yet close proximity to fisheries can have negative consequences for both dolphins and humans (Fertl, 2009; Garrison, 2007; Hung and Jefferson, 2004; Read, 2008). For example, an increased risk of dolphin injury or mortality can result from vessel strike (Dwyer et al., 2014; Parsons and Jefferson, 2000), incidental bycatch (Parsons and Jefferson, 2000; Sutaria and Jefferson, 2004; Wade, 1995), or aggressive interactions with sharks when multi-species aggregations occur (Orams, 2002; Paterson, 1990). Furthermore, dolphins may interfere with commercial fishing operations by taking fish from gear, disturbing prey, or damaging fishing equipment (Fertl, 2009). In some cases, real or perceived damage caused by dolphins elicits hostile responses from fishers including lethal and non-lethal deterrence methods (Fertl, 2009).

3.4 Noise Pollution

High-energy sound source levels associated with modern human activities, such as construction pile driving, explosive detonations, industrial seismic exploration, and military active sonar, have the potential to impair dolphin hearing and equilibrium directly via tissue damage and associated hearing loss (Ketten, 2004; NRC, 2005; Weir et al., 2011). These activities can also elicit behavioural responses by marine mammals such as changes in activity state or exposure avoidance (Southall et al., 2008). For example, Blainville's beaked whales, *Mesoplodon densirostris*, have demonstrated behavioural responses including disrupted foraging bouts and avoidance of impact areas of up to tens of kilometres (Tyack et al., 2011). Active sonar has also been implicated in behavioural responses that lead to injury or death, including altered dive profiles and stranding events of beaked whales (Cox et al., 2006). The effects of lower-energy underwater sound (e.g. chronic engine noise, dredging, or seismic/sonar further from the source) may not directly impair animals physically, although temporary threshold shifts in hearing have been documented to occur in common bottlenose dolphins after noise exposure (Finneran et al., 2005), but shifts in behaviour may occur. In addition to shifts in behaviour, noise associated with human activities can displace prey species, and mask acoustics important in social organization, navigation, and prey acquisition (Jefferson et al., 2009; Jensen et al., 2009; Lesage et al., 1999; Richardson and Würsig, 1997). Masking may occur when sounds of interest (e.g. vocalizations for communication and echolocation for prey detection) are unclear or muted by interfering sounds, in this case, human-generated underwater noise (Würsig and Richardson, 2009).

3.5 Habitat Loss and Degradation

Large-scale ecosystem alterations in marine environments, including direct loss of habitat and degradation of existing habitat, represent a world-wide threat to coastal dolphins. Ecosystem alterations are especially problematic in areas where multiple industrial projects occur in core habitats where marine mammals engage in biologically significant activities such as foraging, mating, and calf rearing (Bejder et al., 2009; Würsig, 1989). Habitat loss can result directly from coastal development (e.g. artificial land formation, port development, and wind farm development), and habitat degradation can result from activities such as dredging, trawling, and mining, which may result in pollution, eutrophication, sedimentation, noise and oil spills (Weir and Pierce, 2013). An extreme example of a habitat alteration impact on a marine mammal species is the baiji, *Lipotes vexillifer*, that was recently declared extinct, due largely to intensive habitat degradation (Turvey et al., 2007).

3.6 Cumulative Effects

Many threats act cumulatively on marine mammal populations, intensifying the overall effects and making it difficult to isolate the most critical threats (Jefferson et al., 2009; Weir et al., 2011). It is inherently difficult to study anthropogenic effects on animal behaviour in natural settings because of confounding factors, both natural and artificial, in dynamic environments where multiple activities occur simultaneously. Some activities may contribute numerous different stressors into the environment. High-speed vessels and marine construction, for example, take up physical space and create a noisy environment. Cumulative effects may threaten population viability at a different rate relative to any single activity acting alone, and care should be taken when interpreting behavioural responses (Bejder et al., 2009). For example, directed killing and habitat degradation cumulatively impacted the population status of short-beaked common dolphins, *Delphinus delphis*, in the Adriatic Sea (Bearzi et al., 2004). In the instance of common bottlenose dolphins in New Zealand, where tourism was implicated in avoidance of certain areas (Lusseau, 2004), other human activities may have played an equal or larger part in this shift. Thermal intolerance has been suggested as a driver of declining calf survival rate, in this case, where an increase in freshwater inflow from a dam, and a subsequent decrease in overall water temperature, may have led to thermal stress in calves (Currey et al., 2009).

4. ANTHROPOGENIC EFFECTS ON HUMPBACK DOLPHINS

Humpback dolphins are coastal animals that use shallow, near-shore habitats and estuarine environments and thus, may frequently be exposed to recreational and commercial activities (Jefferson et al., 2008, 2009; Van Waerebeek et al., 2004; Würsig, 1989). This is especially true for humpback dolphins that live near densely populated urban cities and industrialized coastal areas, such as Hong Kong; Goa, India; and Queensland, Australia. When overlap occurs in time and space, humans infringe on humpback dolphin activities that are biologically important, including foraging behaviours, movement patterns, and social/mating behaviours (Berger-Tal et al., 2011).

4.1 Marine Vessel Traffic

Several examples of avoidance relative to vessel activity are described for *Sousa*. Off South Africa, near-shore motorized boat activity appears to be highly disturbing to Indian Ocean humpback dolphins, *Sousa plumbea*, where dolphins actively avoid fast-moving boats and have longer dive durations, changes in swimming direction, and extended underwater movement away from vessels (Karczmarski et al., 1997). Similar avoidance behaviour by *S. plumbea* was described in the Arabian Gulf, including deep diving, group dispersal, and movement away from vessels (Pilleri and Gihr, 1974). Off South Africa, females were observed moving between approaching boats and their calves, and at times joining other female-calf pairs (Karczmarski et al., 1997). *Sousa plumbea* avoid parts of eastern Maputo Bay, Mozambique, and Port Elizabeth, South Africa, possibly due in part to the considerable human presence and heavy use of powerboats in these areas (Karczmarski et al., 1997, 2000). Furthermore, *S. plumbea* in most areas off South Africa have not been observed to ride the bow or stern of vessels; a behaviour common to many other delphinid species, including bottlenose dolphins, in that area (Karczmarski et al., 1997). However, wave-riding is not common in humpback dolphins (Parra and Ross, 2009). Off India, little is known about the behaviour and ecology of *Sousa* spp., but, avoidance behaviour has been documented in the Indus Delta region relative to boat traffic (Karczmarski et al., 1997). In the Ashtamudi Estuary, off the southwestern coast of India, heavy-motorized boat and trawler traffic, which access fishing harbours on either side of the estuary, has been described as a great threat to *S. plumbea* (Bijukumar and Smrithy, 2012). Avoidance responses to high-speed vessels by Indo-Pacific humpback dolphins, *S. chinensis*, off Sabah, Malaysia, and Hong Kong are described (Hashim and Jaaman, 2011). Off Hong Kong, dolphins flee in response to vessels moving at high speeds, and longer dive durations have been observed during periods of heavy vessel traffic (Ng and Leung, 2003).

Two studies on *Sousa* and vessel traffic found that there were no reactions by dolphins to slow-moving vessels, such as large cargo vessels emitting primarily lower frequency sound (relative to the sounds produced by faster moving vessels). *Sousa chinensis* off Hong Kong and *S. plumbea* in the Indus Delta displayed no observable behavioural changes when vessels operated in the vicinity at slow speeds (Ng and Leung, 2003; Pilleri and Gihr, 1974). Vessels in this context may be more predictable and therefore perceived as less threatening. Off Hong Kong, heavy vessel traffic in *S. chinensis* habitat is typical and dolphins appear accustomed to their presence (Jefferson et al., 2008; and see Würsig et al., 2016). It is possible that the dolphins may be more tolerant of vessels or perhaps more subtle reactions by dolphins were not detected (e.g. heightened vigilance) or not examined (e.g. increased heart rate). Additionally, dolphins spend the majority of time underwater, making detection of small behaviour changes challenging when observing from above-water only.

4.2 Dolphin-Based Tourism

Boat-based tourism activities that focus on, or include, humpback dolphins occur in several regions, including off Zanzibar (Stensland and Berggren, 2007; Stensland et al., 2006), Goa, India (Parsons, 1998; and see Sutaria et al., 2015), Sri Lanka (Nanayakkara et al., 2014), Hong Kong (Ng and Leung, 2003), Sanniang Bay, Guangxi Province, PR China (Wenshi et al., 2006), and there are a few operators in both Moreton Bay and Hervey Bay, Australia (Parra et al., 2004). Several small-scale community-driven operations also exist off the west coast of Africa including Gambia (Van Waerebeek et al., 2003) and Benin (Van Waerebeek et al., 2002).

Off the west coast of Lantau Island, Hong Kong, an important foraging area for *S. chinensis* where female-calf pairs are abundant, independent dolphin-watching tourism vessels operate on a daily basis (Hung, 2014). *Sousa chinensis* behavioural changes have been recorded in the presence of tour boats (Piwetz et al., 2012; Wong et al., 2015), including boat approaches by juvenile dolphins (Ng and Leung, 2003). This perceived positive response toward boats still constitutes a change in natural behaviour and may have adverse consequences. Although behavioural responses to tourism activities may be considered "less critical" than some other human-driven threats (Jefferson and Rosenbaum, 2014), it may be particularly problematic in dolphin "hotspots" (e.g. Tai O), especially those utilized by female-calf pairs (Jefferson et al., 2012). Calves are not fully developed, and therefore are physiologically limited in their ability to swim and dive (Mann and Smuts, 1999), which may result in greater vulnerability to high-speed dolphin-targeted vessel activity.

In Menai Bay, on the southwest side of Zanzibar Island, *S. plumbea* occur year round and are often close to shore. Modern motorized commercial dolphin tourism began in Menai Bay in the 1990s and has grown to receive thousands of guests each month (Amir and Jiddawi, 2001). To our knowledge, research has not been published on tourism effects on *S. plumbea* behaviour off Zanzibar, but Indo-Pacific bottlenose dolphins, *T. aduncus*, in the same area showed reduced foraging, resting, and socializing behaviours, and increased travelling behaviour in the presence of tour boats, indicating that less time was spent engaged in biologically important activities (Christiansen et al., 2010). The perception by some local marine mammal scientists and tourism personnel is that *S. plumbea* are harassed during tourism-based boat approaches where dolphin groups are chased for viewing opportunities (Gautam, 2010). Off Goa, India, *S. plumbea* are described as wary of tourist boats that have also been reported as chasing dolphins (Sutaria and Jefferson, 2004).

In Tin Can Bay, Queensland, Australia, tourism includes hand feeding Australian humpback dolphins, *Sousa sahulensis*, from shore. Wildlife tourism is considered a threat to *S. sahulensis* in some areas of Australia (Department of the Environment and Heritage Protection, DEHP, 2005), but data are lacking.

4.3 Cetacean–Fishery Interactions

When recreational or commercial fishing activities aggregate prey species, humpback dolphins are sometimes attracted to facilitated feeding opportunities. For example, at least some Australian humpback dolphins, *S. sahulensis*, off Australia feed in association with fishing trawlers as a major source of food (Parra and Ross, 2009), and *S. chinensis* off Malaysia have been documented following commercial trawlers (Hashim and Jaaman, 2011). *Sousa chinensis* off Hong Kong fed near commercial pair and hang trawlers for hours at a time when operations were still active (Hung, 2008; Jefferson, 2000). Humpback dolphins actively approached and followed trawlers, in significantly larger schools than those not pursuing trawlers, likely to access prey that were concentrated into a small area or stirred up by fishing nets (Hung and Jefferson, 2004; Jefferson, 2000; Ng and Leung, 2003). There appeared to be inter-individual selection for this type of feeding off Hong Kong, where some dolphins associated with trawlers more than others (Jefferson, 2000). Furthermore, *S. chinensis* groups with calves were more likely to feed near fishing vessels than those without calves (Hung, 2008) which may reflect the energetic needs of breeding females (Fertl and Leatherwood, 1997). A similar distinction occurs in Moreton Bay, Australia, where socially segregated sympatric communities of Indo-Pacific bottlenose dolphins exist and members of only one community obtain food from trawlers (Chilvers and Corkeron, 2001). Although there may be a short-term benefit to feeding near fishing boats, the risks of injury and mortality from net entanglement increase (Jefferson, 2000), which may ultimately be damaging to vulnerable populations, especially if sexually mature females are disproportionately represented (Dungan et al., 2012). Moreover, bottom trawling activities negatively affect the marine environment directly via destruction of benthic communities and post-fishing mortality and indirectly by altering local trophic relationships (Jones, 1992). Trawling activities are currently banned in Hong Kong, but are still active in other areas, including Australia and Malaysia (Hashim and Jaaman, 2011; Parra and Ross, 2009).

Off South Africa and Australia, anti-shark nets near human swimming areas attract fish aggregations, and therefore humpback dolphins such as

S. plumbea and *S. sahulensis* (Atkins et al., 2013; Cockcroft, 1990; Paterson, 1990). Both of these humpback dolphin species are considered vulnerable to incidental mortality in these types of nets. Humpback dolphin mortalities associated with entanglement are well documented and have been problematic off South Africa for decades, emphasizing how human-made structures that attract fishes, and therefore attract dolphins, can be life threatening to humpback dolphins (Atkins et al., 2013; Dudley and Cliff, 1993).

In addition to shifts in behavioural state associated with attraction to fishing-related activities, ranging and movement patterns may also be affected. For example, individual *S. chinensis* ranges off Hong Kong may have been influenced by the distribution of fishing vessels, whereby individuals with greater fishing–boat associations had larger ranges (Hung and Jefferson, 2004). However, it is not clear if larger ranges had an overall positive, neutral, or negative effect, or if the larger ranges actually exceeded natural ranging patterns. Humpback dolphins also avoid certain types of commercial fishing activities in some areas. In the fishing harbours of Neendakara and Sakthikulangara, within the Ashtamudi estuary of India, dolphins have been observed actively avoiding trawlers (Bijukumar and Smrithy, 2012), which is quite different from *S. chinensis* in Hong Kong and Malaysian waters and *S. sahulensis* off Australia (Hung, 2008; Jefferson, 2000; Parra and Ross, 2009; Piwetz et al., 2012). *Sousa chinensis* of the eastern Taiwan Strait also have not been seen feeding in association with trawlers (Wang et al., 2007b). *Sousa plumbea* along the Chennai coast of India are often observed near-shore seiners, likely where fish congregate, but they actively avoid seiners by swimming around them (Muralidharan, 2013). Coastal fisheries in the form of shellfish farming may also alter *Sousa* movement patterns. For example, dense oyster mariculture displaced *S. chinensis taiwanensis* of the Taiwan Strait, in littoral waters of Western Taiwan (Wang et al., 2007a; and see Wang et al., 2016).

4.4 Noise Pollution

Sousa often inhabit murky estuarine waters where vision is limited, making sound production and reception especially important in social communication, navigation, and foraging behaviour (Parra and Ross, 2009). It also has been suggested that *Sousa* in some areas use passive listening to locate sound-producing prey (Barros and Cockcroft, 1999; Parra and Ross, 2009), and such efficient location of prey may be masked by anthropogenic noise.

Throughout their range, the most common noise to which *Sousa* are exposed originates from vessels. Boat noise is largely dependent on the size of the engine and speed of the vessel (Spence, 2007). High-speed vessels, which can produce frequencies within *Sousa* sound production ranges, appear to cause more acoustic disturbance to *S. chinensis* off Hong Kong, due to masking, than vessels travelling at slower speeds with lower frequency output (Ng and Leung, 2003). This may be particularly problematic for *S. chinensis* off Hong Kong (and other parts of coastal south China) where high-speed ferries contribute considerable noise into the environment. Furthermore, fast-moving vessels can approach dolphins quickly, rapidly elevating sound pressure levels, which could elicit a startle response or other behavioural reaction (Sims et al., 2012). Little research has been conducted on *Sousa* behaviour relative to small vessel engine noise, although *S. sahulensis* groups off Australia increased their rate of whistle production directly following motor boat movement through their area, particularly when calves were present (Van Parijs and Corkeron, 2001). This suggests that females with calves attempt to maintain close contact and may have an increased perception of risk associated with vessel noise, and thus may be more sensitive to vessel presence (Van Parijs and Corkeron, 2001). In one region of South Africa, where a major harbour thoroughfare crosses known *S. plumbea* feeding habitat, sounds produced by consistent vessel passage may mask vocalizations important in prey detection and acquisition (Keith et al., 2013).

Species of *Sousa* in several regions are subject to increasing sound exposure associated with large-scale coastal development. Acute noise produced from industrial activities has been suggested as a major threat to *S. chinensis* off Hong Kong (Jefferson et al., 2009; Würsig and Greene, 2002; Würsig et al., 2000) and in the eastern Taiwan Strait (Wang et al., 2007a). *Sousa chinensis* disappeared from an area off Hong Kong when percussive piling was initiated during one construction project (Beasley, 1996) and showed behavioural changes and temporary movement away from another area after percussive pile-driving activity was conducted (Würsig et al., 2000). Off Hong Kong, construction of the Hong Kong–Zhuhai–Macau Bridge (which is designed to be 35 km in total length, including a 6 km underwater tunnel) is underway and extends directly through *S. chinensis* habitat (Chen et al., 2010). Construction involves use of the OCTA-KONG, the largest vibration hammer in the world, and *S. chinensis* are susceptible to complete auditory masking of whistles within several hundred metres of operation (Wang et al., 2014). Current and planned activities in

South Africa include pipeline construction and onshore wind turbine operation. Eight wind turbines are planned for assembly within 3 km of the sea (Koper and Plön, 2012) and are of concern for near-shore *S. plumbea* due to the noise vibrations associated with construction and operation. Construction and development off portions of West Africa (e.g. pile driving, explosives, hydrocarbon exploration, dredging, and ship mooring) also have the potential to acoustically disturb and displace Atlantic humpback dolphins, *Sousa teuszii* (Weir et al., 2011).

4.5 Habitat Loss and Degradation

Humpback dolphins of the genus *Sousa* display limited habitat "flexibility" relative to water depth and distance to shore and are especially vulnerable to the growing threats of coastal habitat loss and degradation (Hung and Jefferson, 2004; Jefferson and Karczmarski, 2001; Jefferson et al., 2009; Karczmarski et al., 2000; Keith et al., 2013; Wang et al., 2007a; Weir and Pierce, 2013). The *Sousa* propensity for inhabiting estuarine water in some areas creates fragmentation and isolation among populations, making it difficult to relocate to suitable habitat (Karczmarski et al., 2000). *Sousa teuszii*, for example, use narrow, specialized niches and loss of near-shore habitat may inhibit group mobility (Weir et al., 2011).

Habitat alteration of inshore areas is considered among the greatest threats to *S. chinensis*, in which energetic opportunities are impacted when historical foraging grounds or nursery grounds of prey species are destroyed (Karczmarski, 2000). In the Pearl River Estuary, Hong Kong, continuous large-scale urban development contributes to loss and degradation of *S. chinensis* habitat (Chen et al., 2011; Hung, 2008; Jefferson and Hung, 2004; Jefferson et al., 2009). Extensive land development from the sea occurs (Jefferson et al., 2009) and, if formally approved, a considerable land-formation project is planned to support a three-runway system at the Hong Kong International Airport (Environmental Investigation Agency, EIA, 2012) which will take years to complete (Wong, 2012). This modification includes dredging activities to supply material for land formation, which not only modifies habitat but also contributes additional sound into the environment. Although some populations of *S. chinensis* (e.g. in Hong Kong Pearl River Estuary) have been somewhat resilient to alterations in habitat, others may be approaching the limits of population viability (Jefferson and Rosenbaum, 2014). The eastern Taiwan Strait subspecies, for example, is a small population and uses shallow near-shore habitat, wherein heavy and rapid coastal degradation is a serious conservation concern listed as one of the top five major threats to this population (Wang et al., 2007a,b).

Habitat destruction and degradation are described as a threat to *S. teuszii* off West Africa, in Senegal (Van Waerebeek et al., 2004) and Angola (Weir et al., 2010). *Sousa sahulensis* off western Australia are vulnerable to habitat degradation, particularly relating to environmental change (Brown et al., 2012; and see Hanf et al., 2016), which has also been suggested as one potential factor in the absence of *S. teuszii* off Ghana (Van Waerebeek et al., 2009). Localized extinction or displacement has not been substantiated, but is an important risk to consider for many populations of *Sousa*. Alternatively, *S. sahulensis* off eastern Australia, in Moreton Bay, are abundant and appear to co-exist with extensive coastal development, though a proactive management approach guided by the precautionary principle is recommended to provide protection to this native species (Chilvers et al., 2005).

4.6 Cumulative Effects

Many humpback dolphin populations occur in dynamic environments that overlap with human activities, often where similar resources, such as prey items, are exploited (Fertl and Leatherwood, 1997). Activities should be identified and assessed collectively in order to minimize overall effects on humpback dolphin populations. Dungan et al. (2011) recently highlighted the need for addressing cumulative effects of human activity on *S. c. taiwanensis* in the eastern Taiwan Strait. A spatiotemporal analysis of data on dolphin distribution and mapped marine impacts dating back to 1996 indicated that the cumulative effects of human activities on *S. chinensis* off Hong Kong did not appear to impact dolphin density distribution on the North Lantau community as a whole; however, localized areas of dolphin density decline correlated with the introduction of high-speed ferry routes (Marcotte et al., 2015). One particularly sensitive social group within the *Sousa* genus is female-calf pairs. Female *Sousa* with offspring have been observed positioning themselves between their calves and boats and increasing whistle rate relative to noise masking (Karczmarski et al., 1997; Van Parijs and Corkeron, 2001). Females with young calves may use smaller nursery areas (relative to prey availability or predator avoidance) due to the energetic requirements of lactation and may increase site fidelity during nursing periods (Karczmarski, 1999; Karczmarski and Cockcroft, 1999). This can become problematic when nursery areas overlap with intensive habitat alteration. Vessel traffic, noise pollution, and habitat degradation may collectively intensify effects of exposure on sexually mature females with offspring, which could be detrimental to small or declining populations.

5. CONCLUSIONS AND FUTURE RESEARCH

To resolve human–wildlife conflicts, we must first understand how animals respond to different types of activity (Beale, 2007). Though several studies have emerged in recent years, there is a paucity of research focusing on anthropogenic effects on humpback dolphins from a behavioural ecology perspective. Overall, qualitative and quantitative data have been lacking on *Sousa* throughout their range. Incorporating variable behaviour data-gathering platforms is needed, especially in coastal areas of high human density that experience a great deal of activity, such as Hong Kong; Goa, India; and parts of Australia, to better address cumulative input. Furthermore, collaborative research that integrates behaviour, physiology, and conservation for an inter-disciplinary approach can better inform management for more successful mitigation outcomes (Cooke et al., 2014).

In many regions where dolphin-based tourism is conducted almost no regulatory or management frameworks exist, and especially lacking are plans with long-term sustainable management in mind (Higham et al., 2009). Even when regulations are in place, management and enforcement are often absent or ineffective (Bejder et al., 2006a), particularly in remote locations. Official codes of conduct for tour operators have been established in Zanzibar, which vary by region, but codes are not mandatory and the majority of operators do not follow best-practise recommendations (Christiansen et al., 2010). To our knowledge, no mandatory regulations exist in India or Hong Kong, two of the regions most heavily populated by humpback dolphin species. In Hong Kong, a voluntary code of conduct is recommended for vessels approaching and viewing marine mammals (Parra and Ross, 2009), but, as with many voluntary codes of conduct, compliance is lacking (Scarpaci et al., 2004). At least some tourism boats in Hong Kong fail to observe recommendations and exceed the maximum number of boats present with a dolphin group, exceed the maximum observation time (Hung, 2008), and enter and depart humpback dolphin habitat abruptly and at high speeds in excess of the recommended 10 knot maximum (Piwetz et al., 2012). National guidelines for commercial dolphin-watching have been established in Australia where state and territory governments regulate local industry, ranging from voluntary codes of conduct to official regulations (Department of the Environment and Heritage Protection, DEHP, 2005). Hand feeding in Tin Can Bay is regulated by the Minister for Environment and Heritage Protection (Department of the Environment and Heritage Protection, DEHP, 2013), with up to hundreds of visitors per day (Parra et al., 2004). Responsible wildlife viewing can stimulate local economies and promote public interest in humpback dolphin conservation in positive ways, but operations should be conducted with appropriate knowledge of *Sousa* behaviour to avoid disturbance (Karczmarski et al., 2000). Dolphin-based tourism should be carefully regulated and closely monitored to ensure that it is sustainable. In most regions, *Sousa* occur close to shore and land-based dolphin viewing has been suggested as a non-invasive alternative to boat-based viewing (Hung, 2008).

Several mitigation efforts have been applied to fisheries practises within humpback dolphin ranges including gear modification and restricted fishing zones. Off Angola, commercial trawlers are required to operate several km from the coast which may reduce interactions with humpback dolphins. However, artisanal fishing practises and coastal gillnet placement are considered a conservation concern in this area (Weir et al., 2011). Off Hong Kong, commercial trawling, including pair, stern, shrimp, and hang trawling, was banned in December 2012 (Hung, 2015). However, illegal trawling occurs sporadically (Agriculture and Conservation Department, AFCD, 2013) often times at night. Although enforcement is an issue, this ban is an excellent example of science informing management for improved humpback dolphin conservation. In the eastern Taiwan Strait, *S. c. taiwanensis* experience substantial interactions with fisheries and a ban on trammel nets, gillnets, and trawling has been suggested in this region (Slooten et al., 2013).

In general, data are lacking on humpback dolphin behavioural responses to both chronic and acute anthropogenic sound sources throughout their range and this is especially concerning in areas with large-scale development and vessel activity. A number of mitigation measures have been developed to minimize impacts of industrial noise in marine mammal environments. However, most approaches for acute sound sources include relatively small safety zones aimed at minimizing risk of direct physical injury with little consideration of behavioural responses at varying distances. Creative techniques have been developed in humpback dolphin habitat aimed to mitigate potential communication interference and behaviour modification. For example, off Hong Kong, a large rubber hose that produced a screen of air bubbles was designed to reduce sound transmission during pile-driving activity in *S. chinensis* habitat. The bubble curtain effectively attenuated sounds in high frequency ranges. During pile driving, concomitant humpback dolphin behavioural data were collected from shore using quantitative methods (Würsig et al., 2000). This type of real-time data collection during noise-intensive construction is an excellent example of future research needs.

One of the biggest challenges to *Sousa* conservation is protecting and maintaining suitable high-quality habitat in developing and densely populated regions that overlap with humpback dolphin coastal ranges (Parra and Ross, 2009). Few data exist on *Sousa* displacement due to habitat loss, and this lack of information is unfortunate, given the increasingly degraded environments many populations occupy. One potentially effective tool for protecting portions of *Sousa* habitat is proper designation and management of Marine Protected Areas (MPAs) in known dolphin "hotspots", especially where site fidelity has been described. Ideally, MPAs should be truly protected with effective enforcement and restriction of development, fisheries, and other vessel activity. MPA boundaries, as well as travel corridors between locations, should be developed and managed with knowledge of humpback dolphin movement patterns and habitat use. Furthermore, research should be ongoing, with adaptive management schemes in place, in order to modify boundaries according to shifts in habitat.

Populations of *Sousa* spp. are regionally endemic with small geographic ranges, some of which show evidence of restricted gene flow (Mendez et al., 2011); thus, research, conservation efforts, and management should focus not only on the species but also on separate populations (Reeves, 2009). The more that local researchers are involved in anthropogenic studies on *Sousa*, particularly in smaller rural areas, the greater the chance for increased local conservation awareness and initiatives that could ultimately reduce human-generated disturbance and sustain individual populations. Creative technological advances and frameworks for behaviour-based data collection and analyses have emerged in recent years (Mann and Würsig, 2014) and should be applied to aid in conservation efforts. Humans will continue to develop in coastal areas, forever altering the natural seascape in which humpback dolphins occur, and we should actively work to mitigate the effects our presence has on wildlife.

ACKNOWLEDGEMENTS

We thank T.A. Jefferson for inviting us to present this information. This chapter benefited greatly from reviews by B.E. Curry, R.W. Davis, D. Fertl, J.M. Packard, D.L. Smee, and an anonymous reviewer. We thank them all.

REFERENCES

Agriculture and Conservation Department (AFCD), 2013. Joint Operations Conducted Against Illegal Fishing. Agriculture, Fisheries and Conservation Department of Hong Kong SAR Government. Press Release [online]. Copyright owned by the Government of the Hong Kong Special Administrative Region. http://www.info.gov.hk/gia/general/201310/11/P201310110607.htm.

Allen, M.C., Read, A.J., 2000. Habitat selection of foraging bottlenose dolphins in relation to boat density near Clearwater, Florida. Mar. Mamm. Sci. 16 (4), 815–824.

Amir, O.A., Jiddawi, N.S., 2001. Dolphin tourism and community participation in Kizimkazi village, Zanzibar. WIOMSA Book Series, 1, pp. 551–560.

Atkins, S., Cliff, G., Pillay, N., 2013. Humpback dolphin bycatch in the shark nets in KwaZulu-Natal, South Africa. Biol. Conserv. 159, 442–449.

Barros, N.B., Cockcroft, V.G., 1999. Prey resource partitioning between Indo-Pacific humpback dolphins (*Sousa chinensis*) and bottlenose dolphins (*Tursiops truncatus*) off South Africa: competitive exclusion or mutual tolerance. In: Abstracts, 13th Biennial Conference on the Biology of Marine Mammals, Wailea, Maui.

Beale, C.M., 2007. The behavioral ecology of disturbance responses. Int. J. Comp. Psychol. 20 (2), 111–120.

Beale, C.M., Monaghan, P., 2004. Human disturbance: people as predation-free predators? J. Appl. Ecol. 41 (2), 335–343.

Bearzi, G., Holcer, D., Notarbartolo di Sciara, G., 2004. The role of historical dolphin takes and habitat degradation in shaping the present status of northern Adriatic cetaceans. Aquat. Conserv. Mar. Freshwat. Ecosyst. 14 (4), 363–379.

Beasley, I., 1996. Land-Based Observations of Indo-Pacific Hump-Backed Dolphins, in Hong Kong Waters. University of Otago, New Zealand.

Bejder, L., Samuels, A., Whitehead, H., Gales, N., 2006a. Interpreting short-term behavioural responses to disturbance within a longitudinal perspective. Anim. Behav. 72 (5), 1149–1158.

Bejder, L., Samuels, A., Whitehead, H., Gales, N., Mann, J., Connor, R., Heithaus, M., Watson-Capps, J., Flaherty, C., Krutzen, M., 2006b. Decline in relative abundance of bottlenose dolphins exposed to long-term disturbance. Conserv. Biol. 20 (6), 1791–1798.

Bejder, L., Samuels, A., Whitehead, H., Finn, H., Allen, S., 2009. Impact assessment research: use and misuse of habituation, sensitisation and tolerance in describing wildlife responses to anthropogenic stimuli. Mar. Ecol. Prog. Ser. 395, 177–185.

Berger-Tal, O., Polak, T., Oron, A., Lubin, Y., Kotler, B.P., Saltz, D., 2011. Integrating animal behavior and conservation biology: a conceptual framework. Behav. Ecol. 22, 236–239.

Bijukumar, A., Smrithy, R., 2012. Behaviour of Indo-Pacific humpback dolphin, *Sousa chinensis* (Osbeck) in the Ashtamudi estuary, southwest coast of India. J. Mar. Biol. Assoc. India 54 (2), 1–10.

Blumstein, D.T., 2006. Developing an evolutionary ecology of fear: how life history and natural history traits affect disturbance tolerance in birds. Anim. Behav. 71 (2), 389–399.

Brown, J.S., 1999. Vigilance, patch use and habitat selection: foraging under predation risk. Evol. Ecol. Res. 1 (1), 49–71.

Brown, A., Bejder, L., Cagnazzi, D., Parra, G.J., Allen, S.J., 2012. The North West Cape, Western Australia: a potential hotspot for Indo-Pacific humpback dolphins *Sousa chinensis*? Pac. Conserv. Biol. 18 (4), 240–246.

Burger, J., Gochfeld, M., 1991. Human distance and birds: tolerance and response distances of resident and migrant species in India. Environ. Conserv. 18 (2), 158–165.

Chen, T., Hung, S.K., Qiu, Y., Jia, X., Jefferson, T.A., 2010. Distribution, abundance, and individual movements of Indo-Pacific humpback dolphins (*Sousa chinensis*) in the Pearl River Estuary, China. Mammalia 74 (2), 117–125.

Chen, T., Qiu, Y., Jia, X., Hung, S.K., Liu, W., 2011. Distribution and group dynamics of Indo-Pacific humpback dolphins (*Sousa chinensis*) in the western Pearl River Estuary, China. Z. Säugetierkd. 76 (1), 93–96.

Chilvers, B.L., Corkeron, P.J., 2001. Trawling and bottlenose dolphins' social structure. Proc. R. Soc. Lond. B Biol. 268 (1479), 1901–1905.

Chilvers, B.L., Lawler, I.R., Macknight, F., Marsh, H., Noad, M., Paterson, R.A., 2005. Moreton Bay, Queensland, Australia: an example of the co-existence of significant marine mammal populations and large-scale coastal development. Biol. Conserv. 122 (4), 559–571.

Christiansen, F., Lusseau, D., 2014. Understanding the ecological effects of whale-watching on cetaceans. In: Higham, J., Bejder, L., Williams, R. (Eds.), Whale-Watching: Sustainable Tourism and Ecological Management. Cambridge University Press, Cambridge, UK, pp. 177–192.

Christiansen, F., Lusseau, D., Stensland, E., Berggren, P., 2010. Effects of tourist boats on the behaviour of Indo-Pacific bottlenose dolphins off the south coast of Zanzibar. Endanger. Species Res. 11 (1), 91–99.

Cockcroft, V.G., 1990. Dolphin catches in the Natal shark nets, 1980 to 1988. S. Afr. J. Wildl. Res. 20 (2), 44–51.

Connor, R.C., Heithaus, M.R., 1996. Approach by great white shark elicits flight response in bottlenose dolphins. Mar. Mamm. Sci. 12 (4), 602–606.

Connor, R.C., Krützen, M., 2015. Male dolphin alliances in Shark Bay: changing perspectives in a 30-year study. Anim. Behav. 103, 223–235.

Constantine, R., 2001. Increased avoidance of swimmers by wild bottlenose dolphins (*Tursiops truncatus*) due to long-term exposure to swim-with-dolphin tourism. Mar. Mamm. Sci. 17 (4), 689–702.

Constantine, R., Brunton, D.H., Dennis, T., 2004. Dolphin-watching tour boats change bottlenose dolphin (*Tursiops truncatus*) behaviour. Biol. Conserv. 117 (3), 299–307.

Cooke, S.J., Blumstein, D.T., Buchholz, R., Caro, T., Fernández-Juricic, E., Franklin, C.E., Metcalfe, J., O'Connor, C.M., St Clair, C.C., Sutherland, W.J., 2014. Physiology, behavior, and conservation. Physiol. Biochem. Zool. 87 (1), 1–14.

Cox, T.M., Ragen, T.J., Read, A.J., Vos, E., Baird, R.W., Balcomb, K., Barlow, J., Caldwell, J., Cranford, T., Crum, L., D'Amico, A., D'Spain, G., Fernández, A., Finneran, J., Gentry, R., Gerth, W., Gulland, F., Hildebrand, J., Houser, D., Hullar, T., Jepson, P.D., Ketten, D., MacLeod, C.D., Miller, P., Moore, S., Mountain, D.C., Palka, D., Ponganis, P., Rommel, S., Rowles, T., Taylor, B., Tyack, P., Wartzok, D., Gisiner, R., Mead, J., Benner, L., 2006. Understanding the impacts of anthropogenic sound on beaked whales. J. Cetac. Res. Manage. 7, 177–187.

Currey, R.J.C., Dawson, S.M., Slooten, E., Schneider, K., Lusseau, D., Boisseau, O.J., Haase, P., Williams, J.A., 2009. Survival rates for a declining population of bottlenose dolphins in Doubtful Sound, New Zealand: an information theoretic approach to assessing the role of human impacts. Aquat. Conserv. Mar. Freshwat. Ecosyst. 19 (6), 658–670.

Dans, S.L., Crespo, E.A., Pedraza, S.N., Degrati, M., Garaffo, G.V., 2008. Dusky dolphin and tourist interaction: effect on diurnal feeding behavior. Mar. Ecol. Prog. Ser. 369, 287–296.

Department of the Environment and Heritage Protection (DEHP), 2005. Australian National Guidelines for Whale and Dolphin Watching. Department of the Environment and Heritage Protection, Queensland, Australia.

Department of the Environment and Heritage Protection (DEHP), 2013. Nature Conservation and Other Legislation Amendment and Repeal Regulation (No. 1). Department of the Environment and Heritage Protection, Queensland, Australia.

Dill, L.M., 1974. The escape response of the zebra danio (*Brachydanio rerio*) I. The stimulus for escape. Anim. Behav. 22 (3), 711–722.

Dudley, S.F.J., Cliff, G., 1993. Some effects of shark nets in the Natal nearshore environment. Environ. Biol. Fish. 36 (3), 243–255.
Dugatkin, L.A., 2009. Principles of Animal Behavior. W.W. Norton, New York.
Dungan, S.Z., Riehl, K.N., Wee, A., Wang, J.Y., 2011. A review of the impacts of anthropogenic activities on the critically endangered eastern Taiwan Strait Indo-Pacific humpback dolphins (*Sousa chinensis*). J. Mar. Anim. Ecol. 4 (2), 3–9.
Dungan, S.Z., Hung, S.K., Wang, J.Y., White, B.N., 2012. Two social communities in the Pearl River Estuary population of Indo-Pacific humpback dolphins (*Sousa chinensis*). Can. J. Zool. 90 (8), 1031–1043.
Dwyer, S.L., Kozmian-Ledward, L., Stockin, K.A., 2014. Short-term survival of severe propeller strike injuries and observations on wound progression in a bottlenose dolphin. N. Z. J. Mar. Freshw. Res. 48 (2), 294–302.
Ellenberg, U., Mattern, T., Seddon, P.J., Jorquera, G.L., 2006. Physiological and reproductive consequences of human disturbance in Humboldt penguins: the need for species-specific visitor management. Biol. Conserv. 133 (1), 95–106.
Environmental Investigation Agency (EIA), 2012. Expansion of Hong Kong International Airport into a Three-Runway System. Environmental Impact Assessment Study Brief No. ESB-250/2012, Environmental Investigation Agency, Washington, DC.
Ewbank, R., 1985. Behavioral responses to stress in farm animals. In: Moberg, G.P. (Ed.), Animal Stress. Springer New York, New York, pp. 71–79.
Fair, P.A., Becker, P.R., 2000. Review of stress in marine mammals. J. Aquat. Ecosyst. Stress. Recover. 7 (4), 335–354.
Fernández-Juricic, E., Tellería, J.L., 2000. Effects of human disturbance on spatial and temporal feeding patterns of Blackbird *Turdus merula* in urban parks in Madrid, Spain. Bird Study 47 (1), 13–21.
Fertl, D., 2009. Interference with fisheries. In: Perrin, W.F., Würsig, B., Thewissen, J.G.M. (Eds.), Encyclopedia of Marine Mammals, second ed. Academic Press, Elsevier, San Diego, California, pp. 439–443.
Fertl, D., Leatherwood, S., 1997. Cetacean interactions with trawls: a preliminary review. J. Northwest Atl. Fish. Sci. 22, 219–248.
Finneran, J.J., Carder, D.A., Schlundt, C.E., Ridgway, S.H., 2005. Temporary threshold shift in bottlenose dolphins (*Tursiops truncatus*) exposed to mid-frequency tonesa). J. Acoust. Soc. Am. 118 (4), 2696–2705.
Foroughirad, V., Mann, J., 2013. Long-term impacts of fish provisioning on the behavior and survival of wild bottlenose dolphins. Biol. Conserv. 160, 242–249.
Frid, A., Dill, L.M., 2002. Human-caused disturbance stimuli as a form of predation risk. Conserv. Ecol. 6 (1), 11. [online] URL: http://www.consecol.org/vol6/iss1/art11/.
Garrison, L.P., 2007. Interactions between marine mammals and pelagic longline fishing gear in the US Atlantic Ocean between 1992 and 2004. Fish. B-NOAA 105 (3), 408–417.
Gautam, A., 2010. Ecotourism in Menai Bay Conservation Area, Zanzibar: Recommendations for Improvement in the Dolphin Tourism Industry, Coastal Resources Center, University of Rhode Island. Coastal Resources Center, University of Rhode Island, Narragansett, p. 32.
Gilman, E., Brothers, N., McPherson, G., Dalzell, P., 2007. A review of cetacean interactions with longline gear. J. Cetac. Res. Manage. 8 (2), 215–223.
Hanf, D.M., Hunt, T., Parra, G.J., 2016. Humpback dolphins of Western Australia: a review of current knowledge and recommendations for future management. Adv. Mar. Biol. 73, (in press).
Hashim, N.A.N., Jaaman, S.A., 2011. Boat effects on the behaviour of Indo-Pacific humpback (*Sousa chinensis*) and Irrawaddy dolphins (*Orcaella brevirostris*) in Cowie Bay, Sabah, Malaysia. Sains Malays. 40 (12), 1383–1392.

Higham, J.E.S., Shelton, E.J., 2011. Tourism and wildlife habituation: reduced population fitness or cessation of impact? Tour. Manage. 32 (6), 1290–1298.
Higham, J.E.S., Bejder, L., Lusseau, D., 2009. An integrated and adaptive management model to address the long-term sustainability of tourist interactions with cetaceans. Environ. Conserv. 35 (4), 294–302.
Humphries, D.A., Driver, P.M., 1970. Protean defence by prey animals. Oecologia 5 (4), 285–302.
Hung, S.K., 2008. Habitat use of Indo-Pacific humpback dolphins (*Sousa chinensis*) in Hong Kong. PhD dissertation, The University of Hong Kong, Pokfulam, Hong Kong. p. 253.
Hung, S.K., 2014. Monitoring of marine mammals in Hong Kong waters (2013–14), final report (1 April 2014 to 31 March 2015): Agriculture, Fisheries and Conservation Department of the Hong Kong SAR Government, Technical Report, Hong Kong. pp. 1–198.
Hung, S.K., 2015. Monitoring of marine mammals in Hong Kong waters (2014–15), final report (1 April 2014 to 31 March 2015): Agriculture, Fisheries and Conservation Department of the Hong Kong SAR Government, Technical Report, Hong Kong. pp. 1–151.
Hung, S.K., Jefferson, T.A., 2004. Ranging patterns of Indo-Pacific humpback dolphins (*Sousa chinensis*) in the Pearl River Estuary, People's Republic of China. Aquat. Mamm. 30 (1), 159–174.
Jefferson, T.A., 2000. Population biology of the Indo-Pacific humpbacked dolphin in Hong Kong waters. Wildl. Monogr. 144, 1–65.
Jefferson, T.A., Curry, B.E., 1994. A global review of porpoise (Cetacea: Phocoenidae) mortality in gillnets. Biol. Conserv. 67 (2), 167–183.
Jefferson, T.A., Hung, S.K., 2004. A review of the status of the Indo-Pacific humpback dolphin (*Sousa chinensis*) in Chinese waters. Aquat. Mamm. 30 (1), 149–158.
Jefferson, T.A., Karczmarski, L., 2001. Sousa chinensis. Mamm. Spec. 655, 1–9.
Jefferson, T.A., Rosenbaum, H.C., 2014. Taxonomic revision of the humpback dolphins (*Sousa* spp.), and description of a new species from Australia. Mar. Mamm. Sci. 30 (4), 1494–1541.
Jefferson, T.A., Webber, M.A., Pitman, R.L., 2008. Marine Mammals of the World: A Comprehensive Guide to Their Identification. Academic Press, London, Elsevier, San Diego, California.
Jefferson, T.A., Hung, S.K., Würsig, B., 2009. Protecting small cetaceans from coastal development: impact assessment and mitigation experience in Hong Kong. Mar. Policy 33 (2), 305–311.
Jefferson, T.A., Hung, S.K., Robertson, K.M., Archer, F.I., 2012. Life history of the Indo-Pacific humpback dolphin in the Pearl River Estuary, southern China. Mar. Mamm. Sci. 28 (1), 84–104.
Jensen, F.H., Bejder, L., Wahlberg, M., Aguilar De Soto, N., Johnson, M.P., Madsen, P.T., 2009. Vessel noise effects on delphinid communication. Mar. Ecol. Prog. Ser. 395, 161–175.
Jones, J.B., 1992. Environmental impact of trawling on the seabed: a review. N. Z. J. Mar. Freshw. Res. 26 (1), 59–67.
Karczmarski, L., 1999. Group dynamics of humpback dolphins (*Sousa chinensis*) in the Algoa Bay region South Africa. J. Zool. 249 (3), 283–293.
Karczmarski, L., 2000. Conservation and management of humpback dolphins: the South African perspective. Oryx 34 (3), 207–216.
Karczmarski, L., Cockcroft, V.G., 1999. Daylight behaviour of humpback dolphins *Sousa chinensis* in Algoa Bay, South Africa. Z. Säugetierkd. 64 (1), 19–29.
Karczmarski, L., Thornton, M., Cockroft, V., 1997. Description of selected behaviours of humpback dolphins, *Sousa chinensis*. Aquat. Mamm. 23, 127–134.

Karczmarski, L., Cockcroft, V.G., McLachlan, A., 2000. Habitat use and preferences of Indo-Pacific humpback dolphins *Sousa chinensis* in Algoa Bay, South Africa. Mar. Mamm. Sci. 16 (1), 65–79.

Kastelein, R.A., Jennings, N., Verboom, W.C., De Haan, D., Schooneman, N.M., 2006. Differences in the response of a striped dolphin (*Stenella coeruleoalba*) and a harbour porpoise (*Phocoena phocoena*) to an acoustic alarm. Mar. Environ. Res. 61 (3), 363–378.

Keith, M., Atkins, S., Johnson, A.E., Karczmarski, L., 2013. Area utilization patterns of humpback dolphins (*Sousa plumbea*) in Richards Bay, KwaZulu-Natal, South Africa. J. Ethol. 31 (3), 261–274.

Ketten, D.R., 2004. Marine mammal auditory systems: a summary of audiometric and anatomical data and implications for underwater acoustic impacts. Polarforschung 72 (2/3), 79–92.

Koper, R.P., Plön, S., 2012. The Potential Impacts of Anthropogenic Noise on Marine Animals and Recommendations for Research in South Africa. EWT Research & Technical Paper No. 1, Endangered Wildlife Trust, South Africa, p. 118.

Lesage, V., Barrette, C., Kingsley, M.C.S., Sjare, B., 1999. The effect of vessel noise on the vocal behavior of belugas in the St. Lawrence River Estuary, Canada. Mar. Mamm. Sci. 15 (1), 65–84.

Lima, S.L., Dill, L.M., 1990. Behavioral decisions made under the risk of predation: a review and prospectus. Can. J. Zool. 68 (4), 619–640.

Lundquist, D., Gemmell, N.J., Würsig, B., 2012. Behavioural responses of dusky dolphin groups (*Lagenorhynchus obscurus*) to tour vessels off Kaikoura, New Zealand. PLoS One 7 (7), e41969. http://dx.doi.org/10.1371/journal.pone.0041969.

Lusseau, D., 2004. The hidden cost of tourism: detecting long-term effects of tourism using behavioral information. Ecol. Soc. 9 (1), 2.

Lusseau, D., 2006. The short-term behavioral reactions of bottlenose dolphins to interactions with boats in Doubtful Sound, New Zealand. Mar. Mamm. Sci. 22 (4), 802–818.

Lusseau, D., Bejder, L., 2007. The long-term consequences of short-term responses to disturbance experiences from whalewatching impact assessment. Int. J. Comp. Psychol. 20 (2), 228–236.

Lusseau, D., Schneider, K., Boisseau, O.J., Haase, P., Slooten, E., Dawson, S.M., 2003. The bottlenose dolphin community of doubtful sound features a large proportion of long-lasting associations. Behav. Ecol. Sociobiol. 54 (4), 396–405.

Lusseau, D., Bain, D.E., Williams, R., Smith, J.C., 2009. Vessel traffic disrupts the foraging behavior of southern resident killer whales *Orcinus orca*. Endanger. Species Res. 6 (3), 211–221.

Mann, J., Kemps, C., 2003. The effects of provisioning on maternal care in wild bottlenose dolphins, Shark Bay, Australia. In: Gales, N., Hindell, M., Kirkwood, R. (Eds.), Marine Mammals: Fisheries, Tourism and Management Issues. CSIRO Publishing, Collingwood, Victoria, pp. 304–317.

Mann, J., Smuts, B., 1999. Behavioral development in wild bottlenose dolphin newborns (*Tursiops* sp.). Behaviour 136 (5), 529–566.

Mann, J., Würsig, B., 2014. Observing and quantifying cetacean behavior in the wild: current problems, limitations, and future directions. In: Yamagiwa, J., Karczmarski, L. (Eds.), Primates and Cetaceans: Field Research and Conservation of Complex Mammalian Societies. Springer, New York, pp. 335–344.

Mann, J., Connor, R.C., Barre, L.M., Heithaus, M.R., 2000. Female reproductive success in bottlenose dolphins (*Tursiops* sp.): life history, habitat, provisioning, and group-size effects. Behav. Ecol. 11 (2), 210–219.

Marcotte, D., Hung, S.K., Caquard, S., 2015. Mapping cumulative impacts on Hong Kong's pink dolphin population. Ocean Coast. Manag. 109, 51–63.

Markowitz, T.M., Harlin, A.D., Würsig, B., McFadden, C.J., 2004. Dusky dolphin foraging habitat: overlap with aquaculture in New Zealand. Aquat. Conserv. Mar. Freshwat. Ecosyst. 14 (2), 133–149.

Mendez, M., Subramaniam, A., Collins, T., Minton, G., Baldwin, R., Berggren, P., Särnblad, A., Amir, O.A., Peddemors, V.M., Karczmarski, L., 2011. Molecular ecology meets remote sensing: environmental drivers to population structure of humpback dolphins in the Western Indian Ocean. Heredity 107 (4), 349–361.

Moberg, G.P., 2000. Biological response to stress: implications for animal welfare. In: Moberg, G.P., Mench, J.A. (Eds.), The Biology of Animal Stress: Basic Principles and Implications for Animal Welfare. CABI Publishing, London, UK, pp. 1–21.

Moore, J.E., Wallace, B.P., Lewison, R.L., Žydelis, R., Cox, T.M., Crowder, L.B., 2009. A review of marine mammal, sea turtle and seabird bycatch in USA fisheries and the role of policy in shaping management. Mar. Policy 33 (3), 435–451.

Muralidharan, R., 2013. Sightings and behavioral observations of Indo-Pacific humpback dolphins *Sousa chinensis* (Osbeck, 1765) along Chennai coast, Bay of Bengal. J. Threat. Taxa 5 (15), 5002–5006.

Nanayakkara, R.P., Kusuminda, T., Jefferson, T.A., 2014. Can the Indian Ocean humpback dolphin (*Sousa plumbea*) survive in Sri Lanka? Occurrence of a relict population in Puttalam Lagoon. Aquat. Mamm. 40 (4), 398–406.

Ng, S.L., Leung, S., 2003. Behavioral response of Indo-Pacific humpback dolphin (*Sousa chinensis*) to vessel traffic. Mar. Environ. Res. 56 (5), 555–567.

Nisbet, I.C.T., 2000. Disturbance, habituation, and management of waterbird colonies. Waterbirds 23 (2), 312–332.

Nowacek, S.M., Wells, R.S., Solow, A.R., 2001. Short-term effects of boat traffic on bottlenose dolphins, *Tursiops truncatus*, in Sarasota Bay, Florida. Mar. Mamm. Sci. 17 (4), 673–688.

Nowacek, D.P., Thorne, L.H., Johnston, D.W., Tyack, P.L., 2007. Responses of cetaceans to anthropogenic noise. Mamm. Rev. 37 (2), 81–115.

NRC, 2005. Marine Mammal Populations and Ocean Noise: Determining When Noise Causes Biologically Significant Effects. National Academies Press, Washington, DC.

O'Connor, S., Campbell, R., Cortez, H., Knowles, T., 2009. Whale watching worldwide: tourism numbers, expenditures and expanding economic benefits: A special report from the International Fund for Animal Welfare, Yarmouth MA, USA. prepared by Economists at Large, p. 295.

Orams, M.B., 2002. Feeding wildlife as a tourism attraction: a review of issues and impacts. Tour. Manage. 23 (3), 281–293.

Parra, G.J., Ross, G.J.B., 2009. Humpback dolphins *Sousa chinensis* and *S. teuszii*. In: Perrin, W.F., Würsig, B., Thewissen, J.G.M. (Eds.), Encyclopedia of Marine Mammals, second ed. Academic Press, Elsevier, San Diego, California, pp. 576–582.

Parra, G.J., Corkeron, P.J., Marsh, H., 2004. The Indo-Pacific humpback dolphin, *Sousa chinensis* (Osbeck, 1765), in Australian waters: a summary of current knowledge. Aquat. Mamm. 30 (1), 197–206.

Parsons, E.C.M., 1998. Observations of Indo-Pacific humpbacked dolphins, *Sousa chinensis*, from Goa. Mar. Mamm. Sci. 14 (1), 166–170.

Parsons, E.C.M., Jefferson, T.A., 2000. Post-mortem investigations on stranded dolphins and porpoises from Hong Kong waters. J. Wildl. Dis. 36 (2), 342–356.

Paterson, R.A., 1990. Effects of long-term anti-shark measures on target and non-target species in Queensland, Australia. Biol. Conserv. 52 (2), 147–159.

Perrin, W.F., Donovan, G.P., Barlow, J., 1994. Gillnets and cetaceans: incorporating the proceedings of the symposium and workshop on the mortality of cetaceans in passive fishing nets and traps. Rep. Int. Whal. Comm. Spec. Iss. 15, 617 pp.

Pilleri, G., Gihr, M., 1974. Contribution to the knowledge of the cetaceans of southwest and monsoon Asia (Persian Gulf, Indus Delta, Malabar, Andaman Sea and Gulf of Siam). Invest. Cet. 5, 95–149.

Piwetz, S., Hung, S., Wang, J., Lundquist, D., Würsig, B., 2012. Influence of vessel traffic on movements of Indo-Pacific humpback dolphins (*Sousa chinensis*) off Lantau Island, Hong Kong. Aquat. Mamm. 38 (3), 325–331.

Preisler, H.K., Ager, A.A., Wisdom, M.J., 2006. Statistical methods for analysing responses of wildlife to human disturbance. J. Appl. Ecol. 43 (1), 164–172.

Read, A.J., 2008. The looming crisis: interactions between marine mammals and fisheries. J. Mammal. 89 (3), 541–548.

Reeves, R.R., 2009. Conservation efforts. In: Perrin, W.F., Würsig, B., Thewissen, J.G.M. (Eds.), Encyclopedia of Marine Mammals, second ed. Academic Press, Elsevier, San Diego, California, pp. 275–289.

Richardson, W.J., Würsig, B., 1997. Influences of man-made noise and other human actions on cetacean behaviour. Mar. Freshw. Behav. Physiol. 29 (1-4), 183–209.

Rodgers Jr., J.A., Smith, H.T., 1997. Buffer zone distances to protect foraging and loafing waterbirds from human disturbance in Florida. Wildl. Soc. Bull. 25 (1), 1–39.

Scarpaci, C., Nugegoda, D., Corkeron, P.J., 2004. No detectable improvement in compliance to regulations by "swim-with-dolphin" operators in Port Phillip Bay, Victoria, Australia. Tour. Mar. Environ. 1 (1), 41–48.

Sims, P.Q., Hung, S.K., Würsig, B., 2012. High-speed vessel noises in West Hong Kong waters and their contributions relative to Indo-Pacific humpback dolphins (*Sousa chinensis*). J. Mar. Biol. 2012, 1–11.

Sirot, E., 2010. Should risk allocation strategies facilitate or hinder habituation to nonlethal disturbance in wildlife? Anim. Behav. 80 (4), 737–743.

Slooten, E., Wang, J.Y., Dungan, S.Z., Forney, K.A., Hung, S.K., Jefferson, T.A., Riehl, K.N., Rojas-Bracho, L., Ross, P.S., Wee, A., Winkler, R., Yang, S.C., Chen, C.A., 2013. Impacts of fisheries on the Critically Endangered humpback dolphin *Sousa chinensis* population in the eastern Taiwan Strait. Endanger. Species Res. 22 (2), 99–114.

Southall, B.L., Bowles, A.E., Ellison, W.T., Finneran, J.J., Gentry, R.L., Greene Jr., C.R., Kastak, D., Ketten, D.R., Miller, J.H., Nachtigall, P.E., 2008. Marine mammal noise-exposure criteria: initial scientific recommendations. Bioacoustics 17 (1-3), 273–275

Spence, J.H., 2007. A summary of existing and future potential treatments for reducing underwater sounds from oil and gas industry activities. In: OCEANS 2007. IEEE, pp. 1–15.

Srinivasan, M., Grant, W.E., Swannack, T.M., Rajan, J., 2010. Behavioral games involving a clever prey avoiding a clever predator: an individual-based model of dusky dolphins and killer whales. Ecol. Model. 221 (22), 2687–2698.

Stankowich, T., 2008. Ungulate flight responses to human disturbance: a review and meta-analysis. Biol. Conserv. 141 (9), 2159–2173.

Stensland, E., Berggren, P., 2007. Behavioural changes in female Indo-Pacific bottlenose dolphins in response to boat-based tourism. Mar. Ecol. Prog. Ser. 332, 225–234.

Stensland, E., Carlen, I., Sarnblad, A., Bignert, A., Berggren, P., 2006. Population size, distribution, and behavior of Indo-Pacific bottlenose (*Tursiops aduncus*) and humpback (*Sousa chinensis*) dolphins off the south coast of Zanzibar. Mar. Mamm. Sci. 22 (3), 667–682.

Stockin, K.A., Lusseau, D., Binedell, V., Wiseman, N., Orams, M.B., 2008. Tourism affects the behavioural budget of the common dolphin *Delphinus* sp. in the Hauraki Gulf, New Zealand. Mar. Ecol. Prog. Ser. 355, 287–295.

Sutaria, D., Jefferson, T.A., 2004. Records of Indo-Pacific humpback dolphins (Sousa chinensis, Osbeck, 1765) along the coasts of India and Sri Lanka: an overview. Aquat. Mamm. 30 (1), 125–136.

Sutaria, D., Panicker, D., Jog, K., Sule, M., Muralidharan, R., Bopardikar, I., 2015. Humpback dolphins (Genus *Sousa*) in India: an overview of status and conservation issues. Adv. Mar. Biol. 72, 229–256.

Thorpe, W.H., 1963. Learning and Instinct in Animals, second ed. Methuen, London.

Turvey, S.T., Pitman, R.L., Taylor, B.L., Barlow, J., Akamatsu, T., Barrett, L.A., Zhao, X., Reeves, R.R., Stewart, B.S., Wang, K., 2007. First human-caused extinction of a cetacean species? Biol. Lett. 3 (5), 537–540.

Tyack, P.L., Zimmer, W.M.X., Moretti, D., Southall, B.L., Claridge, D.E., Durban, J.W., Clark, C.W., D'Amico, A., DiMarzio, N., Jarvis, S., 2011. Beaked whales respond to simulated and actual navy sonar. PLoS One 6 (3), e17009. http://dx.doi.org/10.1371/journal.pone.0017009.

Van Parijs, S.M., Corkeron, P.J., 2001. Boat traffic affects the acoustic behaviour of Pacific humpback dolphins, *Sousa chinensis*. J. Mar. Biol. Assoc. UK 81 (3), 533–538.

Van Waerebeek, K., Nobimé, G., Sohou, Z., Tchibozo, S., Dossou-Bodrènou, J.S., Dossou, C., Dossou-Hountoudou, A., 2002. Introducing whale and dolphin watching to Benin, 2002 exploratory survey: Report to the Netherlands Committee for IUCN, Amsterdam. 9 pp (unpublished).

Van Waerebeek, K., Barnett, L., Camara, A., Cham, A., Diallo, M., Djiba, A., Jallow, A., Ndiaye, E., Samba Ould-Bilal, A.O., Bamy, I.L., 2003. Conservation of Cetaceans in the Gambia and Senegal, 1999–2001, and Status of the Atlantic Humpback Dolphin. UNEP/CMS Secretariat, Bonn, Germany. p. 56.

Van Waerebeek, K., Barnett, L., Camara, A., Cham, A., Diallo, M., Djiba, A., Jallow, A.O., Ndiaye, E., Bilal, A.S.O.O., Bamy, I.L., 2004. Distribution, status, and biology of the Atlantic humpback dolphin, *Sousa teuszii* (Kükenthal, 1892). Aquat. Mamm. 30 (1), 56–83.

Van Waerebeek, K., Baker, A.N., Félix, F., Gedamke, J., Iñiguez, M., Sanino, G.P., Secchi, E., Sutaria, D., Van Helden, A., Wang, Y., 2007. Vessel collisions with small cetaceans worldwide and with large whales in the Southern Hemisphere, an initial assessment. Lat. Am. J. Aquat. Mamm. 6 (1), 43–69.

Van Waerebeek, K., Ofori-Danson, P.K., Debrah, J., 2009. The cetaceans of Ghana, a validated faunal checklist. West Afr J. App. Ecol. 15 (1), 61–90.

Visser, F., Hartman, K.L., Rood, E.J.J., Hendriks, A.J.E., Zult, D.B., Wolff, W.J., Huisman, J., Pierce, G.J., 2011. Risso's dolphins alter daily resting pattern in response to whale watching at the Azores. Mar. Mamm. Sci. 27 (2), 366–381.

Wade, P.R., 1995. Revised estimates of incidental kill of dolphins (Delphinidae) by the purse-seine tuna fishery in the eastern Tropical Pacific, 1959–1972. Fish. B-NOAA 93 (2), 345–354.

Wang, J.Y., Chu Yang, S., Hung, S.K., Jefferson, T.A., 2007a. Distribution, abundance and conservation status of the eastern Taiwan Strait population of Indo-Pacific humpback dolphins, *Sousa chinensis*. Mammalia 71 (4), 157–165.

Wang, J.Y., Riehl, K.N., Klein, M.N., Javdan, S., Hoffman, J.M., Dungan, S.Z., Dares, L.E., Araújo-Wang, C., 2016. Biology and conservation of the Taiwanese humpback dolphin, *Sousa chinensis taiwanensis*. Adv. Mar. Biol. 73, (in press).

Wang, J.Y., Yang, S.C., Reeves, R.R., 2007b. Report of the second international workshop on conservation and research needs of the eastern Taiwan Strait population of Indo-Pacific humpback dolphins, *Sousa chinensis*. National Museum of Marine Biology and Aquarium, Checheng, Pingtung County, Taiwan. p. 62.

Wang, Z., Wu, Y., Duan, G., Cao, H., Liu, J., Wang, K., Wang, D., 2014. Assessing the underwater acoustics of the world's largest vibration hammer (OCTA-KONG) and its potential effects on the Indo-Pacific humpbacked dolphin (*Sousa chinensis*). PLoS One 9 (10), e110590. http://dx.doi.org/10.1371/journal.pone.0110590.

Weir, C.R., Pierce, G.J., 2013. A review of the human activities impacting cetaceans in the eastern tropical Atlantic. Mammal Rev. 43 (4), 258–274.

Weir, C.R., Van Waerebeek, K., Jefferson, T.A., Collins, T., 2010. Challenges and Priorities for the Conservation of the Vulnerable Atlantic Humpback Dolphin (*Sousa teuszii*), with a Case Study from Namibe Province, Angola. IWC Scientific Committee Document SC/62/SMx, Agadir, Morocco. pp. 1–17.

Weir, C.R., Van Waerebeek, K., Jefferson, T.A., Collins, T., 2011. West Africa's Atlantic humpback dolphin (*Sousa teuszii*): endemic, enigmatic and soon endangered? Afr. Zool. 46 (1), 1–17.

Wells, R.S., 2014. Social structure and life history of common bottlenose dolphins near Sarasota Bay, Florida: insights from four decades and five generations. In: Yamagiwa, J., Karczmarski, L. (Eds.), Primates and Cetaceans: Field Research and Conservation of Complex Mammalian Societies. Springer, Tokyo, Japan, pp. 149–172.

Wenshi, P., Porter, L., Yamin, W., Dagong, Q., Yu, L., Tong, J., Zuhong, L., Dezhi, W., Yun, S., Zixing, P., Van Waerebeek, K., 2006. The importance of the Indo-Pacific humpback dolphin (*Sousa chinensis*) population of Sanniang Bay, Guangxi Province, PR China: recommendations for habitat protection. In: Working Paper SC/58/SM18 presented to the Scientific Committee of the International Whaling Commission, June 2006, St. Kitts and Nevis.

Whittaker, D., Knight, R.L., 1998. Understanding wildlife responses to humans. Wildl. Soc. Bull. 26, 312–317.

Williams, R., Trites, A.W., Bain, D.E., 2002. Behavioural responses of killer whales (*Orcinus orca*) to whale-watching boats: opportunistic observations and experimental approaches. J. Zool. 256 (2), 255–270.

Williams, R., Lusseau, D., Hammond, P.S., 2006. Estimating relative energetic costs of human disturbance to killer whales (*Orcinus orca*). Biol. Conserv. 133 (3), 301–311.

Wong, Y., 2012. An analysis of civic engagement for major infrastructure projects in Hong Kong. MSc thesis, The University of Hong Kong, Pokfulam, Hong Kong.

Wong, S.W., Karczmarski, L., Gailey, G., 2015. Impacts of coastal tourism in Indo-Pacific humpback dolphins *Sousa chinensis* in Hong Kong. In: International Conference on Biodiversity, Ecology and Conservation of Marine Ecosystems (BECoME 2015), Hong Kong. p. 234.

Würsig, B., 1989. Cetaceans. Science 244 (4912), 1550–1557.

Würsig, B., Greene Jr., C.R., 2002. Underwater sounds near a fuel receiving facility in western Hong Kong: relevance to dolphins. Mar. Environ. Res. 54 (2), 129–145.

Würsig, B., Richardson, W.J., 2009. Effects of noise. In: Perrin, W.F., Würsig, B., Thewissen, J.G.M. (Eds.), Encyclopedia of Marine Mammals, second ed. Academic Press, Elsevier, San Diego, California, pp. 576–582.

Würsig, B., Würsig, M., 1980. Behavior and ecology of the dusky dolphin, *Lagenorhynchus obscurus*, in the South Atlantic. Fish. B-NOAA 77 (4), 871–890.

Würsig, B., Cipriano, F., Slooten, E., Constantine, R., Barr, K., Yin, S., 1997. Dusky dolphins (*Lagenorhynchus obscurus*) off New Zealand: status of present knowledge. Rep. Int. Whaling Comm. 47, 715–722.

Würsig, B., Greene Jr., C.R., Jefferson, T.A., 2000. Development of an air bubble curtain to reduce underwater noise of percussive piling. Mar. Environ. Res. 49 (1), 79–93.

Würsig, B., Parsons, E.C.M., Piwetz, S., Porter, L., 2016. The behavioural ecology of Indo-Pacific humpback dolphins in Hong Kong. Adv. Mar. Biol. 73, (in press).

CHAPTER THREE

Re-assessment of the Conservation Status of the Atlantic Humpback Dolphin, *Sousa teuszii* (Kükenthal, 1892), Using the IUCN Red List Criteria

Tim Collins[1]

Wildlife Conservation Society, Marine Program, Global Conservation, Bronx, New York, USA
[1]Corresponding author: e-mail address: tcollins@wcs.org

Contents

1. Introduction 48
2. Taxonomic Notes 50
3. Geographic Range 51
   3.1 Range Description 51
   3.2 Countries 53
4. Population 53
   4.1 Abundance 53
   4.2 Population Trend 55
5. Habitat and Ecology 57
   5.1 Habitat 57
   5.2 Food and Feeding 58
   5.3 Reproduction 59
   5.4 Migrations and Movements 62
6. Use and Trade 64
7. Threats 64
   7.1 Major Threats 64
   7.2 Minor Threats 68
8. Conservation Actions 68
9. IUCN Red List Justification 69
Acknowledgements 71
References 71

Abstract

The Atlantic humpback dolphin (*Sousa teuszii*) is an obligate shallow water dolphin that is endemic to the western coasts of Africa, ranging from Western Sahara to Angola. The species occurs exclusively in a limited number of near-shore habitats, a tendency that

Advances in Marine Biology, Volume 72
ISSN 0065-2881
http://dx.doi.org/10.1016/bs.amb.2015.09.001

routinely exposes it to a suite of lethal and deleterious anthropogenic threats. These include habitat degradation, accidental capture in artisanal fishing nets, and hunting for use as food and bait. The species also competes with rapidly expanding human populations for coastal resources in some of the poorest countries on Earth. Data for most aspects of the species' ecology are sparse, but *S. teuszii* is considered by most qualified observers to be rare and greatly threatened. A lack of appropriate survey data precludes a quantitative assessment of population trends and status. Most populations for which any data are available are considered to be extremely small, numbering in the tens or low hundreds of individuals. The available published estimates suggest that the total population likely falls below 3000 individuals. Declines in abundance have been observed or are suspected for each population and will continue, given projected expansions of identified threats that affect most of the species' known range, and a corresponding lack of appropriate management actions. The apparent scale of threats, the presumed isolation of most populations, and a lack of directed conservation efforts in most areas suggest that the species qualifies for a listing of Critically Endangered (under criteria A3cd) on the IUCN Red List.

1. INTRODUCTION

The Atlantic humpback dolphin (*Sousa teuszii*) is endemic to shallow tropical and subtropical waters of the west coast of Africa, ranging from Western Sahara to Angola (Weir and Collins, 2015). It is the only member of the genus *Sousa* that occurs outside of the Indo-Pacific region. The species is dependent on inshore habitats that are routinely exposed to anthropogenic threats, particularly habitat loss, the risk of bycatch in artisanal fishing gear, and directed hunting for "marine bushmeat" (Clapham and Van Waerebeek, 2007; Weir et al., 2011). The species also competes for coastal resources with rapidly expanding human populations in some of the poorest countries on Earth Solutions for these conservation problems are complex. *Sousa teuszii* is considered to be both rare and threatened, and concerns for its long-term survival are significant and justified (Ayissi et al., 2014; Belhabib et al., 2015; Robards and Reeves, 2011; Weir et al., 2011).

The *S. teuszii* holotype (a skull) was collected in Cameroon in 1891 (Kükenthal,1892), and the species was not reported again for over 50 years. A second skull was collected in Senegal in 1925, but went unrecognized until 1965 (Van Bree and Duguy, 1965). Fraser (1949) confirmed the identity of a third skull collected in Senegal in 1943. In doing so, he contributed both a range extension and a prescient first account of bycatch for this species. Other accounts documenting the occurrence of *S. teuszii* were to follow in the 1950s and decades thereafter, although most were exclusive to West Africa[1] (Cadenat, 1956, 1959; Cadenat and Paraiso, 1957). Research efforts have extended to other parts of the range since the 1990s, and its occurrence has now been confirmed in 12 countries, although many areas remain poorly surveyed (Weir and Collins, 2015). The species has never been considered common and several small populations are apparently isolated.

Most aspects of the ecology and biology of *S. teuszii* remain poorly understood. Notable is the absence of data on life history, a continued absence of basic occurrence data from large areas of the presumed range, and an almost total absence of robust abundance estimates. The latter are particularly important for assessing status; available "guesstimates" are subjective, and are based on the experience of few observers and relatively sparse survey data (Van Waerebeek et al., 2004). Most researchers believe that the species is declining in abundance almost everywhere it occurs, a concern repeated emphatically in several recent reviews (Ayissi et al., 2014; IWC, 2011; Perrin and Van Waerebeek, 2007; Van Waerebeek et al., 2004; Weir et al., 2011).

In the absence of directed and sustained conservation efforts, the likelihood that *S. teuszii* will survive in the long-term is poor. The paucity of information reflects several issues that have been noted for years; resources for the study of *S. teuszii* have always been limited, as has technical capacity for initiating and completing surveys in range states. Its persistence in coastal protected areas in some range states offers some hope, and a basis for concerted conservation efforts. Known threats affect many other charismatic species in the region, and conservation efforts for *S. teuszii* could have wider implications for the stewardship of coastal marine resources in general (e.g. Hooker and Gerber, 2004; Hoyt, 2012; Nicholson et al., 2006; Reeves, 2000).

This chapter reviews the taxonomy and ecology of the Atlantic humpback dolphin, based on published literature, unpublished reports, and the observations of qualified observers. The species' conservation status is then evaluated against the International Union for the Conservation of Nature (IUCN) Red List criteria (IUCN, 2001), and recommendations are provided for its Red List status.

[1] "West Africa" is a recognized United Nations subregion that includes all of the Atlantic coast states from Mauritania to Nigeria. "Central" or "Middle Africa" is a different subregion that includes all Atlantic coast states from Cameroon to Angola.

2. TAXONOMIC NOTES

The *S. teuszii* holotype[2] (a skull) was recovered in the "Bucht des Kameruner Kriegsshiffhafens", ("Bay of Warships" or "Man O'War Bay"),[3] in Cameroon by a German agronomist named Eduard Tëusz.[4] The specimen was sent to Jena (Germany), where the zoologist Willy Kükenthal described it as the new species *Sotalia teuszii* (Kükenthal, 1891, 1892; Van Beneden, 1892). Thorough reviews of the species taxonomy are provided by Van Waerebeek et al. (2004) and Jefferson and Rosenbaum (2014).

The collection details for the holotype are imprecise, but were of consequence. The dolphin skull and the shark-mauled carcass of a West African manatee were recovered by Tëusz, who assumed they were from the same animal. This included the latter's stomach containing "grass, weeds and mangrove fruits", which in turn led Kükenthal to hypothesize that the species was perhaps riparian and vegetarian. The Belgian zoologist Van Beneden compounded the misunderstanding in a separate publication (Van Beneden, 1892; see Van Waerebeek et al., 2004). This confusion held until the 1950s, when necropsies of fresh Atlantic humpback dolphins confirmed that they were piscivorous and marine (Cadenat, 1956; Cadenat and Paraiso, 1957).

Recent genetic and morphometric work has served to clarify the taxonomy of the genus *Sousa* (Jefferson and Rosenbaum, 2014; Jefferson and Van Waerebeek, 2004; Mendez et al., 2013). The available data indicate that there is "strong and significant genetic and morphologic differentiation between *S. teuszii* and all other sampling units...with no evidence of exchange or contact" (Mendez et al., 2013). Principal components analysis demonstrated that *S. teuszii* have significantly shorter rostra, wider skulls and lower tooth counts (an average of about 30 teeth per row vs. about 33–37 for other areas) when compared with Southeast African, Arabian/Persian Gulf and Indian *Sousa* specimens (Jefferson and Rosenbaum, 2014; Jefferson and Van Waerebeek, 2004). There are no shared mtDNA haplotypes with other species in the genus, but a phylogenetic assessment of combined nuclear and mtDNA datasets suggests that *S. teuszii* is most closely related to Indian Ocean humpback dolphin, *Sousa plumbea*, from South East Africa

[2] The holotype is held by the British Museum (catalogue number 1893.8.1.1).
[3] Approximately 3°56′54N, 9°13′27E (Ayissi et al., 2014).
[4] The actual year of recovery is never stated, though it is likely to be 1891.

(Mendez et al., 2013). The most plausible mechanism for their isolation is the Benguela upwelling system, an area dominated by cold upwelling that is located within the 2000-km distribution gap between *S. teuszii* and *S. plumbea* (Jefferson and Van Waerebeek, 2004; Mendez et al., 2013). The Benguela has been described as a "species gate", and is inferred in the divergence of other taxa (e.g. Henriques et al., 2014; Perrin and Van Waerebeek, 2007; White et al., 2010).

3. GEOGRAPHIC RANGE

3.1 Range Description

Sousa teuszii is endemic to tropical and subtropical eastern Atlantic nearshore waters (<30 m) of western Africa (Culik, 2004; Jefferson et al., 1997; Van Waerebeek et al., 2004; see Figure 1), ranging from Dakhla Bay (Rio de Oro) in Western Sahara (23°52′N, 15°47′W) to Tômbwa (Namibe Province) in Angola (15°46′S, 11°46′E) (see also Weir and Collins, 2015). Research suggests that these dolphins occur in a series of localized communities with little interchange identified between them; opinions vary on whether their historical distribution was ever continuous (e.g. Jefferson et al., 1997; Maigret, 1980; Ross et al., 1994; Van Waerebeek et al., 2004; Zwart and Weir 2014). Recent work investigating the genetic substructuring of *S. plumbea* in the Western Indian Ocean indicated significant genetic divergence between populations in neighbouring regions, and thus minimal interchange between them over long-time frames. The scale of geographic distribution in West Africa is similar, and suggests that the species' within the genus *Sousa* may have an innate tendency for localized residency (Mendez et al., 2011).

Eight putative "management stocks" (subpopulations) of *S. teuszii* were identified by Van Waerebeek et al. (2004), using "fragmentary evidence". These included: Dakhla Bay (Rio de Oro-Western Sahara), Banc d'Arguin (Mauritania), Saloum Niumi (Senegal The Gambia), Canal do Gêba-Bijagós Archipelago (Guinea and Bissau), southern Guinea and Angola. Gabon and Cameroon Estuary stocks were also included, but were hypothetical, given an absence of recent records; work has since confirmed their occurrence in these states (Ayissi et al., 2014; Collins et al., 2004). Apparent distribution gaps in some areas were attributed in part to insufficient cetacean survey effort, a hypothesis supported by recent sightings in northern Guinea, Benin, Togo and the Republic of Congo (Collins et al., 2010; Weir, 2015; Zwart and Weir, 2014).

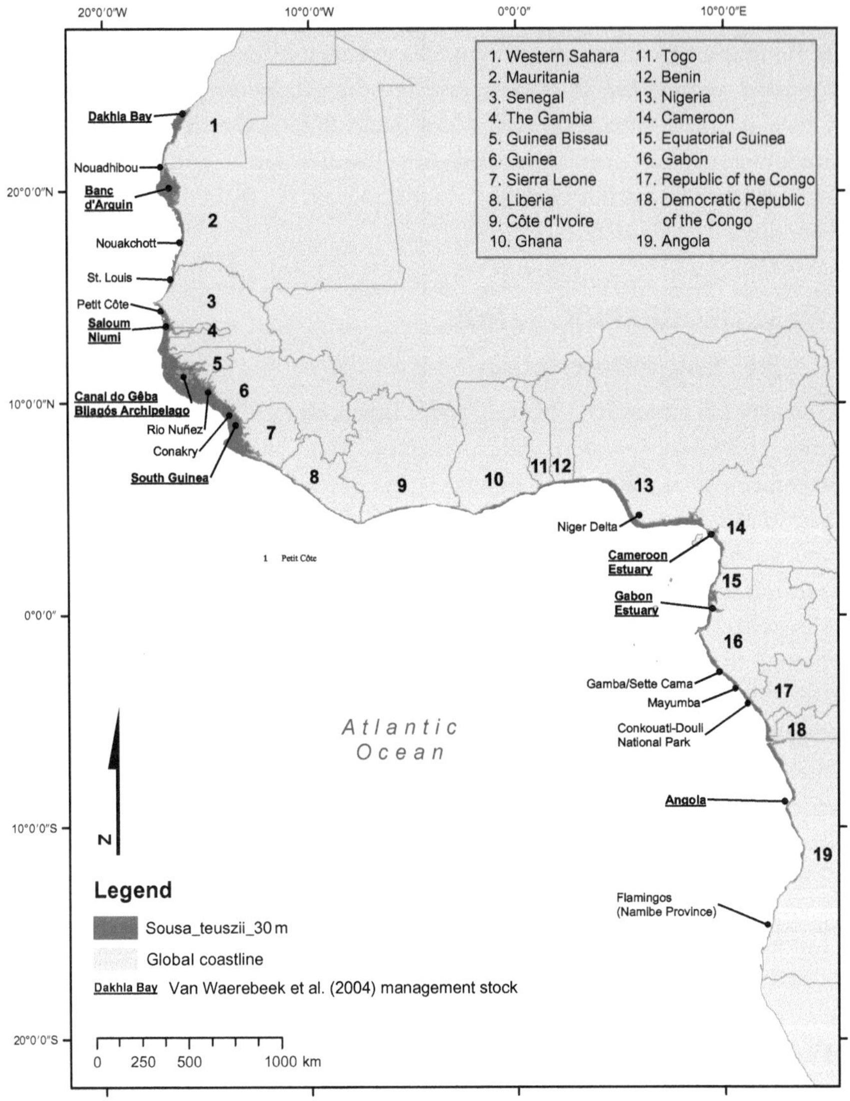

Figure 1 Inferred range of the Atlantic humpback dolphin, *Sousa teuszii*. Underlined localities indicate management stocks identified by Van Waerebeek et al. (2004).

Areas of known distribution may reflect the availability of high-quality *S. teuszii* habitats. The populations at Dakhla Bay, the Banc d'Arguin and in the Saloum-Niumi region are separated from each other by distances exceeding 350 km, and few observations have been recorded between them despite fieldwork (albeit inconsistent) over several decades. This suggests that the distribution of *S. teuszii* may be naturally discontinuous in some areas, with highest densities in optimal habitats and occurrence on intervening coasts reduced (Maigret, 1980; Robineau and Vely, 1998; Van Waerebeek et al., 2004).

Gaps in the distribution of *S. teuszii* may, however, be a relatively recent phenomenon, given mounting human pressures in once pristine regions (Perrin and Van Waerebeek, 2007; Weir et al., 2011). Several sightings and bycatches were reported from the "Petite Côte" region in Senegal (between Dakar and the Saloum Delta) during the 1950s (e.g. Cadenat and Paraiso, 1957), but none were reported between 1975 and 1980 (Maigret, 1980). Van Waerebeek et al. (2004) identified a single catch (of unknown origin) landed at Fadiouth/Joal in 1997. No records of *S. teuszii* have been reported in Ghana despite consistent monitoring of landing sites over many years and the existence of an active dolphin fishery; this suggests that the species has either been extirpated or is extremely rare there (Debrah et al., 2010; Van Waerebeek and Ofori-Danson, 1999; Van Waerebeek et al., 2009). Van Waerebeek et al. (2004) suggested that *S. teuszii* most likely inhabited the Niger Delta before the large-scale oil exploration and extraction altered the coastal environment.

The available data demonstrate that sightings in most areas of known occurrence can be low, and a general absence of records from "gap" areas suggests occurrence in extremely low densities, if at all. For instance, in southern Gabon, where *S. teuszii* occurs in the surf zone on open coastlines, boat-based survey work demonstrates that sightings rates can be very low, even with dedicated effort (see Section 4).

3.2 Countries

The species has been recorded in 12 countries. These include Morocco (Western Sahara), Mauritania, Senegal, The Gambia, Guinea, Guinea Bissau, Togo, Benin, Cameroon, Gabon, Republic of Congo and Angola (Bamy et al., 2010; Collins et al., 2010; Van Waerebeek et al., 2004; Weir et al., 2011; Weir and Collins, 2015; Zwart and Weir, 2014). Interview surveys with fishermen in Sierra Leone and Nigeria yielded unverified accounts of *S. teuszii* bycatches, and more work is needed in these areas to confirm the species' occurrence (Lewison and Moore, 2012; Moore et al., 2010).

4. POPULATION

4.1 Abundance

There are no global abundance estimates for *S. teuszii*. Assessments for the stocks described by Van Waerebeek et al. (2004) were derived from survey work that was typically localized, sporadic and of limited duration. Varied survey methods were used that do not facilitate abundance estimation or comparison. The few associated "guesstimates" are subjective and based on the experience of a limited number of researchers; these provide a crude index of abundance at best.

Populations at the northern (Dakhla Bay, Western Sahara) and southern (Namibe, Angola) extremities of the range appear to be very small, which may indicate the limits of habitat suitability (Weir, 2009). Beaubrun (1990) described the Dakhla Bay population as "miniscule", and sightings in the same area by Notarbartolo di Sciara et al. (1998) reinforce this impression. This population is likely limited to "a few tens of individuals" (Van Waerebeek et al., 2004). Verifiable sightings by locals and birders suggest that the subpopulation is extant (Weir and Collins, 2015).

The Banc d'Arguin and Saloum-Niumi populations have been estimated repeatedly at ~100 animals since the mid-1970s (Maigret, 1980; Van Waerebeek et al., 2003, 2004). Incidental sightings from the southern Banc d'Arguin suggest that the species is sighted relatively frequently (Antonio Araujo, Fondation Internationale du Banc d'Arguin, personal communication, January 2014). However, the stock has never been considered large by those who have completed assessments (Maigret, 1980; Robineau and Vely, 1998). In the Saloum-Niumi region, encounter rates and group sizes recorded during more recent surveys indicate a small population "unlikely [to] exceed low hundreds, maybe less" (Van Waerebeek et al., 2000, 2004).

Data and sighting records for Guinea Bissau suggested the continued occurrence of a significant population of *S. teuszii* into at least the late 1990s (Powell et al., 1996; Spaans, 1990; Van Waerebeek et al., 2000, 2004). Van Waerebeek et al. (2004) ventured a guess of "at least several hundred, if not more". A more recent review of sightings records indicates that *S. teuszii* is still widely distributed in Guinea Bissau (Leeney et al., 2015b), but sightings appear to be declining in regularity (P. Campredon, IUCN country program for Guinea Bissau, personal communication, 11 May 2015). Recent sightings in the Río Nuñez (northern Guinea) may indicate that distribution across the Guinea Bissau/Guinea border is contiguous (Weir, in press; K. Van Waerebeek, Peruvian Centre for Cetacean Research (CEPEC), personal communication, 2 June 2015).

Recent surveys in Cameroon yielded a single humpback dolphin sighting, despite extensive beach- and boat-based survey effort. A recorded encounter rate of 0.386 sightings per 100 km suggests that abundance there may be very low (Ayissi et al., 2014). Boat surveys in Gabon have also yielded low sightings rates. Sporadic surveys between 2003 and 2006 yielded five sightings at an average rate of 0.15 sightings per 100 km. Recent surveys in the Gamba/Sette Cama region of Gabon yielded three sightings of *S. teuszii* with an encounter rate of 0.13 sightings per 100 km (Collins et al., 2010; Minton, 2015). However, sightings rates during shore-based work in Congo are much higher (though not directly comparable), and suggest that the coasts of southern Gabon and a limited area in the adjacent Republic of Congo may harbour a total population in the low hundreds (Collins et al., 2013). Most of the Angolan coast is unsurveyed, but intensive survey effort demonstrated that a small group of 10 individuals was resident in the Flamingos area (Weir, 2009).

The 2008 IUCN Red List status assessment concluded that "considering the relatively small numbers observed, and even taking account of the many areas of the species' range where there has been little or no assessment, the total population probably numbers only a few thousand" (Reeves et al., 2008). The renewed assessment completed here further reinforces the inference of small total population size (see Table 1). Based on the sparse information available, it appears highly plausible that the species' total abundance may be below 3000 individuals.

4.2 Population Trend

The absence of robust estimates precludes any quantitative assessments of trend. However, the evidence of recent work in some areas and a consensus of expert opinion indicate that most described populations of *S. teuszii* are small and that all populations have experienced significant declines in recent decades (IWC, 2011). These declines are likely linked to the rapid expansion of coastal human populations and associated coastal fisheries effort (Ayissi et al., 2014; IWC, 2011; Perrin and Van Waerebeek, 2007; Van Waerebeek et al., 2004; Weir et al., 2011). Of additional concern are recent dolphin bycatches and observed or suspected declines in Guinea Bissau, which is believed to harbour the largest population of the species (Leeney et al. 2015b).

Limited research effort for each putative *S. teuszii* population has either identified significant mortality or yielded strong evidence to infer it. Artisanal fishing bycatch and directed takes are the principal causes of these declines, although habitat loss is also likely a contributing factor. Recent work in Conkouati Douli National Park (Republic of Congo) provides

Table 1 A Summary of Available Information Reported for *Sousa teuszii* Abundance and Population Sizes

Putative Population	Estimated Size	Source(s)
Dakhla Bay	"Miniscule"	Beaubrun (1990)
	"Low tens"	Van Waerebeek et al. (2004)
Banc d'Arguin	"Probably does not exceed 100 animals"	Maigret (1980)
	"Stock is apparently fairly small"	Van Waerebeek et al. (2004)
Saloum-Niumi	"Low hundreds, maybe less"	Van Waerebeek et al. (2004)
Guinea Bissau	56 Sightings recorded during a two month period (1986/1987)	Spaans (1990)
	"Several hundred, maybe more…until at least 1998"	Van Waerebeek et al. (2004)
	"Widespread"	Leeney et al. (2015a,b)
	"Still regular in the area, but in lower numbers"	P. Campredon (personal communication, May 2015)
Guinea	Eight sightings in the Río Nuñez Estuary, for a minimum population estimate of 47 animals	Weir (2015)
	Single sighting south of Conakry	Bamy et al. (2010)
Togo	4	Zwart and Weir (2014)
Benin	27	L. Henry (personal communication, April 2015)
Cameroon	"Abundance may be very low"	Ayissi et al. (2014)
Gabon	"Low hundreds"	Collins et al. (2013)
Congo	"Low hundreds"	Collins et al. (2013)
Angola	10	Weir (2009)

some indication of the substantial bycatch risk (see Figure 2). A programme of intensive monitoring, enforcement, and cooperative (incentivized) reporting, identified 19 dolphin bycatches across all artisanal landing sites ($n=14$) on a 60-km stretch of protected beach over 5 years of work. Of

Figure 2 Work in Conkouati Douli National Park (Republic of the Congo) provides some indication of the potential scale of *Sousa teuszii* bycatches; fishermen voluntarily report captures, which are first examined by park researchers and officials. The meat is distributed thereafter to villagers. Humpback dolphins are protected in Congo, and all captures that are not reported are considered poached and the fishermen responsible are prosecuted.

these, 10 were *S. teuszii*, and the testimony of fishermen showed that all were caught in gillnets less than 1 km from shore (Collins et al., 2013). Although alarming mortality figures have been reported for other areas such as the Banc d'Arguin (Campredon and Cuq, 2001; Robineau and Vely, 1998) and the Saloum Delta (Van Waerebeek et al., 2004), the monitoring of bycatches in other areas is either non-existent or limited to very few landing sites (e.g. Debrah et al., 2010; Moore et al., 2010). Those reported are considered unrepresentative of the true toll (Clapham and Van Waerebeek, 2007; Collins et al., 2010; Debrah et al., 2010; Lewison and Moore, 2012; Reeves et al., 2003; Van Waerebeek et al., 2004).

5. HABITAT AND ECOLOGY

5.1 Habitat

Atlantic humpback dolphins occur exclusively in shallow (<30 m) nearshore waters. Preferred habitats appear to be strongly influenced by tides and include areas of sandbanks, deltas, estuaries and mangrove systems (Maigret, 1980; Ross et al., 1994; Van Waerebeek et al., 2004; Weir et al., 2011). They do occur on exposed coastlines (Figure 3), although there is some evidence that densities in these areas are lower and possibly linked to the availability of shelter (Collins et al. 2010; Weir et al., 2011). Recorded

Figure 3 Atlantic humpback dolphins, *Sousa teuszii*, sighted off a coastline exposed to surf in Gabon.

sightings are typically coastal, but the species may also occur some distance from shore when suitable habitats are present, up to at least 13 km (Van Waerebeek et al., 2004; Weir and Collins, 2015). They have been recorded some distance uprivers but there is no evidence that they ascend beyond the influence of marine waters (Maigret, 1980; Van Bree and Duguy, 1965; Van Waerebeek et al., 2004; Weir and Collins, 2015). Maigret (1980) described a sighting at Foundiougne, in the main channel of the Saloum, some 33 miles (53 km) from the estuary mouth, an area within the influence of tides (Van Waerebeek et al., 2004). In Guinea Bissau, they have been recorded 30 km upstream from the mouth of the Rio Grande de Buba and 20 km upstream from the mouth of the Rio Cacine (Leeney et al., 2015b; Van Waerebeek et al., 2000). A skull was collected in Gabon ca. 1949 at Kango, a village 95 km from the Gabon Estuary mouth and at the limit of saline intrusion. Collection details are scant and the skull's true origin is unknown (Van Waerebeek et al., 2004; Vande weghe, 2005). Recent surveys in the Gabon Estuary, including as far as Kango, located no humpback dolphins (Collins et al., 2004). They are not known to enter the coastal lagoons that are a prevalent feature of equatorial West African coasts. For more detailed discussion of apparent habitat preferences, see Weir and Collins (2015).

5.2 Food and Feeding

Knowledge of the feeding ecology of *S. teuszii* is incomplete, as few stomach samples have been examined and observations of feeding are rare (Van Waerebeek et al., 2004). It is, however, clear that *S. teuszii* diets consist largely of coastal, estuarine and reef-associated fishes, many of which are characteristically vocal (Cadenat and Paraiso, 1957; Cadenat, 1959; Ladich, 2000; Van Waerebeek et al., 2004; Weir, 2009). Recorded prey items are listed in Table 2.

Foraging has been linked to rising (flood) tides (Weir, 2009; Van Waerebeek et al., 2004). In the Saloum Delta, high tides were thought to provide access to inner reaches of "bolons"[5] and mangrove edges (Maigret, 1980). In the Banc d'Arguin, the link between tides and foraging was less evident, although daily movements into channels inshore were coupled with flood tides (Maigret, 1980), and Duguy (1976) reported *S. teuszii* chasing fish in tidal channels between the Tidra and Nair islets. In the Bijagós Archipelago (Guinea Bissau), *S. teuszii* were most frequently observed during low tide, suggesting that they fed when fish were concentrated in "gullies and creeks" (Spaans, 1990; Weir, 2009). Feeding has been observed at river confluences within the Rio Grande de Buba (Van Waerebeek et al., 2000), and in the Saloum Delta the waters around Sangomar Island appear to be a preferred foraging area (Van Waerebeek et al., 2004, and references therein). Groups observed foraging in Saloum were also seen to be widely dispersed, a tendency also noted in Gabon (Collins et al., 2004; Maigret, 1980; Van Waerebeek et al., 2004). Feeding activity in some areas is also coincident with observations of larger aggregations, including the largest group sizes (20–40) recorded (Collins et al., 2004; Maigret, 1980; Van Waerebeek et al., 2004, and references therein).

Sousa teuszii observed off Flamingos, Angola, spent approximately half of their daylight time engaged in travel and foraging and foraged preferentially around rocks and reefs, as well as at the mouths of rivers, including the typically dry Flamingo River (Weir, 2009). These tendencies appear to be shared in Gabon, although data are limited (Collins et al., 2004). Animals performing longer dives (>1 min) were presumed to be in pursuit of demersal fishes and other benthic prey (Weir, 2009).

5.3 Reproduction

Information on reproduction of this species is extremely scant. In the Saloum Delta, births were thought to occur in March and April, based upon observations of juveniles (Maigret, 1980). This pattern was also suggested for Guinea Bissau (Krömer et al., 1994, as cited in Van Waerebeek et al., 2004). No neonates have been examined, but lengths at birth may be similar to *S. plumbea* (Van Waerebeek et al., 2004). In Congo, a lactating female

[5] "*bolon*" translates to "mangrove channel" in Wolof (e.g. Maigret, 1980; Van Waerebeek et al., 2003).

Table 2 A Summary of *Sousa teuszii* Prey Items Identified Through the Examination of Specimens or Direct Observations of Feeding

Family	Genus	Species	Common Name	Country	Described Locality	Sample Type	Source(s)
Haemulidae	*Pomadasys*	*jubelini*	Sompat grunt	Senegal	Joal, M'bour	Bycatch	Cadenat (1956), Cadenat and Paraiso (1957) and Maigret (1980)
Clupeidae	*Ethmalosa*	*fimbriata*	Bongo shad	Senegal	Joal, M'bour	Bycatch	Cadenat (1956), Cadenat and Paraiso (1957) and Maigret (1980)
Mugilidae	*Mugil*	sp.	Mullet	Senegal	Joal, M'bour	Bycatch	Cadenat (1956), Cadenat and Paraiso (1957) and Maigret (1980)
Lutjanidae	*Lutjanus*	*goreensis*	Gorean snapper	Guinea Bissau	Ilha Canhabaque	Bycatch	Sequeira and Reiner (1992)
Lethrinidae	*Lethrinus*	*atlanticus*	Atlantic emperor	Guinea Bissau	Ilha Canhabaque	Bycatch	Sequeira and Reiner (1992)
Ephippidae	*Chaetodipterus*	*lippei*	West African spadefish	Guinea Bissau	Ilha Canhabaque	Bycatch	Sequeira and Reiner (1992)
Scianidae	*Pseudotolithus*	*typus*	Long-neck croaker	Republic of Congo	Conkouati	Bycatch	T. Collins and M. Smale (unpublished data)
Scianidae	*Pseudotolithus*	*senegalensis*	Cassava croaker	Republic of Congo	Conkouati	Bycatch	T. Collins and M. Smale (unpublished data)
Paralichthyidae	*Bothidae*	sp.	Flounder	Republic of Congo	Conkouati	Bycatch	T. Collins and M. Smale (unpublished data)

Paralichthyidae	*Pseudorhombus*	sp.	Flounder	Republic of Congo	Conkouati	Bycatch	T. Collins and M. Smale (unpublished data)
Haemulidae	*Pomadasys*	sp.	Unidentified grunt	Republic of Congo	Conkouati	Bycatch	T. Collins and M. Smale (unpublished data)
Polynemidae	*Pentanemus*	*quinquarius*	Royal threadfin	Republic of Congo	Conkouati	Bycatch	T. Collins and M. Smale (unpublished data)
Squillidae	*Squilla*	*mantis*	Mantis shrimp	Republic of Congo	Conkouati	Bycatch	T. Collins and M. Smale (unpublished data)
Mugilidae	*Mugil*	*cephalus*	Flathead grey mullet	Mauritania	Nouamghar	Observation	Busnel (1973)
Mugilidae	*Liza*	*aurata*	Golden grey mullet	Mauritania	Nouamghar	Observation	Busnel (1973)
Mugilidae	*Liza*	*richardsonii*	South African mullet	Angola	Flamingos	Observation	Weir (2009)
Scombridae	*Sarda*	*sarda*	Atlantic bonito	Angola	Flamingos	Observation	Weir (2009)
Clupeidae	*Sardinella*	sp.	Sardine	Angola	Flamingos	Observation	Weir (2009)

Figure 4 The prominent dorsal hump of a presumed mature *Sousa teuszii* male.

caught in a gillnet during May (2013) measured 2.50 m (T. Collins, unpublished data). The species is suspected to be sexually dimorphic (males larger at maturity and with a more prominent dorsal hump—Figure 4), but the current sample size (~20 animals) is too small to assess statistically (Jefferson and Rosenbaum, 2014). Taylor et al. (2007) estimated a generation length of 18.4 years for *Sousa teuszii*, although Moore (2015) provided a figure closer to 25 years for the Indo-Pacific humpback dolphin, *S. chinensis*, and *S. plumbea*. The data required to estimate generation length and other *S. teuszii* vital rates remain unavailable.

5.4 Migrations and Movements

Migrations and movements are poorly understood. Localized movements have been linked to feeding opportunities facilitated by tides (e.g. Dupuy, 1983; Maigret, 1980; Van Waerebeek et al., 2003; Weir, 2009). Cadenat (1949) stated that in the Saloum they are "observed daily on the dropping tide between Pointe Jackonsa and the mouth of the Djiombos" (Maigret, 1980; Dupuy, 1983), a tendency that helped to optimize encounters in later surveys (Van Waerebeek et al., 2003).

Movements on larger scales are rarely documented, but have been inferred. Fishermen's accounts from the Baie du Lévrier and sightings at the Banc de Corbine suggested movement north of the Banc d'Arguin (Maigret, 1980); sightings between Nouamghar and Nouakchott[6] may indicate occasional movements south (Robineau and Vely, 1998). Maigret

[6] A sighting of *S. teuszii* by a trustworthy observer was recorded in 2005 "halfway between Nouakchott and Nouamghar". (M. Nicoll, World Wildlife Fund, *in litt.* to T. Collins, April 2015).

(1980) mused that animals from the Saloum Delta moved[7] as far north as "Yenne" (Yene Kao, ~100 km north of Saloum), and as far south as the Gambia River Estuary (~15 to 25 km). Landings of *S. teuszii* on the Petit Côte of Senegal provided some support for this idea (Cadenat, 1956; Cadenat and Paraiso, 1957), although capture locations were unknown (Van Waerebeek et al., 2004). More recent observations of *S. teuszii* groups passing between Barra and Buniada Points indicate routine movement between Senegal and Gambia (Murphy et al., 1997), and a dolphin photographed crossing this border at Djinack (October 2000) provided the rationale for the species' Convention on Migratory Species (CMS) listing (Van Waerebeek et al., 2004).

Larger migrations between the populations at Dakhla Bay, the Banc d'Arguin, and the Saloum-Niumi region are likely to be rare, if they occur at all, given the distances involved (Maigret, 1980; Van Waerebeek et al., 2003), although Robineau and Vely (1998) believed it "inconceivable" that the Dakhla animals were isolated. Maigret (1980) hypothesized that Saloum Delta animals swam north to Mauritania during the May–December period based on available records. A more recent stranding "midway" between Saloum and the Banc d'Arguin provided some support (Robineau and Vely, 1998), but this idea was rejected by Van Waerebeek et al. (2003), given the availability of a larger dataset. These include bycatch and interview data, indicating year round occurrence in the Saloum Delta. Moreover, there are no records between Dakar and St Louis, and very few from south of Nouakchott (Maigret, 1980; Van Waerebeek et al., 2004). Van Waerebeek et al. (2004) suggested that movements between Saloum-Niumi and Bijagos are probable, "considering the relatively limited distance (ca. 280 km)". Recent sightings in the Rio Nunez suggest this connectivity extends into Guinea (Weir and Collins, 2015).

Beach-based observations indicate routine movements of *S. teuszii* across the Gabon/Congo border within the Mayumba-Conkouati transboundary protected area; it remains unclear if these animals range further afield (Collins et al., 2010). The very small population at Flamingos (Angola) appears to be both resident and highly localized in its occurrence (Weir et al., 2011). Records elsewhere suggest opportunities for transboundary movements (e.g. Ayissi et al., 2014; Bamy et al., 2010; Zwart and Weir, 2014).

Inferring demographic exchange (or lack of) between populations through analyses of genetics and/or photographs is not possible at this point,

[7] Maigret used the phrase "petits déplacements le long des côtes".

given a lack of samples and identification images from almost all areas. If their distribution was once continuous (or less fragmented than it now appears to be), localized movements of the kind observed at Saloum-Niumi and between Congo and Gabon may have been sufficient for ensuring step-wise exchanges (and thus gene flow) between adjacent sub-groups (see Mendez et al., 2011).

6. USE AND TRADE

The sale of dolphin meat for either human consumption or bait has been documented or suspected from almost every *S. teuszii* range state. Evidence for the consumption and sale of *S. teuszii* specifically has been reported from Mauritania, Senegal, Gambia, Guinea Bissau, Guinea, Sierra Leone (unverified interview data), Togo (suspected), Nigeria (unverified interview data), Cameroon and Republic of the Congo. Many instances involve the secondary (i.e. non-targeted) use of dolphin bycatch, but it is clear that the species has been, and is directly and increasingly, targeted for food in many areas (Ayissi et al., 2014; Bamy et al., 2010; Cadenat, 1956; Clapham and Van Waerebeek, 2007; Collins et al., 2013; Lewison and Moore, 2012; Moore et al., 2010; Van Waerebeek et al., 2004; Weir and Pierce, 2013).

7. THREATS

7.1 Major Threats

The nearshore habitats of *S. teuszii* greatly increase its susceptibility to many anthropogenic threats. High human population densities, widespread poverty and habitat degradation (and destruction) associated with development are important determinants of the species' distribution and status (Perrin and Van Waerebeek, 2007; Weir et al., 2011). Fisheries are extensive and are a major source of income and protein (Belhabib et al., 2015; Brashares et al., 2004; Kaczynski and Fluharty, 2002; Weir et al., 2011). Bycatches of *S. teuszii* in artisanal gillnets have been reported from most areas and for as long as the species has been studied (see Figures 5 and 6), including the first records for some areas (e.g. Busnel, 1973; Cadenat, 1949; Campredon and Cuq, 2001; Collins et al., 2010; Murphy et al., 1997; Robineau and Vely, 1998; Van Waerebeek et al., 2000, 2004). The scale of mortality is unknown, but bycatches are considered the principal cause for their range-wide decline and the greatest immediate threat for most populations (Ayissi et al., 2014; Collins et al., 2010, 2013;

Figure 5 Captures in coastal gillnets are considered the principal cause of the *Sousa teuszii.*

Figure 6 Accidental captures of *Sousa teuszii* are used opportunistically by fishermen across the range for either food or bait.

Jefferson et al., 1997; Northridge, 1984; Perrin and Van Waerebeek, 2007; Weir et al., 2011).

Accounts of the deliberate capture of *S. teuszii* are rare (e.g. Cadenat, 1949, 1956, 1957; Duguy, 1976; Maigret, 1980; Van Waerebeek et al., 2004), but directed captures likely occur in some areas (Perrin and Van Waerebeek, 2007; Van Waerebeek et al., 2000). Maigret (1994) believed that there were no traditional marine mammal fisheries in West Africa, but there is evidence that the opportunistic use of bycatches may develop into "directed entanglement" or "non-target-deliberate acquisition" (respectively, Clapham and Van Waerebeek, 2007; Robards and Reeves, 2011). The scale of this practice is unknown, but appears to be part of a developing global trend in the utilization of "marine bushmeat" and reflective of general fisheries declines (Alfaro-Shigueto and Van Waerebeek, 2001; Brashares et al., 2004; Clapham and Van Waerebeek, 2007; Leeney et al., 2015a; Robards and Reeves, 2011; Uwagbae and Van Waerebeek, 2010; Van Waerebeek et al., 2004). Debrah et al. (2010) noted that the sale of cetaceans and sea turtles contributes to the economic viability of gillnet fisheries in Ghana. This included the killing of live entangled animals, and the supply of dolphin meat to markets for shark bait (Van Waerebeek et al., 2004; Weir and Pierce, 2013). Captures may be concealed, given legal prohibitions (Reeves et al., 2003), an issue for surveys in some areas (Moore et al., 2010; Van Waerebeek et al., 2004).

A related concern is the pervasive spread of migrant fishermen across western Africa over the past few decades (e.g. Bakhayokho and Kébé, 1991; Binet et al., 2012; Campredon and Cuq, 2001; Cormier-Salem, 2000; Duffy-Tumasz, 2012; Leeney et al., 2015b). Migrant fishermen (including those who move within states) may not abide by local injunctions, taboos or laws, and are often better resourced and more aggressive in their exploitation of local resources (Campredon and Cuq, 2001; Binet et al., 2012; Leeney et al., 2015b). They have been implicated in the captures of *S. teuszii* in areas adjacent to the Banc d'Arguin (Campredon and Cuq, 2001). Fishermen from Senegal, Guinea (Conakry), and Sierra Leone exploit the waters of Guinea Bissau, which has a limited artisanal fishing tradition; captures of dolphins and manatees, and declines of nesting sea turtles have been reported, raising concern for *S. teuszii* (Almeida e Silva et al., 1999; Campredon and Cuq, 2001; Leeney et al., 2015b; Maigret, 1994; Powell, 1990).

Fisheries declines in general may affect populations of *S. teuszii* in a variety of ways, including reductions in the prey base and increased artisanal fishing effort, leading to greater risk of capture (Weir and Pierce, 2013). Industrial fisheries compound this problem by competing for resources, as well as fishing in illegal zones. For instance, trawlers fishing illegally within Conkouati Douli National Park (Republic of Congo) influence artisanal fishermen to set their nets closer to shore (for fear of losing their nets in trawls), raising bycatch risks for coastal species (Collins et al., 2013). Upstream water diversion and dams have also been blamed for declines in fisheries and mangroves in Senegal (Campredon and Cuq, 2001).

Figure 7 Coastal developments necessarily affect the limited inshore habitats of *Sousa teuszii*; ports and other large urban areas may present barriers to longshore movement.

Coastal developments and their varied cumulative effects represent largely understated, unquantified and poorly recognized threats for *S. teuszii* (Weir et al., 2011). The species is innately timid, and often noted avoidance of research boats is seemingly prognostic of their susceptibility to disturbance (Van Waerebeek et al., 2004). Port developments are particularly prevalent, and preferred sites frequently overlap with the habitats of coastal dolphins; the scale of some ports suggests that they present effective physical barriers, and thus have potential for disrupting longshore movements (see Figure 7). Indirect or "non-lethal" disturbances are likely during construction, and may become more permanent if maintenance (e.g. dredging) and urban development occurs at port sites (Jefferson et al., 2009). Work in Hong Kong has demonstrated that vessel collisions and disturbance are of major concern for *S. chinensis* (Jefferson et al., 2006). Animals seeking to avoid disturbance might also be forced into less optimal areas (Ayissi et al., 2014). Work on other species, e.g. bottlenose dolphins (*Tursiops* spp.), has indicated that short-term disturbances to individuals may translate to longer term consequences for population health and fecundity (Lusseau et al., 2007; Nowacek et al., 2001; Pirotta et al., 2013; and see New et al., 2013). Moreover, population-level consequences are more likely to arise when populations are already facing other ecological constraints.

Anthropogenic noise is an identified threat for cetaceans in general (e.g. Weilgart, 2007), and for humpback dolphins specifically; noise from shipping and other industrial activities are described threats for *S. chinensis* in Taiwan (Wang et al., 2007) and Hong Kong (Jefferson et al., 2009). Work by Van Parijs and Corkeron (2001) indicates that Australian humpback dolphins (*Sousa sahulensis*) may increase their vocalization rates following the passage of motor boats, suggesting that motor boats affect group cohesion in the short term. Noise associated with development may also lead to disturbance of prey species, many of which are also vocal (Slabbekoorn et al., 2010).

7.2 Minor Threats

Environmental contamination is a potential threat in some areas and may increase as development accelerates. The habitat preferences of *S. teuszii* increase the potential for exposure to terrestrial sources of pathogens and pollution (Weir et al., 2011). Phosphorite mining in coastal Togo contaminates inshore sediments with trace elements, including cadmium (Segniagbeto and Van Waerebeek, 2010). Oil exploration and development are underway in many marine areas within the species' range. Spills and chronic pollution associated with some of these facilities may be detrimental to dolphin health but remain completely uninvestigated in *S. teuszii*.

8. CONSERVATION ACTIONS

Conserving the Atlantic humpback dolphin requires the proactive and urgent implementation of practical conservation measures (Ayissi et al., 2014; IWC, 2011; Maigret, 1994). Of utmost concern is improving the understanding of their decline, and confronting its primary causes, particularly fisheries bycatch, directed hunts and habitat loss (Perrin and Van Waerebeek, 2007; Weir et al., 2011). Inshore gillnet fisheries are considered the major threat, and explicit fisheries management measures to reduce bycatch need to be sought and implemented. Addressing fundamental ecological and biological data gaps is also vital, particularly for improving conservation and management decisions. This needs to include efforts to generate robust and comparable abundance estimates for areas of described occurrence, as well as areas that are under sampled. Surveys for this species can be challenging, but the biggest limitation to date has been an absence of sufficient and targeted resources (Ayissi et al., 2014; Collins et al., 2010; Maigret, 1980; Reeves et al., 2003; Van Waerebeek et al., 2004; Weir et al., 2011).

The scale of coastal development and associated anthropogenic pressures makes it imperative that the identification and protection of critical habitats (hotspots, migratory pathways, foraging sites) becomes a cornerstone of conservation management efforts. Nine range states (Western Sahara, Mauritania, Senegal, Gambia, Guinea Bissau, Guinea, Cameroon, Gabon and Republic of Congo) and two states within the wider range (Equatorial Guinea and Democratic Republic of Congo), have Marine Protected Areas (MPAs) that incorporate proven or potential *S. teuszii* habitats (Hoyt, 2012; Weir et al., 2011). It is clear that their effectiveness varies; artisanal fishing and associated bycatches within these MPAs are known issues (Collins et al., 2013; Hoyt, 2012; Van Waerebeek et al., 2003; Weir and Pierce, 2013), and financing and governance are habitual problems (e.g. Gallegos et al., 2005; Bennett and Dearden, 2014). States that have committed to MPA targets under the Convention on Biological Diversity (COP10, 2010) should include consideration of critical habitats for *S. teuszii* as part of spatial planning efforts toward MPA creation. Explicit consideration of *S. teuszii* (and other threatened taxa) must also be included in offset programmes associated with large-scale coastal developments (e.g. DEFRA, 2011; BBOP, 2012). If appropriately directed, these could help kick-start the effective management of important species and habitats (e.g. Bull et al., 2013; Pilgrim and Bennun, 2014).

Significant conservation concerns for *S. teuszii* have been raised for decades (e.g. Cadenat and Paraiso, 1957; Maigret, 1980; Van Waerebeek, 2003; Van Waerebeek et al., 2004; Weir et al., 2011). In 2003, the IUCN Species Survival Commission's Cetacean Specialist Group identified *S. teuszii* as a high priority for research and conservation (Reeves et al., 2003). A range of specific research and conservation objectives for Atlantic humpback dolphins were also identified by the Small Cetacean Sub-Committee of the International Whaling Commission (IWC) Scientific Committee (IWC, 2011). They represent a balanced consensus of scientific opinion and included deliberate consideration of earlier strategies, including the CMS *Action Plan for the Conservation of Small Cetaceans of Western Africa and Macaronesia* (CMS 2008; Perrin and Van Waerebeek, 2007). Immediate action is clearly required if additional losses, local extirpations, and potentially even the extinction of the species, are to be avoided in the next few decades.

9. IUCN RED LIST JUSTIFICATION

The Atlantic humpback dolphin has been described as rare for most of the period it has been known. Although there are no empirically based abundance estimates for either the species, or any of the putative populations, the consensus of expert opinion is that the species occurs in very low and declining numbers throughout most of its range.

Similar to other species in the genus, *S. teuszii* has very precise habitat requirements, limiting its resilience and its ability to escape environmental and anthropogenic threats (Davidson et al., 2011). Remnant populations are faced with a wide suite of "largely intractable threats" (Ayissi et al., 2014) that either affect them directly, or contribute to the rapid and widening deterioration of their inshore habitats (Perrin and Van Waerebeek, 2007; Van Waerebeek et al., 2004; Weir et al., 2011). Bycatch is the greatest immediate concern and available data indicate high rates of fisheries mortality in some areas, a trend that has also been suspected in areas where data are lacking (Perrin and Van Waerebeek, 2007; Van Waerebeek et al., 2004; Collins et al., 2013; Leeney et al., 2015a). Other biological factors, including a loss of genetic diversity through inbreeding depression (Lacy, 1997) and low immigration rates (Pimm et al., 1988) are likely, and would increase their susceptibility to environmental stochasticity (e.g. Frankham, 2005; Weir et al., 2011).

It is likely that *S. teuszii* has a low reproductive rate (Jefferson and Rosenbaum, 2014; Moore, 2015; Taylor et al. 2007). Given small apparent population sizes, any mortality over and above natural rates is likely to lead to significant declines in abundance (*sensu* Pimm et al., 1988). Moore (2015) estimated that given an inferred generation time of 25 years (as estimated for *S. plumbea* and *S. chinensis*), an annual average adult mortality rate of ~4% across the range would lead to a 50% decline over 75 years (three generations), qualifying *S. teuszii* for Endangered. However, a slightly higher adult mortality rate of 5.3% per year (equivalent to one or two additional deaths per year per 100 mature animals) would qualify the same population for Critically Endangered status (80% decline after 75 years). Other less evident factors, including a lack of immigration, density dependent resource limitation and ecosystem impacts that compromise vital rates could also have deleterious effects on population growth rates (Gilpin, 1986; Moore, 2015; Thompson et al., 2000).

Most populations for which any data are available are extremely small; the available published estimates suggest that the total species' population size plausibly falls well below 3000 individuals, and that the number of mature individuals is less than 1500 (Taylor et al., 2007). Declines have been observed or are suspected for each population, and continued declines are strongly likely given projected expansions of identified threats that affect most of the species' known range. Bycatch in fisheries, the principal cause of these declines, have not ceased and may even be increasing. Appropriate management interventions that limit habitat loss and mortalities from bycatch and hunting are non-existent in much of the range, and in the absence of directed and sustained conservation management efforts, long-term prospects are grim. The available information, much of it characterized by uncertainty, suggests that the Atlantic humpback dolphin merits consideration of an IUCN Red List classification of Critically Endangered (CR) under Criterion A3cd. This is based on a suspected and observed decline of $\geq$80% over three *S. teuszii* generations ($\sim$75 years), with declines likely to have begun with the expansion of coastal fisheries during the 1980s, and bycatches likely to increase as new areas are targeted and fishery pressures increase.

ACKNOWLEDGEMENTS

The best recent information on *S. teuszii* is to be found in the comprehensive reports of the UNEP/CMS-WAFCET projects on which this review relies heavily. The initiative, led by Koen Van Waerebeek, provided a foundation on which further conservation efforts for this species have been built. The work inspired research and conservation efforts in Gabon and Congo, and all other recent work uses the resulting and seminal publication by Van Waerebeek et al. (2004) as a benchmark reference. T.C. is grateful to the staff of the WCS Library for their excellent help, particularly Debra Levinson and Kerry Prendergast.

REFERENCES

Alfaro-Shigueto, J., Van Waerebeek, K., 2001. Drowning in the sea of silence: the bushmeat concept applied for marine fauna. In: Fourth Biennial Zoos and Aquariums: Committing to Conservation Conference, Cocoa Beach, Florida, 28 November–2 December 2001.

Almeida e Silva, M.A., Araujo, A., Djedjo, F., Gomes, L., Monteiro, H., 1999. Plano Nacional de Conservaçao do Manatim Africano (*Trichechus senegalensis*) na Guiné Bissau. UICN-Instituto para a Conservaçao da Natureza, Bissau.

Ayissi, I., Segniagbeto, G.H., Van Waerebeek, K., 2014. Rediscovery of Cameroon dolphin, the Gulf of Guinea Population of *Sousa teuszii* (Kükenthal, 1892). Int. Sch. Res. Notices Biodiv. 2014, 819827.

Bakhayokho, M., Kébé, M., 1991. Réactions des pêcheurs face aux variations d'abondance et de disponibilité des ressources: approche méthodologique. In: Durand, J.R., Lemalle, J., Weber, J. (Eds.), La Recherche Face à la Pêche Artisanale. Symposium International. vol. 2. ORSTOM, France, pp. 943–955.

Bamy, I.L., Van Waerebeek, K., Bah, S.S., Dia, M., Kaba, B., Keita, N., Konate, S., 2010. Species occurrence of cetaceans in Guinea, including humpback whales with southern hemisphere seasonality. Mar. Biodivers. Rec. 3, e48.

BBOP (Business and Biodiversity Offsets Programme), 2012. Biodiversity Offsets: Principles. Criteria and Indicators, Forest Trends, Washington, DC, USA.

Beaubrun, P.C., 1990. Un Cétacé nouveau pour les côtes sud-marocaines: *Sousa teuszii* (Kukenthal, 1892). Mammalia 54, 162–164.

Belhabib, D., Sumaila, U.R., Pauly, D., 2015. Feeding the poor: contribution of West African fisheries to employment and food security. Ocean Coast. Manag. 111, 72–81.

Bennett, N.J., Dearden, P., 2014. From measuring outcomes to providing inputs: governance, management, and local development for more effective marine protected areas. Mar. Policy 50, 96–110.

Binet, T., Failler, P., Thorpe, A., 2012. Migration of Senegalese fishers: a case for regional approach to management. Marit. Stud. 11 (1), 1–14.
Brashares, J.S., Arcese, P., Sam, M.K., Coppolillo, P.B., Sinclair, A.R.E., Balmford, A., 2004. Bushmeat hunting, wildlife declines, and fish supply in West Africa. Science 306, 1180–1183.
Bull, J.W., Suttle, K.B., Gordon, A., Singh, N.J., Milner-Gulland, E.J., 2013. Biodiversity offsets in theory and practice. Oryx 47, 369–380.
Busnel, R.G., 1973. Symbiotic relationship between man and dolphins. Trans. N. Y. Acad. Sci. 35 (Series II), 112–131.
Cadenat, J., 1949. Notes sur les cétacés observés sur les côtes du sénégal de 1941 à 1948. Bulletin del 'Institut Français d'Afrique Noire 11, 1–15.
Cadenat, J., 1956. Un delphinidae encore mal connu de la côte occidentale d'Afrique: *Sotalia teuszii* Kükenthal 1892. Bull. IFAN 18A, 555–566.
Cadenat, J., 1957. Observations de cétacés, siréniens, chéloniens et sauriens en 1955–1956. Bulletin de l'IFAN 19A (4), 1358–1369.
Cadenat, J., 1959. Rapport sur les petits Cétacés Ouest-Africains. Résultats des recherches entreprises sur ces animaux jusqu'au mois de mars 1959. Bull. IFAN 21A, 1367–1409.
Cadenat, J., Paraiso, F., 1957. Nouvelle observation de *Sotalia teuszii* (Cétacé, Delphinidé) sur les côtes du Sénégal. Bull. IFAN 19A, 324–332.
Campredon, P., Cuq, F., 2001. Artisanal fishing and coastal conservation in West Africa. J. Coast. Conserv. 7, 91–100.
Clapham, P., Van Waerebeek, K., 2007. Bushmeat and bycatch: the sum of the parts. Mol. Ecol. 16, 2607–2609.
CMS, 2008. Memorandum of understanding concerning the conservation of the manatee and small cetaceans of western Africa and Macaronesia. In: Convention on Migratory Species. http://www.cms.int/en/legalinstrument/western-african-aquatic-mammals. (accessed 28.05.2015).
Collins, T., Ngouessono, S., Rosenbaum, H.C., 2004. A note on recent surveys for Atlantic humpback dolphins, *Sousa teuszii* (Kukenthal, 1892) in the coastal waters of Gabon. In: Paper SC/56/SM23 Presented to the IWC Scientific Committee, June 2004. (unpublished; available from the office of this journal).
Collins, T., Boumba, R., Thonio, J., Parnell, R., Vanleeuwe, H., Ngouessono, S., Rosenbaum, H.C., 2010. The Atlantic humpback dolphin (*Sousa teuszii*) in Gabon and Congo: cause for optimism or concern. In: Paper SC/62/SM9 Presented to the IWC Scientific Committee, June 2010. (unpublished; available from the office of this journal).
Collins, T., Stindberg, S., Boumba, R., Dilambaka, E., Thonio, J., Mouissou, C., Boukaka, R., Saffou, G.K., Buckland, L., Leeney, R., Antunes, R., Rosenbaum, H.C., 2013. Progress on Atlantic humpback dolphin conservation and research efforts in Congo and Gabon. In: Paper SC/65a/SM16rev Presented to the IWC Scientific Committee. (unpublished; available from the office of this journal).
COP10, 2010. Decision X/2, Strategic Plan for Biodiversity 2011–2020. In: Conference of the Parties to the Convention on Biological Diversity. Convention on Biological Diversity, Montreal, Canada.
Cormier-Salem, M.C., 2000. Appropriation des ressources, enjeu foncier et espace halieutique sur le littoral ouest-africain. In: Chauveau, J.-P., Jul-Larsen, E., Chaboud, C. (Eds.), Les Pêches Piroguières en Afrique de l'Ouest: Pouvoirs, Mobilités, Marchés. Karthala, Paris, pp. 205–229.
Culik, B.M., 2004. Review of small cetaceans. Distribution, behaviour, migration and threats: Compiled for CMS/UNEP. Marine Mammal Action Plan/Regional Seas reports and studies, 177. 343 pp.

Davidson, A.D., Boyer, A.G., Kim, H., Pompa-Mansilla, S., Hamilton, M.J., Costa, D.P., Ceballos, G., Brown, J.H., 2011. Drivers and hotspots of extinction risk in marine mammals. Proc. Natl. Acad. Sci. 109, 3395–3400. http://dx.doi.org/10.1073/pnas.1121469109.

Debrah, J.S., Ofori-Danson, P.K., Van Waerebeek, K., 2010. An update on the catch composition and other aspects of cetacean exploitation in Ghana. In: Paper SC/62/SM10 Presented to the IWC Scientific Committee, June 2010. (unpublished; available from the office of this journal).

DEFRA, 2011. Biodiversity Offsets: Guiding Principles for Biodiversity Offsetting. Department for the Environment, Food and Rural Affairs, London, UK.

Duffy-Tumasz, A., 2012. Migrant fishers in West Africa: roving bandits? Afr. Geogr. Rev. 31 (1), 50–62.

Duguy, R., 1976. Contribution à l'étude des mammifères marins de la côte nord-ouest Afrique. Rev. Trav. l'Inst Pêches Marit. 39, 321–332.

Dupuy, A.R., 1983. Données complémentaires concernant le statut des mammifères marins du Sénégal. Bull. IFAN 45A, 380–391.

Frankham, R., 2005. Genetics and extinction. Biol. Conserv. 126, 131–140.

Fraser, F.C., 1949. A specimen of *Sotalia teuszii* Kukenthal from the coast of Senegal. J. Mammal 30, 274–276.

Gallegos, V.L., Vaahtera, A., Wolfs, E., 2005. Sustainable Financing for Marine Protected Areas: Lessons from Indonesian MPAs Case Studies: Komodo and Ujung Kulon National Parks. IVM, Vrije Universiteit Amsterdam, Amsterdam.

Gilpin, M.E., 1986. Minimum viable populations: the processes of species extinctions. In: Soulé, M. (Ed.), Conservation Biology: The Science of Scarcity and Diversity. Sinauer Associates, Sunderland, MA, pp. 13–34.

Henriques, R., Potts, W.M., Santos, C.V., Sauer, W.H., Shaw, P.W., 2014. Population connectivity and phylogeography of a coastal fish, *Atractoscion aequidens* (Sciaenidae), across the Benguela Current region: evidence of an ancient vicariant event. PLoS One 9 (2), e87907.

Hooker, S.K., Gerber, L.R., 2004. Marine reserves as a tool for ecosystem-based management: the potential importance of megafauna. Bioscience 54 (1), 27–39.

Hoyt, E., 2012. Marine Protected Areas for Whales, Dolphins and Porpoises: A World Handbook for Cetacean Habitat Conservation and Planning. Routledge, Earthscan/ Taylor and Francis, London and New York, 479 pp.

IUCN Species Survival Commission, 2001. IUCN red list categories and criteria. IUCN.

IWC (International Whaling Commission), 2011. Report of the Sub Committee on Small Cetaceans. J. Cetac. Res. Manage. 12, 272–295.

Jefferson, T.A., Rosenbaum, H.C., 2014. Taxonomic revision of the humpback dolphins (*Sousa* spp.), and description of a new species from Australia. Mar. Mamm. Sci. 30, 1494–1541.

Jefferson, T.A., Van Waerebeek, K., 2004. Geographic variation in skull morphology of humpback dolphins (*Sousa* spp.). Aquat. Mamm. 30, 3–17.

Jefferson, T.A., Curry, B.E., Leatherwood, S., Powell, J.A., 1997. Dolphins and porpoises of West Africa: a review of records (Cetacea: Delphinidae, Phocoenidae). Mammalia 61, 87–108.

Jefferson, T.A., Hung, S.K., Lam, P.K.S., 2006. Strandings, mortality and morbidity of Indo-Pacific humpback dolphins in Hong Kong, with emphasis on the role of organochlorine contaminants. J. Cetac. Res. Manage. 8, 181–193.

Jefferson, T.A., Hung, S.K., Würsig, B., 2009. Protecting small cetaceans from coastal development: impact assessment and mitigation experience in Hong Kong. Mar. Policy 33, 305–311.

Kaczynski, V.M., Fluharty, D.L., 2002. European policies in West Africa: who benefits from fisheries agreements? Mar. Policy 26, 75–93.

Krömer, J.L., Insali, P., Gomes, M., 1994. Projet de développement durable de la Pêche Artisanale dans le Rio Grande de Buba. Bio-écologie et paramètres environnementaux. UICN, Ministère des Pêches de Guinée-Bissau. 119 pp. (Unpublished).

Kükenthal, W., 1891. Porpoises in African rivers. Nature 44, 175.

Kükenthal, W., 1892. *Sotalia teuszii* n. sp. ein pflanzenfressender (?) Delphin aus Kamerun. Zool. Jahrb. Abt. Syst. 6 (3), 442–446.

Lacy, R.C., 1997. Importance of genetic variation to the viability of mammalian populations. J. Mammal 78, 320–335.

Ladich, F., 2000. Acoustic communication and the evolution of hearing in fishes. Philos. Trans. R. Soc. Lond. 355B, 1285–1288.

Leeney, R.H., Dia, I.M., Dia, M., 2015a. Food, pharmacy, friend? Bycatch, direct take and consumption of dolphins in West Africa. Hum. Ecol. 43, 105–118.

Leeney, R.H., Weir, C.R., Campredon, P., Regalla, A., Foster, J., 2015b. Occurrence of Atlantic humpback (*Sousa teuszii*) and bottlenose (*Tursiops truncatus*) dolphins in the coastal waters of Guinea-Bissau, with an updated cetacean species checklist. J. Mar. Biol. Assoc. U.K. 1–9.

Lewison, R.L., Moore, J.E., 2012. Improving interview-based assessments of sea turtle and marine mammal bycatch in West Africa: putting fishing activity into a socio-economic context: Project report prepared for NOAA Office of International Affairs, September 2012.

Lusseau, D., Bejder, L., 2007. The long-term consequences of short-term responses to disturbance experiences from whalewatching impact assessment. Int. J. Comp. Psychol. 20, 228–236. Available at: http://escholarship.org/uc/item/42m224qc.

Maigret, J., 1980. Données nouvelles sur l'écologie du *Sousa teuszii* (Cetacea, Delphinidae) de la côte Ouest Africaine. Bull. IFAN 42A, 619–633.

Maigret, J., 1994. Marine mammals and fisheries along the West African coast. Rep. Int. Whal. Comm. 15 (special issue 15), 307–316.

Mendez, M., Subramaniam, A., Collins, T., Minton, G., Baldwin, R., Berggren, P., Särnblad, A., Amir, O.A., Peddemors, V.M., Karczmarski, L., Guissamulo, A., Rosenbaum, H.C., 2011. Molecular ecology meets remote sensing: environmental drivers to population structure of humpback dolphins in the Western Indian Ocean. Heredity 107 (4), 349–361.

Mendez, M., Jefferson, T.A., Kolokotronis, S.-O., Krützen, M., Parra, G.J., Collins, T., Minton, G., Baldwin, R., Berggren, P., Särnblad, A., Amir, O.A., Peddemors, V.M., Karczmarski, L., Guissamulo, A., Smith, B., Sutaria, D., Amato, G., Rosenbaum, H.C., 2013. Integrating multiple lines of evidence to better understand the evolutionary divergence of humpback dolphins along their entire distribution range: a new dolphin species in Australian waters? Mol. Ecol. 22, 5936–5948.

Minton, G., 2015. Report on gamba area marine surveys: July 2105. WWF Gabon. 21pp.

Moore, J.E., 2015. Intrinsic growth (r_{max}) and generation time (T) estimates for the cetacean genera *Sousa, Orcaella,* and *Neophocaena*, in support of IUCN Red List assessments. In: NOAA Technical Memorandum NOAA-TM-NMFS-SWFSC-550.

Moore, J.E., Cox, T.M., Lewison, R.L., Read, A.J., Bjordland, R., Mcdonald, S.L., Crowder, L.B., Aruna, E., Ayissi, I., Espeut, P., Joynson-Hicks, C., Pilvher, N., Poonian, C.N.S., Solarin, B., Kiszka, J., 2010. An interview-based approach to assess marine mammal and sea turtle captures in artisanal fisheries. Biol. Conserv. 143, 795–805.

Murphy, P., Van Waerebeek, K., Jallow, A., 1997. Cetaceans from Gambian coastal waters. In: Paper SC/49/ SM11 Presented to the Scientific Committee of the International Whaling Commission, Bournemouth, UK. (unpublished; available from the office of this journal).

New, L.F., Harwood, J., Thomas, L., Donovan, C., Clark, J.S., Hastie, G., Thompson, P.M., Cheney, B., Scott-Hayward, L., Lusseau, D., 2013. Modelling the biological significance of behavioural change in coastal bottlenose dolphins in response to disturbance. Funct. Ecol. 27, 314–322.

Nicholson, E., Westphal, M.I., Frank, K., Rochester, W.A., Pressey, R.L., Lindenmayer, D.B., Possingham, H.P., 2006. A new method for conservation planning for the persistence of multiple species. Ecol. Lett. 9, 1049–1060.

Northridge, S.P., 1984. World review of interactions between marine mammals and fisheries. FAO Fish. Tech. Pap. 251. 190 pp.

Notarbartolo di Sciara, G., Politi, E., Bayed, A., Beaubrun, P.C., Knowlton, A., 1998. A winter cetacean survey off southern Morocco, with a special emphasis on right whales. Rep. Int. Whal. Comm. 48, 547–550.

Nowacek, S. M., Wells, R.S., Solow, A. R., 2001. Short-term effects of boat traffic on bottlenose dolphins, *Tursiops truncatus*, in Sarasota Bay, Florida. Mar. Mamm. Sci. 17, 673-688.

Perrin, W.F., Van Waerebeek, K., 2007. The small-cetacean fauna of the west coast of Africa and Macaronesia: diversity and distribution. In: Western African Talks on Cetaceans and Their Habitats. UNEP/CMS-WATCH-Inf. 6, Convention on the Conservation of Migratory Species of Wild Animals, Adeje, Tenerife. Available at: www.cms.int/species/waam/watch1_docs/Inf06_Small_Cetacean_ Fauna_E.pdf.

Pilgrim, J.D., Bennun, L., 2014. Will biodiversity offsets save or sink protected areas? Conserv. Lett. 7, 423–424.

Pimm, S.L., Jones, H.L., Diamond, J., 1988. On the risk of extinction. Am. Nat. 132, 757–785.

Pirotta, E., Laesser, B.E., Hardaker, A., Riddoch, N., Marcoux, M., Lusseau, D., 2013. Dredging displaces bottlenose dolphins from an urbanised foraging patch. Mar. Pollut. Bull. 74, 396–402.

Powell, J.A., 1990. Manatees in the Bijagós Archipelago: recommendations for their conservation. IUCN Wetland Programme.

Powell, J.A., Jefferson, T.A., Curry, B.E., Leatherwood, S., Pankhurst, S., 1996. Marine mammals of West and Central Africa: a report on distribution and status: Report to UNEP, Nairobi, Kenya. (unpublished).

Reeves, R.R., 2000. The value of sanctuaries, parks, and reserves (protected areas) as tools for conserving marine mammals: Final report to the Marine Mammal Commission, contract number T74465385. Marine Mammal Commission, Bethesda, MD. 50 pp.

Reeves, R.R., Smith, B.D., Crespo, E.A., di Sciara, Notarbartolo, 2003. Dolphins, Whales and Porpoises: 2002–2010 Conservation Action Plan for the World's Cetaceans. IUCN/SSC Cetacean Specialist Group, IUCN, Gland, Switzerland and Cambridge, UK. ix + 139 pp.

Reeves, R.R., Dalebout, M.L., Jefferson, T.A., Karczmarski, L., Laidre, K., O'Corry-Crowe, G., Rojas-Bracho, L., Secchi, E.R., Slooten, E., Smith, B.D., Wang, J.Y., Zhou, K., 2008. *Sousa chinensis*. The IUCN Red List of Threatened Species. Version 2014.3. www.iucnredlist.org. Downloaded on 12 December 2014.

Robards, M.D., Reeves, R.R., 2011. The global extent and character of marine mammal consumption by humans: 1970–2009. Biol. Conserv. 144 (12), 2770–2786.

Robineau, D., Vely, M., 1998. Cetaceans of coastal Mauritania (NW Africa). Particularities and spatio-temporal variations of their distribution: the role of oceanographical factors. Rev. Ecol. (Terre Vie) 53, 123–152.

Ross, G.J., Heinsohn, G.E., Cockcroft, V.G., 1994. Humpback dolphins *Sousa chinensis* (Osbeck, 1765), *Sousa plumbea* (G. Cuvier, 1829) and *Sousa teuszii* (Kukenthal, 1892). In: Ridgway, S.H., Harrison, R.J. (Eds.), Handbook of Marine Mammals vol. 5. Academic Press, London, pp. 23–42.

Segniagbeto, G., Van Waerebeek, K., 2010. A note on the occurrence and status of cetaceans in Togo. In: Paper SC/62/SM11 Presented to the IWC Scientific Committee, June 2010. (unpublished; available from the office of this journal).

Sequeira, M., Reiner, F., 1992. First record of an Atlantic humpback dolphin, *Sousa teuszii* Kukenthal, 1892 (Cetacea; Delphinidae) in Guinea-Bissau. Mammalia 56, 311–313.

Slabbekoorn, H., Bouton, N., van Opzeeland, I., Coers, A., ten Cate, C., Popper, A.N., 2010. A noisy spring: the impact of globally rising underwater sound levels on fish. Trends Ecol. Evol. 25, 419–427.

Spaans, B., 1990. Dolphins in the coastal area of Guiné Bissau. Lutra 33, 126–133.

Taylor, B.L., Chivers, S.J., Larese, J., Perrin, W.F., 2007. Generation length and percent mature estimates for IUCN assessments of cetaceans: Southwest Fish. Sci. Cent. Admin. Rep. LJ-07-01. 18 pp.

Thompson, P.M., Wilson, B., Grellier, K., Hammond, P.S., 2000. Combining power analysis and population viability analysis to compare traditional and precautionary approaches to conservation of coastal cetaceans. Conserv. Biol. 14, 1253–1263.

Uwagbae, M., Van Waerebeek, K., 2010. Initial evidence of dolphin takes in the Niger Delta region and a review of Nigerian cetaceans. In: Paper SC/62/SM1 Presented to the IWC Scientific Committee, June 2010. (unpublished; available from the office of this journal).

Van Beneden, P.J.H., 1892. Un cétacé fluviatile d'Afrique. Bull. Acad. Roy. Sci. Belg. 3, 350–355.

van Bree, P., Duguy, R., 1965. Sur un crâne de *Sotalia teuszii* Kükenthal, 1892 (Cetacea, Delphinidae). Z. Säugetierkd. 30, 311–314.

Van Parijs, S.M., Corkeron, P.J., 2001. Boat traffic affects the acoustic behaviour of Pacific humpback dolphins, *Sousa chinensis*. J. Mar. Biol. Assoc. U.K. 81, 533–538.

Van Waerebeek, K., 2003. The Atlantic humpback dolphin: in retreat. CMS Bull. 17, 10–11.

Van Waerebeek, K., Ndiaye, E., Djiba, A., Diallo, M., Murphy, P., Jallow, A., Camara, A., Ndiaye, P., Tous, P., 2000. A Survey of the Conservation Status of Cetaceans in Senegal, The Gambia and Guinea-Bissau. UNEP/CMS Secretariat, Bonn, Germany. (unpublished), 80 pp.

Van Waerebeek, K., Barnett, L., Camara, A., Cham, A., Diallo, M., Djiba, A., Jallow, A., Ndiave, E., Ould-Bilal, A.O., Bamy, I.L., 2003. Conservation of cetaceans in the Gambia and Senegal 1999–2001, and status of the Atlantic humpback dolphin: WAFCET-2 report. UNEP/CMS, Bonn, Germany. (unpublished).

Van Waerebeek, K., Barnett, L., Camara, A., Cham, A., Diallo, M., Djiba, A., Jallow, A.O., Ndiaye, E., Samba Ould Bilal, A.O., Bamy, I.L., 2004. Distribution, status, and biology of the Atlantic humpback dolphin, *Sousa teuszii* (Kükenthal, 1892). Aquat. Mamm. 30, 56–83.

Van Waerebeek, K., Ofori-Danson, P.K., 1999. A first checklist of cetaceans of Ghana, Gulf of Guinea, and a shore-based survey of interactions with coastal fisheries. Paper SC/51/SM35 Presented to the IWC Scientific Committee. (unpublished; available from the office of this journal).

Van Waerebeek, K., Ofori-Danson, P.K., Debrah, J., 2009. The cetaceans of Ghana, a validated faunal checklist. West Afr. J. Appl. Ecol. 15, 61–90.

Vande weghe, J.P., 2005. Akanda et Pongara: plages et mangroves. In: Les Parc Nationaux du Gabon, The Wildlife Conservation Society, New York.

Wang, J.Y., Yang, S.C., Reeves, R.R. (Eds.), 2007. Conservation Action Plan for the Eastern Taiwan Strait Population of Indo-Pacific Humpback Dolphins. National Museum of Marine Biology and Aquarium, Checheng, Pingtung County, Taiwan. 4 pp. (English) + 4 pp. (Chinese).

Weilgart, L., 2007. The impacts of anthropogenic ocean noise on cetaceans and implications for management. Can. J. Zool. 85, 1091–1116.

Weir, C.R., 2009. Distribution, behaviour and photo-identification of Atlantic humpback dolphins *Sousa teuszii* off Flamingos, Angola. Afr. J. Mar. Sci. 31, 319–331.

Weir, C.R., 2015. Photo-identification and habitat use of Atlantic humpback dolphins *Sousa teuszii* around the Río Nuñez Estuary in Guinea. African Journal of Marine Science, West Africa. http://dx.doi.org/10.2989/1814232X.2015.1069757.

Weir, C.R., Collins, T., 2015. A review of the geographical distribution and habitat of the Atlantic humpback dolphin (*Sousa teuszii*). Adv. Mar. Biol. 72, 79–118.

Weir, C.R., Pierce, G.J., 2013. A review of the human activities impacting cetaceans in the eastern tropical Atlantic. Mamm. Rev. 43, 258–274.

Weir, C.R., Van Waerebeek, K., Jefferson, T.A., Collins, T., 2011. West Africa's Atlantic humpback dolphin (*Sousa teuszii*): endemic, enigmatic and soon endangered? Afr. Zool. 46, 1–17.

White, C., Selkoe, K.A., Watson, J., Siegel, D.A., Zacherl, D.C., Toonen, R.J., 2010. Ocean currents help explain population genetic structure. Proc. R. Soc. B. 277, 1685–1694. rspb20092214.

Zwart, S.J., Weir, C.R., 2014. Filling in the gaps: first record of *Sousa teuszii* in Benin (Gulf of Guinea: Africa). Mar. Biodivers. Rec. 7, e59.

CHAPTER FOUR

A Review of the Geographical Distribution and Habitat of the Atlantic Humpback Dolphin (*Sousa teuszii*)

Caroline R. Weir*[,1], Tim Collins†
*Ketos Ecology, Kingsbridge, Devon, United Kingdom
†Wildlife Conservation Society, Marine Program, Global Conservation, Bronx, New York, USA
[1]Corresponding author: e-mail address: caroline.weir@ketosecology.co.uk

Contents

1. Introduction 80
2. Methods 81
3. Distribution 85
   3.1 Temporal Trend in Records 85
   3.2 Global Distribution 86
   3.3 Distribution by Country 86
   3.4 Distribution by Ecoregion 97
   3.5 Distribution Gaps? 98
4. Group Size 98
5. Habitat 101
   5.1 Water Depth 102
   5.2 Distance from Shore 102
   5.3 Temperature 104
   5.4 Similarities and Differences 104
6. Conclusions and Future Work 105
Acknowledgements 106
Appendix A 106
   Note 1 106
   References to Tables A1 and A2 113
References 116

Abstract

Understanding of the distributional ecology of the Atlantic humpback dolphin (*Sousa teuszii*) has been hampered by a lack of systematic and consistent sampling effort. The only comprehensive species distribution review was published in 2004; since then a considerable amount of novel information has emerged. We compiled 853 sighting, capture and specimen records of the species, and produced global and regional

Advances in Marine Biology, Volume 72
ISSN 0065-2881
http://dx.doi.org/10.1016/bs.amb.2015.08.001

distribution maps. Of the 830 records where year was available, 63.1% dated from ≥2005 and confirm a contemporary occurrence in six marine ecoregions and 11 countries: Western Sahara, Mauritania, Senegal, Gambia, Guinea-Bissau, Guinea, Benin, Cameroon, Gabon, Congo Republic and Angola. Additionally, Togo is a recently confirmed range state. Group sizes ranged from 1 to 45 animals, with small groups of 1 to 10 animals comprising 65% of the sightings. Similarities were noted in the regions inhabited by Atlantic humpback dolphins across their range, particularly an occurrence in relatively shallow (predominantly ≤20 m) depths, in warm waters (average SSTs of 15.8–31.8 °C) and in dynamic habitat strongly influenced by tidal patterns. These conditions occur in various habitats occupied by the species, including estuarine systems, open coasts, archipelagos, tidal mud-flats and sheltered bays. Sightings were recorded at distances of 13 m to 12.8 km (mean of 573 m) from land, indicating that the species occurs several kilometres from shore when suitable shallow habitat is present. The Atlantic humpback dolphin may be a 'nearshore' species based on oceanographic definitions incorporating water depth, wave action and sedimentation rather than on spatial distance from the coast.

1. INTRODUCTION

Cetacean species inhabit dynamic marine systems, and a range of biotic and abiotic environmental parameters and intra- and inter-specific interactions may define their ecological niche and govern their spatio-temporal distribution. Some of the key environmental parameters affecting cetacean occurrence include water depth, seabed topography and sea-surface temperature (SST; e.g. Hamazaki, 2002; Kaschner et al., 2006). Understanding the habitat preferences of, and ecological niche inhabited by, a species is fundamental to informing conservation and management efforts, with particular relevance for species of high conservation concern.

The Atlantic humpback dolphin (*Sousa teuszii*) is one of only two cetacean species known to be endemic to the African continent, and the only member of the *Sousa* genus inhabiting the tropical Atlantic Ocean. Little information has been published on its ecology (Maigret, 1980; Van Waerebeek et al., 2004; Weir, 2009). Although considered a vulnerable species, robust evaluations of its global and regional distribution have been hampered by a lack of targeted systematic and consistent sampling effort. In a major review of its status and distribution, Van Waerebeek et al. (2004) described a discontinuous occurrence from Western Sahara to Angola.

There has been no previous attempt to plot all known records of the species worldwide, and the emergence of many new records in the decade following the review of Van Waerebeek et al. (2004) warrants a revised distribution assessment. Thus, this review has several objectives: (1) to map all available records of the Atlantic humpback dolphin to provide a comprehensive distribution and range state assessment; (2) to assess distribution, seasonality and group size according to marine ecoregion and (3) to carry out a preliminary assessment of at-sea distribution with respect to some core parameters known to influence cetacean habitat, specifically water depth, distance from shore and SST.

2. METHODS

A literature review was carried out to locate published Atlantic humpback dolphin records. Unpublished records were acquired from colleagues working along the African coast on marine mammals or other fauna. A summary of the data sources used in this chapter is provided in Appendix A. Species validation was carried out whenever photographs were available. However, the majority of records were lacking in supporting information and have been accepted as identified by the authors. The dorsal profile of Atlantic humpback dolphins is very distinctive (and obviously different from the sympatric common bottlenose dolphin, *Tursiops truncatus*) and misidentification by biologists is likely uncommon. 'Probable' records were not included.

Only individual records for which a locality (GPS or written description) was specified were entered into the resulting database ($n=853$). Records were allocated an accuracy code of 1–7 based on the quality of the positional data provided (Table 1). For written locality descriptions, the latitude and longitude were estimated using GoogleEarth. Eight specimen records where the only locality was the country of origin were omitted. Additionally, the locality of a recent sighting off Togo has not yet been published (see Appendix A). Consequently, Togo is acknowledged as a range state, but that record is not included in this analysis. For most records, an individual date (at least year; $n=830$) and number of animals ($n=807$) were available. Group size was standardised as described in Leeney et al. (2015). Papers that documented overall Atlantic humpback dolphin occurrence without providing these specific record details (e.g. Dupuy, 1983) were omitted. Authors of summarised Atlantic humpback dolphin data (e.g. Notarbartolo di Sciara et al., 1998; Spaans, 1990) were contacted to request individual records. However, in most cases these were no longer available (see Appendix A).

Table 1 Definition of Accuracy Code (AC) Assigned to the Geographic Positions of 853 Atlantic Humpback Dolphin (*Sousa teuszii*) Records

Accuracy Code	km	Definition	*n* ASRs	*n* ORs
1	1	GPS position accurate to ≤1 km	275	7
2	2	GPS position accurate to ≤2 km or position estimated from symbol location on a published figure	116	1
3	2	Position calculated from place name and specified distance from shore (or known to be ≤1 km offshore or within a narrow waterway)	282	13
4	5	Position calculated from place name (no distance information) or location provided as waterway ≤5 km in extent	72	6
5	10	Only waterway or vague locality (≤10 km in extent) provided and position plotted centrally	24	3
6	10	Only waterway or vague locality (≥10 km in extent) provided and position plotted centrally	31	7
7	N/A	Landing site, actual position of capture unknown	0	16

Records were assigned to two categories: (1) at-sea records (ASRs) including sightings, captures and bycatch records for which the given locality reflected a position at sea and therefore a correct representation of at-sea distribution ($n=800$) and (2) other records (ORs) comprising all records where the reported locality related to terrestrial sites of strandings, landings (where actual capture position was unclear), carcasses and skeletal remains ($n=53$). Due to uncertainty regarding their origin, a few other miscellaneous records were categorised as ORs, for example, Kükenthal's (1892) skull, the origin of which was seemingly confused with that of the carcass of a West African manatee (*Trichechus senegalensis*) retrieved from Man O'War Bay in Cameroon. The OR category may provide an inaccurate representation of species distribution (e.g. if animals died at sea and drifted or were landed some distance from their capture position).

Quantum Geographic Information System (QGIS) (v2.6.1, www.qgis.org) was used to map all records. Records were assigned to country using an exclusive economic zone (EEZ) shapefile (v8, 28 February 2014; www.marineregions.org). Although allocating cetacean records to country is valuable when considering national legislation and management, the maritime borders designating national EEZs are of little ecological significance.

Consequently, distribution, seasonality and group size were instead analysed according to marine ecoregion (Spalding et al., 2007; Table 2) using a GIS shapefile (http://data.unep-wcmc.org). Provinces are large areas defined by the presence of distinct biotas which have arisen as a result of distinctive abiotic features that circumscribe their boundaries. Ecoregions are areas of relatively homogeneous species composition, clearly distinct from adjacent systems and determined by oceanographic or topographic features. The four Gulf of Guinea ecoregions in Table 2 are a relatively homogeneous grouping and are considered as a single ecoregion by some, while others consider them to represent distinct units (M.D. Spalding, The Nature Conservancy, personal communication, 5 January 2015).

Only ASRs of AC 1–3 ($n=673$) were used for habitat analysis, resulting in a positional accuracy of ≤2 km. A GEBCO (General Bathymetric Chart

Table 2 Marine Ecoregions (Spalding et al., 2007) Within the Atlantic Humpback Dolphin (*Sousa teuszii*) Global Distribution Range. A brief summary of characteristics defining each ecoregion (M.D. Spalding, The Nature Conservancy, personal communication, 5 January 2015) is provided

Province—Ecoregion	Boundaries	Characteristics	*n* ASRs	*n* ORs
Lusitanian—Saharan Upwelling	Strait of Gibraltar south to the Banc d'Arguin in Mauritania	Warm temperate region influenced by the Canary Current. Characterised by coastal upwelling. The southern boundary, close to Cap Blanc, is a major biogeographic boundary.	16	1
West African Transition—Sahelian Upwelling	Banc d'Arguin to Senegal	Tropical region with mangroves throughout, from extensive, diverse forests in the southern estuaries to remnant dwarfed populations in northern Mauritania. The area between Cape Vert and Cap Blanc experiences seasonal winter upwelling.	186	37
Gulf of Guinea—Gulf of Guinea West	Guinea-Bissau to Liberia	Tropical region. Largest continental shelf in West Africa with large riverine input producing thermal stability. Tropical mangrove and estuarine communities.	119	3

Continued

Table 2 Marine Ecoregions (Spalding et al., 2007) Within the Atlantic Humpback Dolphin (*Sousa teuszii*) Global Distribution Range. A brief summary of characteristics defining each ecoregion (M.D. Spalding, The Nature Conservancy, personal communication, 5 January 2015) is provided—cont'd

Province—Ecoregion	Boundaries	Characteristics	*n*	
			ASRs	**ORs**
Gulf of Guinea—Gulf of Guinea Upwelling	Côte d'Ivoire to Accra, Ghana	Tropical region. Thermally unstable subsystem characterised by seasonal upwelling of cold, nutrient-rich, subthermocline water with fluctuating productivity.	0	0
Gulf of Guinea—Gulf of Guinea Central	Accra, Ghana to Cap Lopez, Gabon	Tropical region. Characterised by thermal stability and a strong pycnocline, with variable freshwater input from the Niger Delta and Cameroon Estuary depending on monsoons.	19	3
Gulf of Guinea—Gulf of Guinea South	Cap Lopez, Gabon to northern Angola	Tropical region. Extensive coastal wetlands, high riverine input and expansive mangrove areas. Characterised by the southward-flowing nutrient-poor Angola Current.	385	9
Gulf of Guinea—Angolan	Angola	Sub-tropical transitional region influenced by the Angola Current and extending south to the Angola-Benguela Frontal Zone (14–17°S). Characterised by coastal wetlands and mangrove in the north, but by arid coastline south of 12.5°S.	75	0

of the Oceans) 2014 global 30 arc-second grid of water depth data and a raster file (4 km^2 grid) of mean annual global SST averaged for 2003–2007 (see Kershaw, 2008) were imported into QGIS and values extracted for each ASR. The initial distance of each ASR was measured (using a QGIS plugin) from a high-resolution coast shapefile downloaded from the GADM database of Global Administrative Areas (v2.0, January 2012; www.gadm.org/). Statistical tests to analyse variation in water depth, distance from shore, SST and group size were carried out using Minitab 16 statistical software (Minitab Inc., 2010).

3. DISTRIBUTION

3.1 Temporal Trend in Records

Although the Atlantic humpback dolphin was first described off Cameroon in 1892 (Kükenthal, 1892), only a handful of further specimens were collected prior to the 1950s: one from Senegal in 1925 (van Bree and Duguy, 1965), three from Senegal in the 1940s (Cadenat, 1947, 1949) and one from Gabon in the 1940s (Van Waerebeek et al., 2004). The first at-sea sighting (off Guinea) was not until 1953 (Cadenat, 1956), and the next did not occur until 20 years later off Mauritania (Busnel, 1973). Consequently, species occurrence has been poorly documented until relatively recently.

An assessment of the individual Atlantic humpback dolphin records compiled for this chapter (Table 3) indicates that of the 830 records where

Table 3 Summary by Range State[a] of 830 Atlantic Humpback Dolphin (*Sousa teuszii*) Records for Which Year Was Available (1892–2015)

		At-Sea Records (ASRs)/Other Records (ORs)			
Country	**Total Records**	**<1970**	**1970–1989**	**1990–2004**	**≥2005**
Western Sahara	10	–	2/0	4/1	3/0
Mauritania	121[b]	0/1	102/5	3/5	5/0
Senegal	76	8/7	29/1	12/6	13/0
Gambia	10	–	–	5/1	2/2
Guinea-Bissau	112	–	55/0	37/1	18/1
Guinea	10	1/0	–	0/1	8/0
Benin	1	–	–	–	1/0
Cameroon	3	0/1	–	–	1/1
Gabon	119	0/1	–	12/1	105/0
Congo Republic	293	–	–	–	285/8
Angola	75	–	–	4/0	71/0
Total	830	9/10	188/6	77/16	512/12

[a]Togo is also a recently confirmed range state but is not included in this analysis (see Section 2).
[b]The 23 records that lacked year information were all from Mauritania.

year was available, 63.1% dated from ≥2005; i.e. within the most recent decade and after the review of Van Waerebeek et al. (2004). Additionally, the proportion of ASRs has increased from 89.5% ≤2004 to 97.7% ≥2005. These trends reflect both increased logging of opportunistic records and the occurrence of several systematic scientific surveys focussed on Atlantic humpback dolphins within the last decade, for example, off Gabon, Congo Republic and Angola (e.g. Collins et al., 2010; Weir, 2009).

3.2 Global Distribution

The distribution of both ASRs (Figure 1) and ORs (Figure 2) indicates a global occurrence limited to warm Atlantic African waters. All records fell within previously documented northern and southern latitudinal limits of 23°55′N (Beaubrun, 1990) and 15°38′S (Weir, 2009), respectively. Unusually, both known latitudinal limits are marked by at-sea sightings of apparently resident populations (in Dakhla Bay, Western Sahara, and Namibe Province, southern Angola) rather than by vagrant records or strandings as is commonly the case for cetaceans. No Atlantic humpback dolphin records are yet confirmed for Sierra Leone, Liberia, Côte d'Ivoire, Ghana, Nigeria, Equatorial Guinea, or the Democratic Republic of the Congo. However, no systematic at-sea cetacean monitoring programmes exist in those countries. We emphasise caution in interpretation because the illustrated data are 'presence only'. Consequently, the apparent distribution 'gaps' and the concentrations of sightings shown in Figures 1–3 may represent areas lacking at-sea survey effort or areas of greater reporting, respectively.

3.3 Distribution by Country

Atlantic humpback dolphins are currently confirmed from 12 range states (Table 3). Movements between countries occur, for example, between Senegal and Gambia (Van Waerebeek et al., 2004) and between Gabon and the Congo Republic (Collins et al., 2010).

3.3.1 Western Sahara

The northernmost location where the Atlantic humpback dolphin has been recorded is approximately 23°55′N in Western Sahara. All known records from Western Sahara are currently limited to the innermost portion of Dakhla Bay (Figure 3A), and its occurrence in the remainder of Dakhla Bay or elsewhere in the country is unknown.

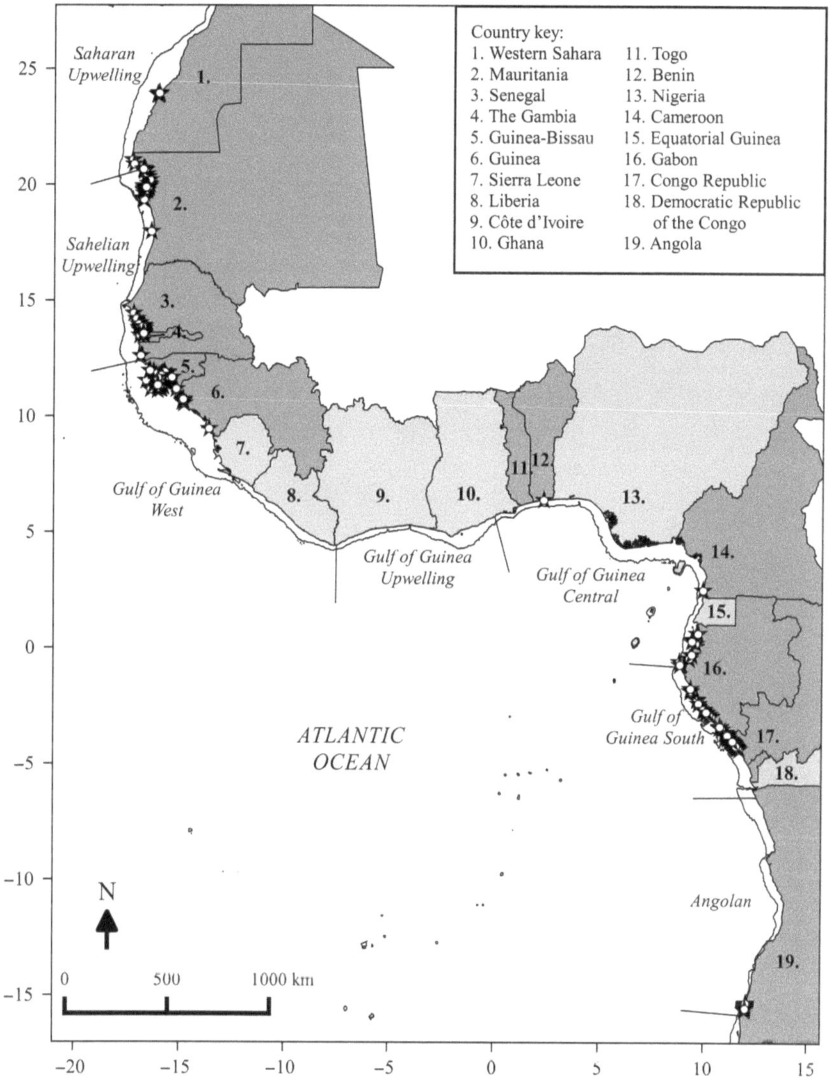

Figure 1 Location of 800 at-sea records (ASRs; sightings, captures and bycatch) of Atlantic humpback dolphins (*Sousa teuszii*), 1942–2015. The 200-m isobath and the boundaries between the marine ecoregions (Spalding et al., 2007) are shown. Confirmed range states are shown in darker grey.

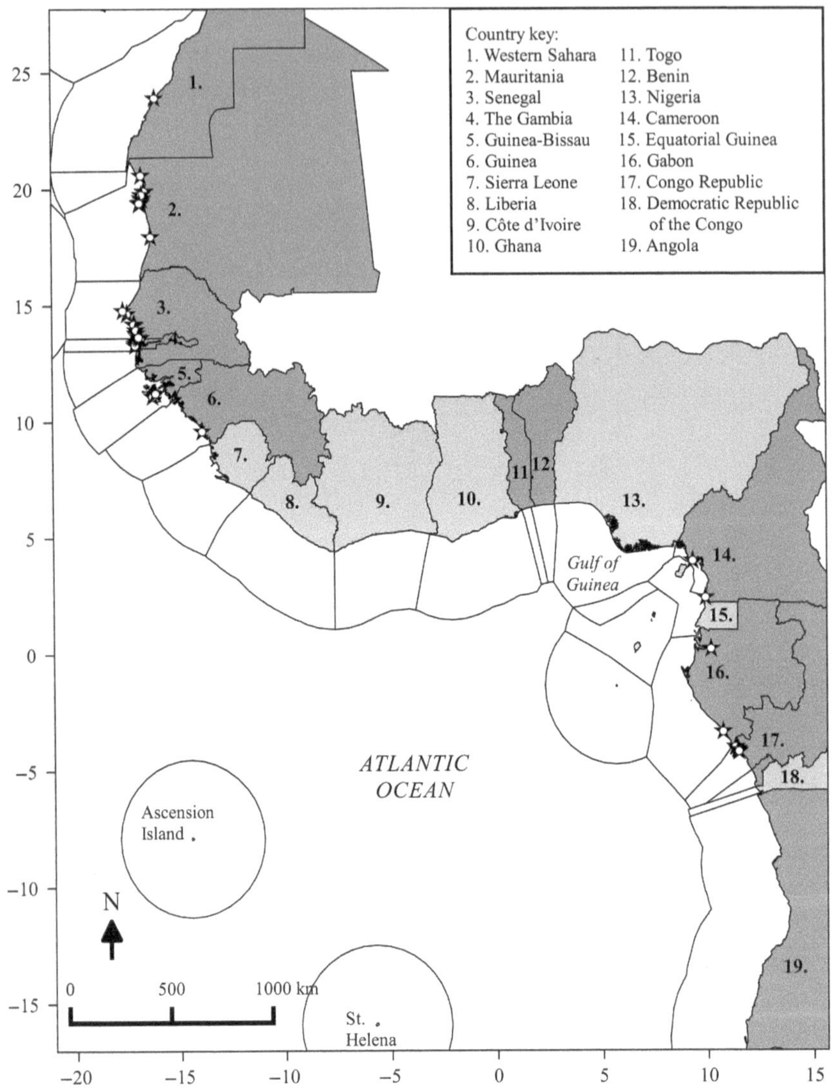

Figure 2 Location of 53 other records (ORs; i.e. strandings, landings, carcasses and skeletal remains) of Atlantic humpback dolphins (*Sousa teuszii*), 1891–2014. The boundaries of the exclusive economic zones (EEZs) are shown.

3.3.2 Mauritania

In northern Mauritania near Nouadhibou, there are four sightings from Baie de l'Étoile and three from the Banc de la Corbine in the Baie du Lévrier (Figure 3B). However, the majority of Mauritanian records are from the

Parc National du Banc d'Arguin (PNBA), a World Heritage site located between Nouadhibou and Nouamghar. In the northern part of the PNBA, sightings are clustered in Baie d'Arguin both along the mainland coast and around the islands. There are sporadic records southwards along the coast,

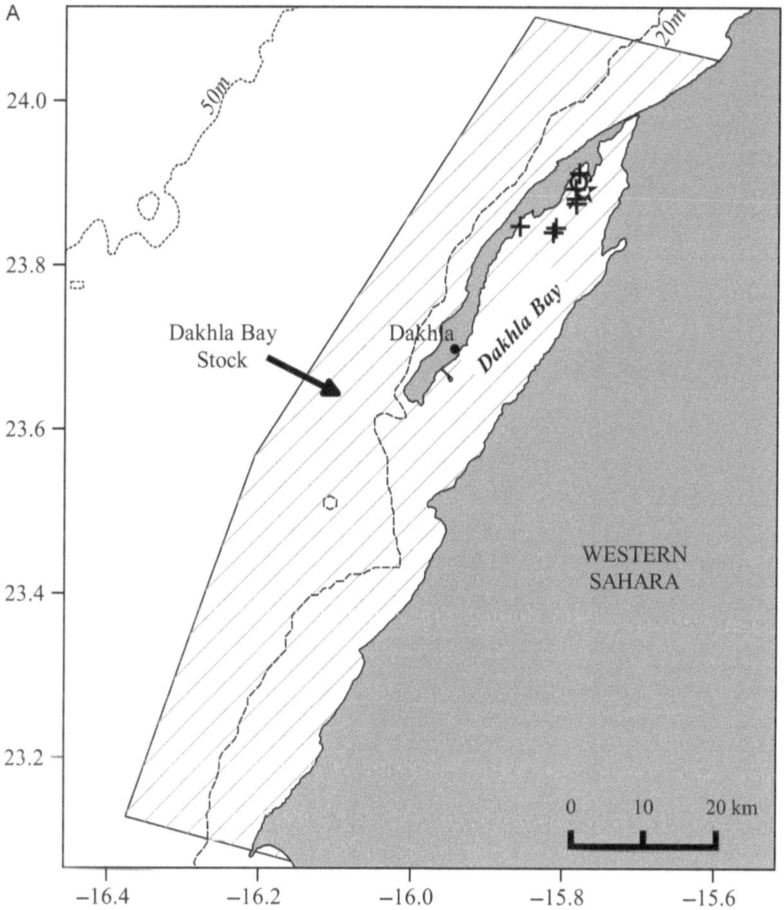

Figure 3A Fine-scale distribution of 853 Atlantic humpback dolphin records (cross symbol: at-sea records (ASRs), with Accuracy Code 1–3; circle symbol: ASRs with Accuracy Code 4–6; star symbol: Other records, ORs) in: (A) Western Sahara, (B) Mauritania (enlargement: Banc d'Arguin), (C) Senegal and Gambia (enlargement: Saloum Delta), (D) Guinea-Bissau, (E) Guinea, (F) Cameroon, (G) Gabon and Congo Republic and (H) Angola. The stocks of Van Waerebeek et al. (2004) are shown. The coast shapefile was accessed from the GADM database of Global Administrative Areas (v2.0, January 2012; www.gadm.org/); the inland extent of waterways is not fully portrayed.

(Continued)

including off Cap El Zass, Baie de Tanoudert, Ten Alloul and Îles Kiaones. However, most records originate from a relatively small area of the PNBA comprising the waters between Iwik and Baie Aouatif on the mainland and the islands of Arel, Niroumi and Nair to the north of Île Tidra. The PNBA base is located at Iwik, which may contribute to explaining the greater number of records in this region. In the southern part of the PNBA, there are scattered records from around Îles Tidra, Kiji and Cheddid and on the mainland at Teichott, R'Gueiba and along the Cap Timiris peninsula to the Grande Plage south of Nouamghar. There are only two records reported along the approximately 380 km stretch of Mauritanian coast south of the PNBA: a sighting and a stranding south of Nouakchott.

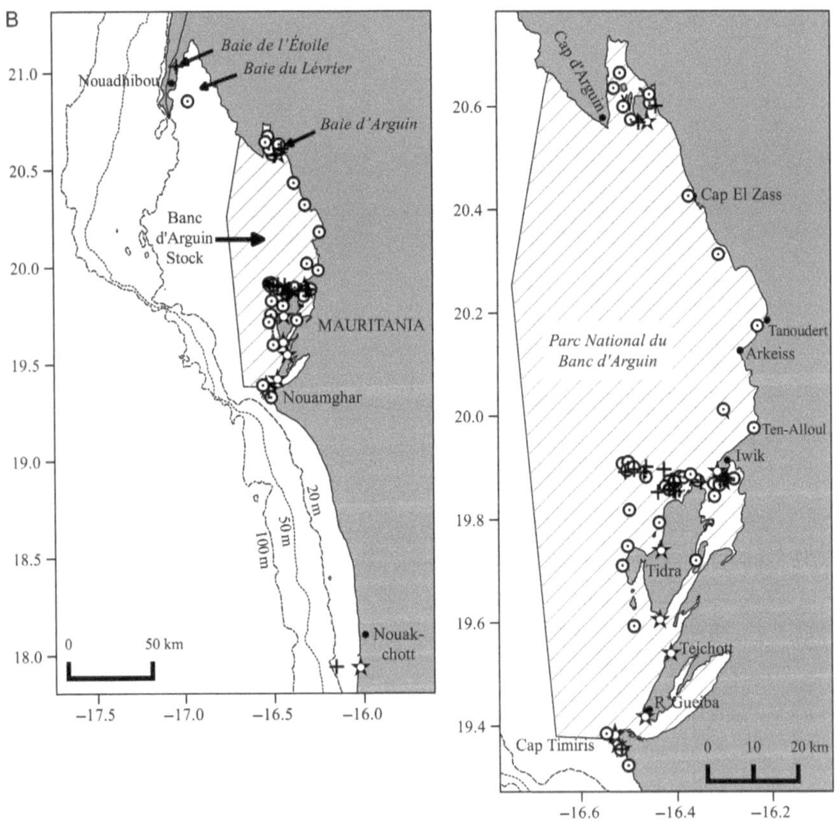

Figure 3B—Cont'd

3.3.3 Senegal

No records are available north of Dakar. Between Dakar and the Saloum Delta all records relate to captures and landings (mostly landed at Joal with unknown capture locations) and are predominantly those documented ≤1950s by Jean Cadenat. The majority of records in Senegal originate from the Delta du Saloum Biosphere Reserve (Figure 3C). Atlantic humpback dolphins inhabit the Saloum at least as far upstream as Djiffer and the Marigot de Ndangane mouth. They occur throughout the lower and central reaches of the Diombos from Pointe de Jackonsa and the mouth of the Bôlon de Bétanti to the channels around the Île de Poutaké. Sightings occur at least 20 km upstream of the Diombos mouth, in the Bôlon Labor and at the confluence with the upper Bandiala River. There are records from Île Diamanio and Île aux Oiseaux, the Bandiala and Djinack mouths and along the

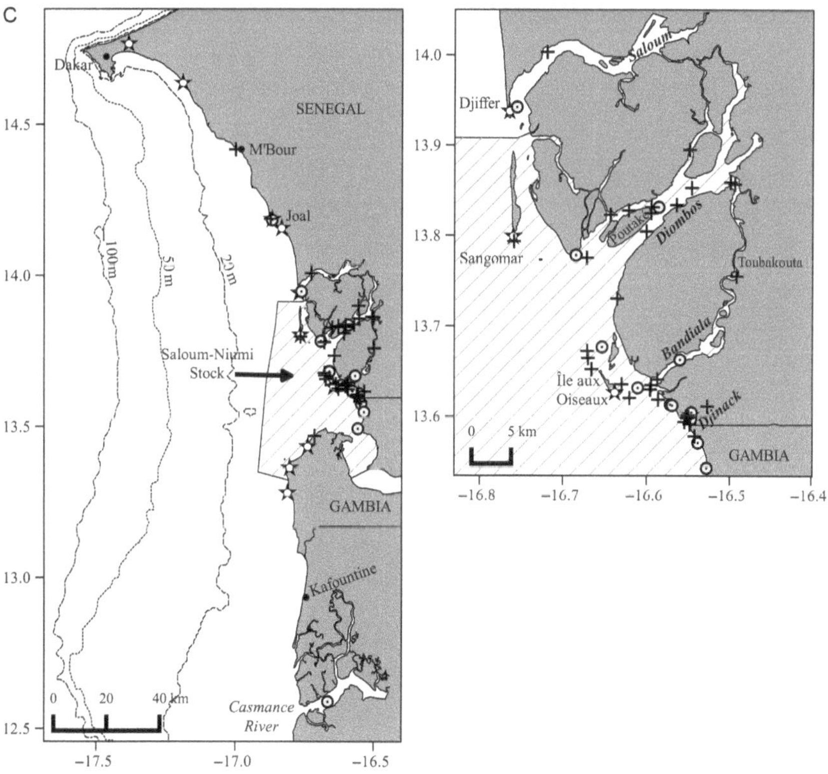

Figure 3C—Cont'd

intermediate coast. Additionally, there is one 1979 sighting of 10 animals approximately 21 km upstream from the Bandiala mouth to the south of Toubakouta (Maigret, 1980). South of Gambia only a single Senegalese record could be located; a sighting in the lower part of the Casamance River (Dupuy and Maigret, 1979).

3.3.4 Gambia

There are relatively few confirmed records, with six sightings occurring in the Niumi National Park located between Barra Point at the Gambia River and the Delta du Saloum Biosphere Reserve (Figure 3C). One sighting, two carcasses and one skull are known from the southern coast between Kotu and Sanyang Point.

3.3.5 Guinea-Bissau

The occurrence of the Atlantic humpback dolphin off Guinea-Bissau was recently reviewed by Leeney et al. (2015). The species is widely distributed from the Canal de Jeta and the Canal de Pecixe in the north, to the Rio Cacine in the south. Records are most numerous along the north coast of the outer Rio Gêba, around the island of Arelas (located south of the

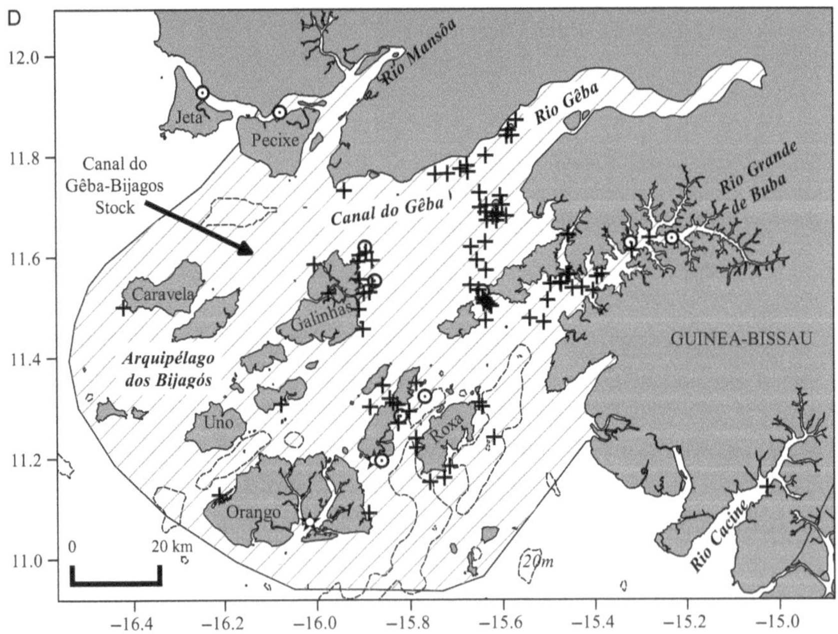

Figure 3D—Cont'd

Rio Gêba mouth), around the western tip of Bolama and in the Rio Grande de Buba (Figure 3D). It also occurs in the Arquipélago dos Bijagós, especially around the eastern islands of Roxa, Bubaque, Galhinas and Maio. It has been recorded 30 km upstream from the Rio Grande de Buba mouth, and 20 km upstream from the Rio Cacine mouth.

3.3.6 Guinea

The species is documented from a sighting (Cadenat, 1956) and capture (Van Waerebeek et al., 2004) in the Conakry region of southern Guinea (Figure 3E). Weir (2015) recently reported eight sightings from the wider Río Nuñez region of northern Guinea.

3.3.7 Togo

A group of 27 animals was photographed off Togo during March 2015 (see Appendix A). While mentioned here for completeness, details have not yet been published and consequently the record is not included in this analysis.

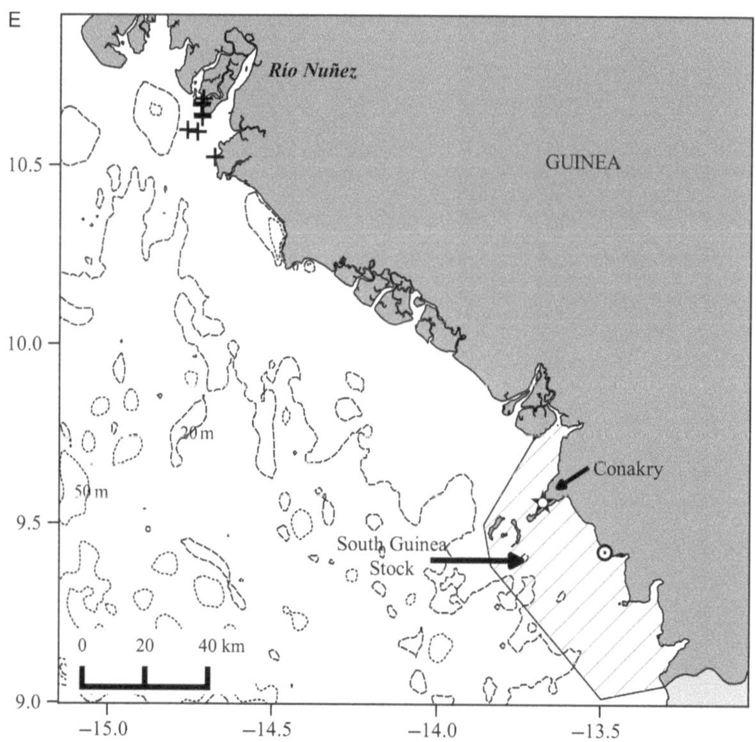

Figure 3E—Cont'd

3.3.8 Benin

The only record is a single sighting recorded west of Cotonou in 2013 (Zwart and Weir, 2014).

3.3.9 Cameroon

The species holotype originates from the Douala region of Cameroon (Kükenthal, 1892). Recently, a sighting and bycatch (landed) were documented from the far south of Cameroon close to Equatorial Guinea (Ayissi et al., 2014; Figure 3F).

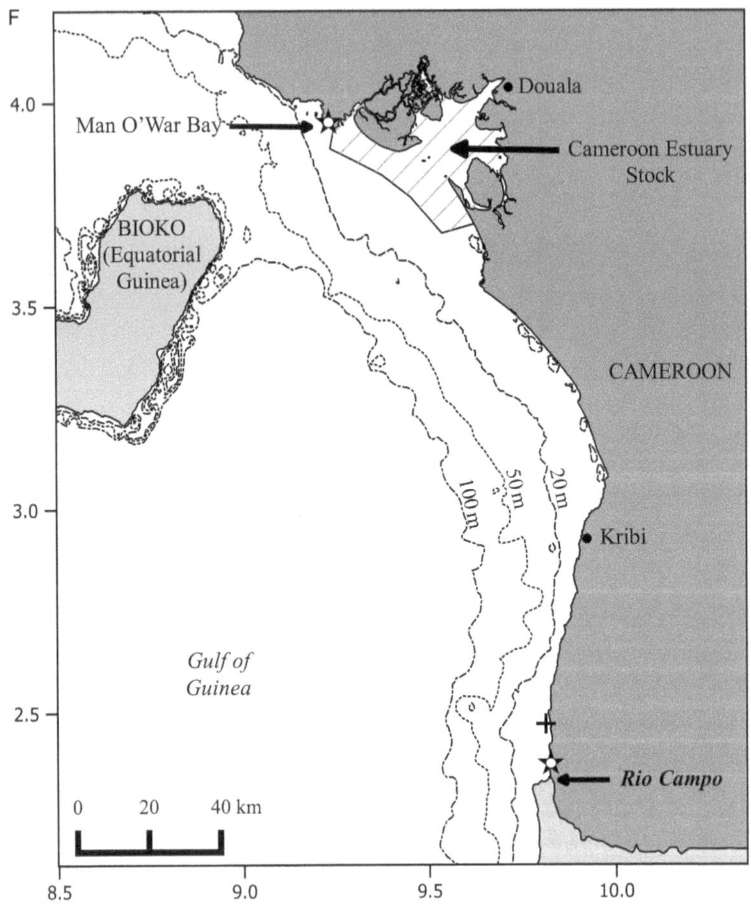

Figure 3F—Cont'd

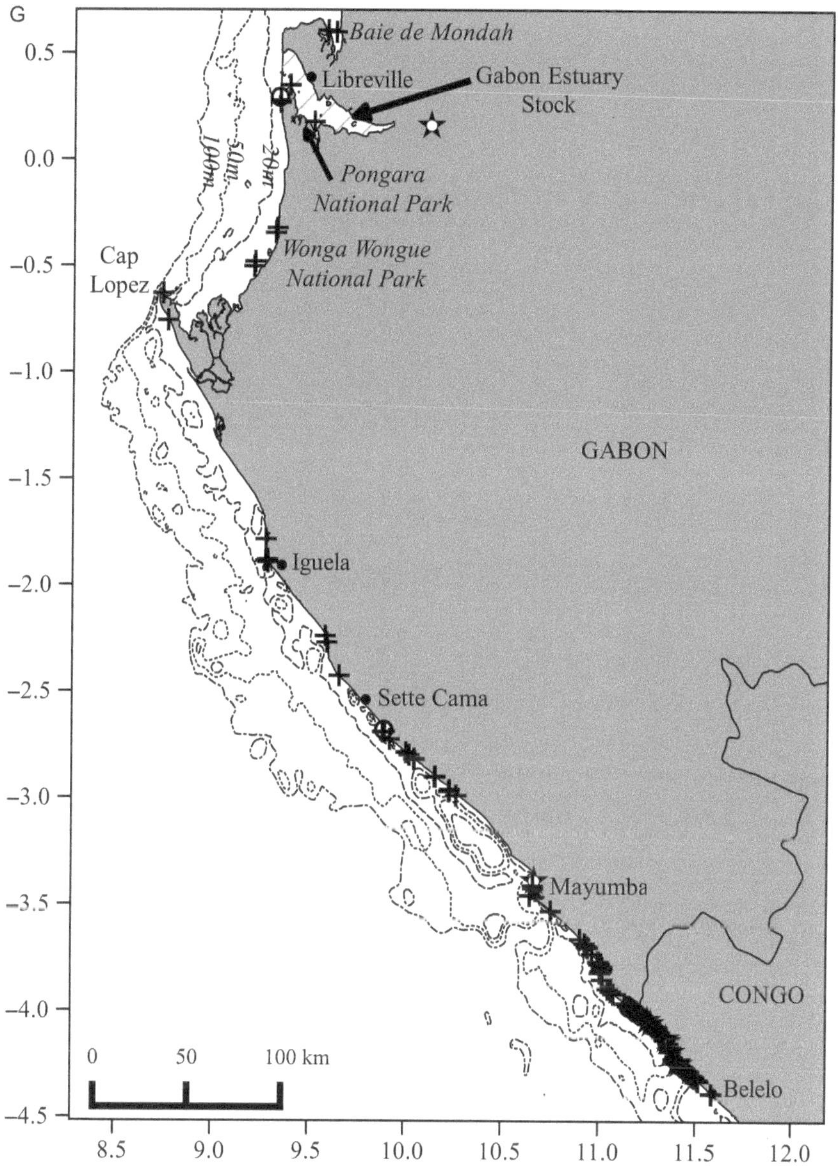

Figure 3G—Cont'd

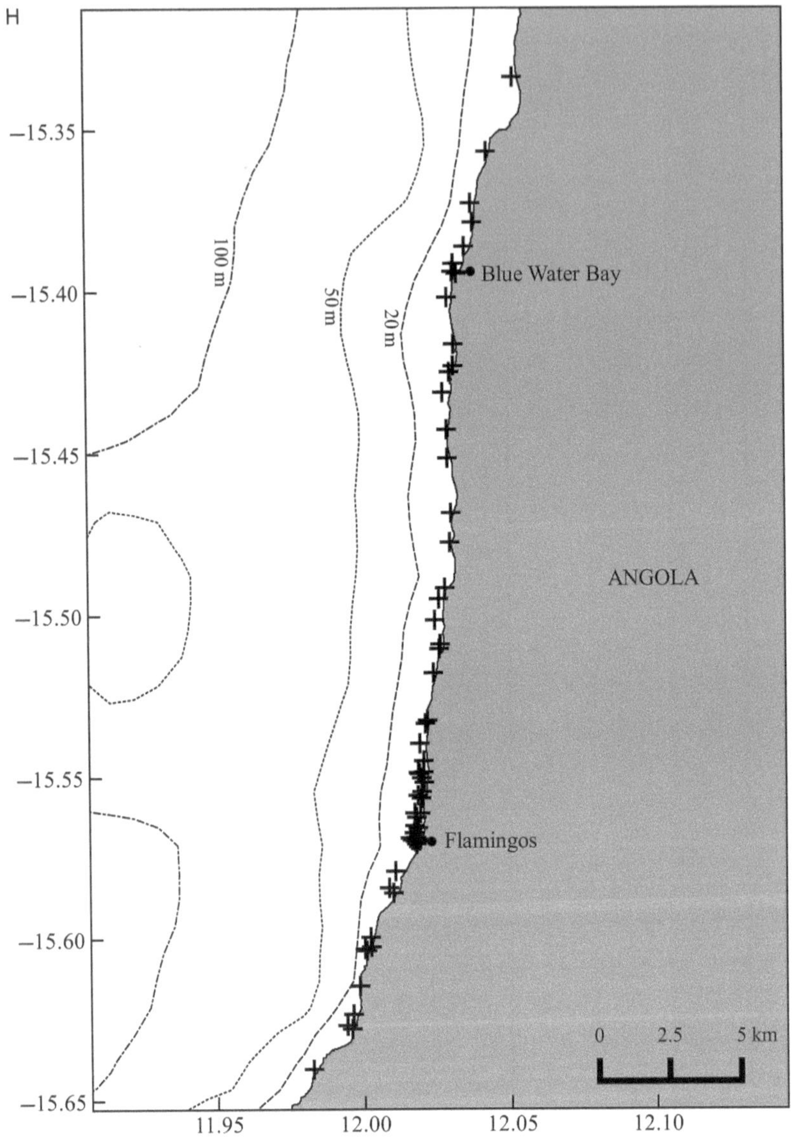

Figure 3H—Cont'd.

3.3.10 Gabon

There are 16 Atlantic humpback dolphin records in northern Gabon from Baie de Mondah, the Gabon Estuary (including the first Gabonese specimen record from Kango), off the coasts of the Pongara and Wonga Wongue National Parks, and off Cap Lopez. However, the majority of records originate from the southern portion of Gabon (Figure 3G), where fieldwork has produced 102 sightings (and one specimen) between Iguela (1°46′S) and the border with the Congo Republic, particularly along the coast of the Mayumba National Park.

3.3.11 Congo Republic

All records in Congo originate from the northern part of the country (Figure 3G), with the vast majority ($n=202$) occurring within the Conkouati-Douli National Park. Records are distributed throughout monitored sections of the coast, with clusters (sightings and bycatch) at the fishing camps of Niandji, Lifelo, Mvandji, Bondi and Longobondi. These relate to beach patrols along the coast and reports from a network of fishermen. The southernmost records for Congo are currently documented from the fishing village of Belelo (~4°22′S).

3.3.12 Angola

Although Angola has over 1600 km of coastline, confirmed sightings of Atlantic humpback dolphins have only been reported from a 35 km stretch of coast in Namibe Province (~15°30′S; Weir, 2009; Figure 3H). This region currently marks the southern known distribution of the species.

3.4 Distribution by Ecoregion

While global Atlantic humpback dolphin distribution spans seven marine ecoregions, no records have been documented from the Gulf of Guinea Upwelling ecoregion in the centre of their range (see Table 2). The majority of records are from the Sahelian Upwelling (26.1%) and Gulf of Guinea South (46.2%) ecoregions. However, the paucity of systematic survey effort throughout the species range precludes interpretation of the relative importance of these regions. With the exception of the population inhabiting Dakhla Bay in Western Sahara, species distribution falls within two shelf provinces: the West African Transition in the north and the Gulf of Guinea in the south, divided at the Senegal–Guinea-Bissau border.

3.5 Distribution Gaps?

Van Waerebeek et al. (2004) provisionally discerned eight Atlantic humpback dolphin 'stocks' for practical reasons and conservation purposes: Dakhla Bay, Banc d'Arguin, Saloum-Niumi, Canal do Gêba-Bijagos, South Guinea, Cameroon Estuary, Gabon Estuary and Angola (see Figure 3). The Cameroon Estuary and Gabon stocks were considered historical, and the remaining six extant. Recent occurrence (i.e. within the last decade; see Table 3) is confirmed in four of those, with no contemporary records from South Guinea, the Cameroon Estuary or the Gabon Estuary. The Angola stock was illustrated by Van Waerebeek et al. (2004) as occurring around Luanda (9°S), but we surmise that was for scale convenience since Angolan occurrence has been confirmed only in Namibe Province (15°30′S: Weir, 2009).

With the large increase in records during the last decade, evidence has emerged for the presence of the species in additional areas, including northern Guinea, Togo, Benin, southern Cameroon, most of Gabon and northern Congo. This generally corresponds with the onset of survey effort, suggesting that stocks defined according to perceived distribution gaps merit a revision, particularly since rigorous systematic at-sea survey effort has been lacking in the vast majority of the species' potential geographic range. Large gaps in the location of confirmed records certainly exist (see Figures 1 and 2), most notably the 945 km between records in Benin and Cameroon, 1500 km between northern Congo and southern Angola and approximately 2000 km between southern Guinea and Togo/Benin. However, species occurrence in intermediate waters requires thorough assessment if population structure is to be definitively described. This is aptly illustrated by recent records of the species during surveys of the Río Nuñez in northern Guinea (Weir, 2015) located in an area between Van Waerebeek et al.'s (2004) Canal do Gêba-Bijagos and South Guinea stocks. The resulting gaps of <100 and 170 km between the Río Nuñez sightings and records in Casamance and Conakry, respectively, are well within the described ranging habits of *Sousa* dolphins. For example, Indian Ocean humpback dolphins, *Sousa plumbea* off South Africa may make linear long-range movements approximating at least a few hundred kilometres (Karczmarski, 2000).

4. GROUP SIZE

Atlantic humpback dolphins were reported (acknowledging that observers may vary in their definition of group) in groups ranging from 1 to 45 animals (Table 4). This excludes bottlenose dolphins recorded during

Table 4 Group Size, Water Depth, Distance from Shore and Sea-Surface Temperature (SST) of Atlantic Humpback Dolphin (*Sousa teuszii*) Records in Six Marine Ecoregions (Spalding et al., 2007)

	Marine Ecoregion						
Parameter	Saharan Upwelling	Sahelian Upwelling	Gulf of Guinea West	Gulf of Guinea Central	Gulf of Guinea South	Angolan	Total
Group size (*n*)	15	151	117	13	363	75	734
Mean	6.3	7.4	6.5	10.2	13.1	3.7	9.7
SE	1.52	0.64	0.51	0.43	0.43	0.33	0.30
Median	4	4	4	10	12	2	8
Range	1–17	1–40	1–25	2–45	1–43	1–8	1–45
Habitat (*n*)	12	83	104	16	383	75	673
Water depth (m)							
Mean	1.0	1.2	5.3	5.4	2.4	2.1	2.7
SE	0.00	0.18	0.55	1.91	0.22	0.47	0.173
Median	1	1	3	3.5	2	1	1
Range	1	1–16	1–30	1–33	1–40	1–24	1–40
Distance from shore (m)							
Mean	1027	826	1658	1115	269	158	573
SE	251.0	173.0	154.0	333.0	10.8	10.8	39.1

Continued

Table 4 Group Size, Water Depth, Distance from Shore and Sea-Surface Temperature (SST) of Atlantic Humpback Dolphin (*Sousa teuszii*) Records in Six Marine Ecoregions (Spalding et al., 2007)—cont'd

	Marine Ecoregion						
Parameter	**Saharan Upwelling**	**Sahelian Upwelling**	**Gulf of Guinea West**	**Gulf of Guinea Central**	**Gulf of Guinea South**	**Angolan**	**Total**
Median	785	342	1197	469	236.1	138.6	245.1
Range	392–2947	13–12766	136–8873	38–4480	34–1979	14–385.4	13–12766
SST (°C)							
Mean	21.8	25.4	27.7	28.7	26.9	20.3	26.1
SE	0.33	0.26	0.09	0.21	0.04	0.17	0.10
Median	21.8	25.1	27.3	28.5	26.8	20.5	26.8
Range	18.7–22.8	21.6–30.6	26.7–31.8	27.0–30.1	26.1–29.6	15.8–21.0	15.8–31.8

Only sightings ($n=734$) were used for group size analysis, and only at-sea records with localities accurate to <2 km ($n=673$) were included in the habitat analysis.

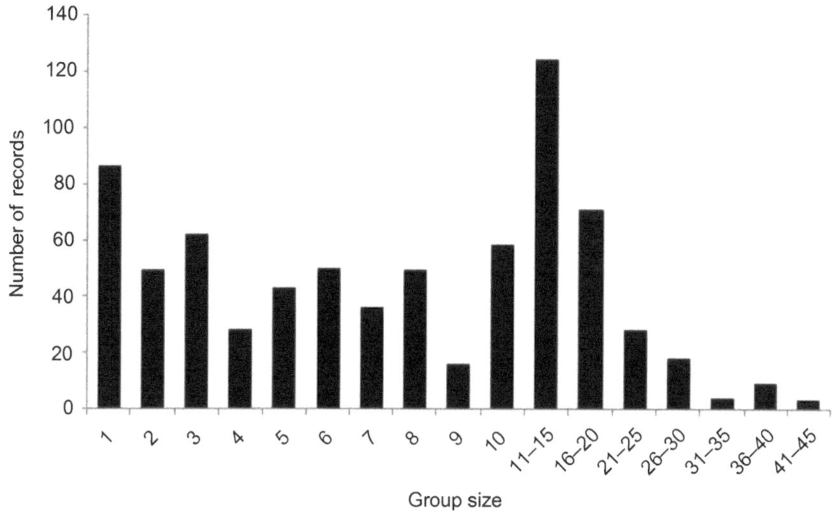

Figure 4 Group sizes for 734 Atlantic humpback dolphin (*Sousa teuszii*) sightings.

mixed-species associations in Western Sahara, Mauritania, Senegal, Guinea-Bissau, Gabon, Congo Republic and Angola.

The species typically occurs in small schools (Figure 4), with groups of 1–10 animals comprising 65% of the sightings. In contrast, groups of over 20 animals constituted only 8.4% of the sightings.

There was a significant difference between the median group sizes according to marine ecoregion (Kruskal–Wallis test: $H=194.1$, d.f. $=5$, $p<0.001$; Table 4). Mann–Whitney U-tests revealed highly significant ($p<0.001$) differences between Angolan and all other ecoregions except the Saharan Upwelling ($p=0.974$), and between Gulf of Guinea South and all other ecoregions except for a weak difference ($p=0.036$) with the Gulf of Guinea Central. Consequently, the Saharan Upwelling and the Angolan ecoregions, both located at the species distribution limits, had the smallest groups, while those inhabiting the Gulf of Guinea South (comprising southern Gabon and Congo) tended to be larger than elsewhere.

5. HABITAT

Atlantic humpback dolphin distribution in relation to bathymetry, distance from land, and SST is examined here at only a relatively broad (several kilometres) scale. Limitations include the reliance on a presence-only dataset and low accuracy arising from: (1) the lack of available high-resolution bathymetry and shoreline data for the study area, (2) the 4-km resolution of available SST data and (3) the 2-km accuracy of the sighting positions. Nevertheless, the results provide an overall indication of typical habitat.

5.1 Water Depth

The Atlantic humpback dolphin has a distribution clearly limited to the continental shelf (<200 m depth). No records are known seaward of the shelf break despite extensive offshore effort in some areas (e.g. northern Angola and Gabon; Weir, 2011). Furthermore, sightings occurred in grid cells with central depths at mean low water of 1–40 m (Table 4; including 273 records where the centre of the grid cell was on land and an arbitrary value of 1 m depth was assigned), indicating that the species inhabits the shallowest portion of the shelf. Only 13 of the 673 records had values exceeding 16 m, and those were due to the cell being centred on pockets of deeper water (e.g. southern Bijagos Archipelago in Guinea-Bissau and off Mayumba in Gabon) or areas where the shelf edge occurred close to shore (e.g. Cap Lopez in Gabon). The mean and median water depths of records in all ecoregions were ≤5.4 m (Table 4). The distribution maps indicate an occurrence predominantly shoreward of the 20 m isobath (see Figure 3). Water depths of approximately 20 m seem to limit the seaward distribution of other *Sousa* species (e.g. Karczmarski, 2000; Wang et al., 2007), and a similar depth may be important for *S. teuszii*.

5.2 Distance from Shore

Reported sightings of Atlantic humpback dolphins are inevitably biased towards waters in immediate proximity (i.e. <1 km) to land, since most available data originate from shore-based observers, coastal pleasurecraft, ornithological surveys targeting mudflats and dedicated surveys close to shore aimed at maximising dolphin encounters (e.g. Collins et al., 2010; Weir, 2009). Consequently, the species is frequently described as having 'nearshore' or 'inshore' distribution, despite those terms rarely being defined.

Reported sightings occurred at distances of 13 m to 12.8 km from shore, with a mean of 573 m (Table 4). The median distance differed significantly between ecoregions (Kruskal–Wallis test: $H=252.3$, d.f. = 5, $p<0.001$) and Mann–Whitney tests revealed significant differences between all pairs except for Gulf of Guinea Central vs. Saharan ($p=0.416$) and Sahelian ($p=0.604$)

Upwellings, and between Gulf of Guinea West and Saharan Upwelling ($p=0.127$). In particular, sightings in the Gulf of Guinea West tended to be at the greatest distance from shore (mean of 1.7 km), while those in the Gulf of Guinea South (mean of 269 m) and Angolan (mean of 158 m) ecoregions were closest to shore (Table 4). Accuracy errors in sighting locations and resolution of the coast file are acknowledged, but it is clear that in all regions the Atlantic humpback dolphin is regularly sighted within a kilometre of the coast. However, we note 12 sightings for which a GPS position accurate to 1 km (AC1) was provided for which the calculated distances from shore exceeded 3 km. They include two sightings off Wonga Wongue National Park in Gabon (3.3 and 4.5 km) and two sightings off the Río Nuñez in Guinea (5.1 and 6.8 km). Dolphins during one of the latter sightings were tracked to a locality 12.1 km from shore (Weir, 2015). In Guinea-Bissau, four AC1 sightings in the Canal do Gêba and outer Rio Grande de Buba were at distances of 3.3–8.9 km from shore (a further 10 sightings of AC2 were at distances of 3.0–6.1 km). Finally, in Mauritania there were three sightings at distances of 3.5–4.5 km from shore between the islands of Arel and Nair on the Banc d'Arguin, and an additional sighting 12.8 km off the coast south of Nouakchott. Consequently, while the Atlantic humpback dolphin appears to be limited to shallow (~20 m) water depths, it can occur several kilometres from shore when suitable shallow habitat is present.

As noted by Weir (2015), factors including spatial distance from land, water depth, seabed topography and oceanographic processes comprising wave motion and sedimentation variously (and inconsistently) contribute to scientific definitions of terms such as 'nearshore'. With regards to Atlantic humpback dolphin habitat, Weir (2015) adopted Davidson-Arnott and Greenwood's (2003) oceanographic definition of the nearshore zone as the area of sea located from the shoreline to the water depth where wave motion ceases to affect the sea floor (i.e. the depth at which the largest waves no longer move sediment on the bottom). In other words, combined water depth, wave action and sedimentation may be more relevant factors than spatial distance from the coast. By Davidson-Arnott and Greenwood's (2003) definition the nearshore zone is most often 0.5–2.5 km in width, but in geographic regions with particularly shallow-sloping seabeds (e.g. off Guinea) it can extend for tens of kilometres from the shoreline (R. Davidson-Arnott, University of Guelph, personal communication, 22 November 2014). The area of coast from southern Senegal to Sierra Leone has the widest shelf within the species range and might be expected to offer suitable habitat away from immediate proximity to the coast, which corresponds with the greater mean distance recorded for sightings in the Gulf of Guinea West ecoregion. Similarly, the shallow (≤20 m) extent of the central Banc d'Arguin in Mauritania extends over 70 km from shore and potentially could provide 'nearshore' habitat for the species at a considerable spatial distance from land.

5.3 Temperature

Little information has been published on the temperature range tolerated by the Atlantic humpback dolphin. MacLeod (2009) described it as a tropical and sub-tropical species that is generally warm water-limited. Available records occurred in grid cells with averaged SSTs ranging from 15.8 to 31.8 °C, with a mean of 26.1 °C (Table 4). As expected, mean temperatures were coolest at either end of the species range, at 21.8 and 20.3 °C in the Saharan and Angolan ecoregions, respectively. While these data reflect the broad-scale temperature range inhabited, the lowest temperatures that will be tolerated by the species remain unclear. For example, averaged SSTs for grid cells in southern Angola ranged from 15.8 to 20.9 °C, while the actual *in situ* SSTs recorded for 19 sightings in Namibe Province during January 2008 (i.e. in summer) ranged from 18.8 to 24.5 °C (C. R.Weir, unpublished data). It is possible that winter temperatures lower than the minimum averaged 15.8 °C grid cell value are experienced by dolphins in that region.

5.4 Similarities and Differences

There are similarities in the areas occupied by Atlantic humpback dolphins across their range, particularly an occurrence in relatively shallow (usually ≤20 m) depths, in warm waters (average SSTs probably exceeding 15 °C) and in dynamic habitat strongly influenced by tidal patterns (e.g. mudflats, sandbars, channels, surf zones and estuaries). Van Waerebeek et al. (2004) noted an occurrence over soft-sediment bottoms, though it remains unclear whether this is a specific preference (of dolphins or of their prey) or simply reflects sand and mud being the dominant bottom types across much of their geographic range. For example, Weir (2009) reported that dolphins sometimes foraged over rocky reefs in southern Angola.

When comparing Atlantic humpback dolphin occurrence in the Saloum Delta and the Banc d'Arguin, Maigret (1980) observed that while those areas shared common ecological conditions (e.g. emergent sandbars and tidal channels), their landscapes were different. We similarly note that the apparent ecological preferences of the species are provided by a range of very different habitat types, including: (1) large estuarine systems, often bordered by mangrove, comprising main and inner channels, and the estuary-influenced water in outer regions further from the coast (e.g. Saloum Delta, Río Nuñez); (2) open coasts bordered by sandy beach exposed to breaking swell (e.g. Angola, Congo Republic, southern Gabon and Benin); (3) coastal archipelagos (e.g. Arquipélago dos Bijagós); (4) wide expanses of open shallow waters characterised by extensive mud-flats and seagrass (e.g. Banc d'Arguin); and (5) large, sheltered enclosed shallow bays (e.g. Dakhla Bay). Nevertheless, all of them may be embraced as 'nearshore' habitat as defined by Davidson-Arnott and Greenwood (2003). The relative importance of these different habitat types in terms of species abundance is unknown.

6. CONCLUSIONS AND FUTURE WORK

We emphasise the reliance on presence-only data for most of this review, and interpretation must therefore consider both the accuracy margin and the fact that the species is poorly documented in almost every part of its geographic range. Allowing for that, we provide the first range-wide quantitative habitat analysis for the Atlantic humpback dolphin and demonstrate that although the species occurs in several different habitat types there are consistent features. Future work should incorporate novel datasets from undersampled regions and examine additional habitat parameters. For example, turbidity, salinity and bottom sediment have been suggested to influence Atlantic humpback dolphin occurrence (Maigret, 1980; Van Waerebeek et al., 2004), and distribution relative to specific habitat features (e.g. river mouths, seagrass beds and coral reefs) also warrants investigation. A more comprehensive evaluation of the species' ecological niche may facilitate predictive habitat modelling that would help to identify areas of suitable habitat and potential species strongholds that could be prioritised during future survey work and may help to inform management assessments.

In spatial terms, and disregarding anthropogenic influences, the Atlantic humpback dolphin would be considered widely distributed, with distributional limits encompassing the entire west coast of the tropical African continent. However, within that geographic range it inhabits only nearshore (as defined earlier) waters and seemingly has a relatively narrow habitat niche limited by maximum depths of around 20 m. The continental shelf along the west coast of Africa is relatively narrow (exceptions being Mauritania and southern Senegal–Guinea), and the 20 m isobath usually occurs well within 15 km of the coastline. Consequently, the overall amount of potential habitat available to the Atlantic humpback dolphin appears limited.

The paucity of long-term systematic scientific datasets (i.e. representative survey coverage providing both presence and absence data) for the Atlantic humpback dolphin continues to represent the largest limitation in understanding its distribution and ecology, and consequently in implementing appropriate management measures. This updated distribution review highlights the lack of information for several potential range states including Sierra Leone, Liberia and Côte d'Ivoire, and a continued reliance on predominantly opportunistic records in some areas of apparent importance such as the Banc d'Arguin, the Saloum Delta and Guinea-Bissau. Until this is addressed, it will remain uncertain whether perceived gaps in species distribution are genuine or simply the result of inadequate survey effort.

ACKNOWLEDGEMENTS

The datasets used in this review are listed in Appendix A; we are extremely grateful to all contributors who supplied unpublished dolphin data or responded to queries regarding published datasets, and to the authors of all of the published work cited. Sincere thanks to Colin MacLeod (GIS in Ecology) for GIS advice and to Mark Spalding (The Nature Conservancy) for unpublished information on marine ecoregions.

APPENDIX A

A review of the geographical distribution and habitat of the Atlantic humpback dolphin (*Sousa teuszii*).

This appendix consists of two tables listing the published (Table A1) and unpublished (Table A2) sources of information used in the data analysis, and one note (Note 1) describing an additional recent record that is relevant to the paper, but was not included in the data analysis.

Note 1

A group totalling 27 Atlantic humpback dolphins (in small subgroups of 2–5 animals) was photographed off Togo during March 2015, by operators of the tour group 1001 Pistes (see "Un beau groupe de 27 Dauphins à bosse (*Sousa teuszii*) cet après-midi, le long de la cote togolaise", available from: www.facebook.com/1001Pistes/photos/a.601867513273803.1073741828.585686184891936/677485095712044 [Accessed 1 July 2015]). However, the exact locality has not been published, and consequently it was not possible to include that record in this analysis. Since it represents the first verified sighting of Atlantic humpback dolphins off Togo, the sighting is mentioned here in the context of providing a twelfth confirmed range state for the species.

Table A1 Sources of published information from which individual Atlantic humpback dolphin (*Sousa teuszii*) records (i.e. with position/locations, years and group sizes) were extracted for analysis

Source	Country	Sousa Record Date Range	Study Type*	Notes on Source/Data
Altenburg and Piersma (1982)	Mauritania	1980	Ornithological	
Ayissi et al. (2014)	Cameroon	2011–2012	Cetacean: survey (boat- and shore-based); interview	
Beaubrun (1990)	Western Sahara	1989	Ornithological	
Busnel (1970)	Mauritania	N/A	N/A	Source not seen—cited in Van Waerebeek et al. (2004), but not included in their reference list.
Busnel (1973)	Mauritania	1967–1972/73	Cetacean: survey (shore-based)	
Cadenat (1947)	Senegal	1942–1943	Cetacean: opportunistic	
Cadenat (1949)	Senegal	1942–1949	Cetacean: opportunistic	
Cadenat (1956)	Senegal, Guinea	1953–1955	Cetacean: opportunistic	
Cadenat (1957)	Senegal	1955–1956	Cetacean: opportunistic	
Cadenat (1959)	Senegal	1958	Cetacean: opportunistic	
Cadenat and Paraiso (1957)	Senegal	1955–1956	Cetacean: opportunistic	
Collins et al. (2004)	Gabon	2001–2003	Cetacean: survey (boat-and shore-based); opportunistic	
Collins et al. (2010)	Gabon, Republic of Congo	2001–2009	Cetacean: survey (boat-and shore-based); opportunistic	
Duguy (1976)	Mauritania	1973–1974	Cetacean: opportunistic	
Dupuy and Maigret (1979)	Senegal	1978	Cetacean: opportunistic	

Continued

Table A1 Sources of published information from which individual Atlantic humpback dolphin (*Sousa teuszii*) records (i.e. with position/locations, years and group sizes) were extracted for analysis—cont'd

Source	Country	Sousa Record Date Range	Study Type	Notes on Source/Data
Fraser (1949)	Senegal	1943	Cetacean: opportunistic	
Fraser (1973)	Mauritania	1970	Cetacean: opportunistic	
Fulling et al. (2008)	Guinea Bissau	2008	Cetacean: survey (boat-based)	Available individual data were provided to Leeney et al. (2015).
Kükenthal (1892)	Cameroon	1891?	Cetacean: opportunistic	Location coded as Man O'War Bay, although there is doubt over the actual origin of this holotype specimen due to apparent confusion with a manatee (*Trichechus senegalensis*) carcass from this location.
Leeney et al. (2015)	Guinea Bissau	1993–2014	Cetacean: survey (boat-based); opportunistic; interview	This review compiled published records from several other sources in this table (e.g. Spaans, 1990; Wolff et al., 1992), but also included unpublished novel data and individual records from previously published compilations (e.g. Fulling et al., 2008) for which a date range is listed here.
Maigret (1980)	Mauritania, Senegal	1972–1980	Cetacean: opportunistic	
Maigret (1981)	Mauritania	1981	Cetacean: opportunistic	
Maigret et al. (1976)	Mauritania	1972–1974	Cetacean: opportunistic	
Marchessaux and Campredon (1988)	Mauritania	1984–1988	Cetacean: opportunistic	Individual data records from this source were provided by P. Campredon (IUCN, personal communication, 11 January 2015; also see Table A2).

Murphy et al. (1997)	Gambia	1996–1997	Cetacean: opportunistic	
Notarbartolo di Sciara et al. (1998)	Western Sahara	1996	Cetacean: survey (boat-based); interview	Individual sighting data are no longer available (G. Notarbartolo di Sciara, Tethys Research Institute, personal communication, 18 October 2014). Approximate locations for 4 sightings were extracted from Figure 2. Mean group size for the 4 records was 6.9 animals (Notarbartolo di Sciara et al., 1998), and these records were coded in the database as two groups each of 6 and 7 animals (as approximate indications).
Robineau and Vély (1998)	Mauritania	N/A	Cetacean: opportunistic	Individual sighting data may still be available (M. Vély, Megaptera, personal communication, 4 February 2015) but unfortunately were not provided in time for inclusion in this analysis. Consequently, only the 14 sightings and 8 strandings for which approximate locations were mentioned by Robineau and Vély (1998) were included in analysis.
Sequeira and Reiner (1992)	Guinea Bissau	1989	Cetacean: opportunistic	
Spaans (1990)	Guinea Bissau	1986/87	Ornithological	Individual sighting data are no longer available (B. Spaans, Royal Netherlands Institute for Sea Research, personal communication, 16 October 2014). The locations of 53 sightings (with approximate group sizes) were extracted from Figure 2.
van Bree and Duguy (1965)	Senegal	1925	Cetacean: opportunistic	
	Senegal	1996		

Continued

Table A1 Sources of published information from which individual Atlantic humpback dolphin (*Sousa teuszii*) records (i.e. with position/locations, years and group sizes) were extracted for analysis—cont'd

Source	Country	Sousa Record Date Range	Study Type	Notes on Source/Data
Van Waerebeek et al. (1997)			Cetacean: survey (shore-based); interview; opportunistic	
Van Waerebeek et al. (2000)	Senegal, Guinea Bissau	1952–1998	Cetacean: survey (shore-based); interview; opportunistic	
Van Waerebeek et al. (2003)	Senegal, Gambia	1999–2000	Cetacean: survey (boat- and shore-based); interview; opportunistic	
Van Waerebeek et al. (2004)	Mauritania, Senegal, Guinea Bissau, Guinea	1949–2002	Cetacean: survey (boat- and shore-based); interview; opportunistic	This comprehensive review compiled many published records from the other sources in this table. However, only the previously unpublished data are summarised here.
Weir (2007)	Angola	2004	Cetacean: survey (boat-based); opportunistic	
Weir (2009)	Angola	2008	Cetacean: survey (boat- and shore-based); opportunistic	
Weir (2015)	Guinea	2013	Cetacean: survey (boat-based); interview; opportunistic	
Wolff (1998)	Guinea Bissau	1992	Ornithological	
Zwart and Weir (2014)	Benin	2013	N/A: incidental	

**Note on study type*. Published studies have been broadly assigned to categories defined as follows: (1) Cetacean—survey: targeted survey work aimed at recording cetaceans (i.e. including systematic field survey effort); (2) Cetacean—opportunistic: study of (or including) cetaceans but without systematic survey effort (e.g. compilations of specimen records, strandings by-catch and/or sightings); (3) Cetacean—interview: study of (or including) cetaceans using interviews/questionnaires of local people to collect information; (4) Ornithological: survey primarily focussed on birds, with cetaceans recorded incidentally (either opportunistically or systematically) and (5) Incidental: records unrelated to scientific studies.

We provide the first source to mention each particular record (and occasionally also an additional source that expands on the details provided by the initial source); many records were extensively re-cited in subsequent publications.

Table A2 Sources of unpublished Atlantic humpback dolphin (*Sousa teuszii*) records used in the analysis

Source (and affiliation, where available)	Country	Sousa Record Date Range	Study Type*	Notes on Source/Data
Anon.	Western Sahara	2008	N/A: incidental	Photograph available online at: http://ozone-family-maroc.blogit.fr/page_29.html (accessed on 12 February 2014).
Armstrong, Roy (University of Cumbria)	Gambia	2010–2011	Natural history	
Brown, Dan (BiOME Consulting Ltd, United Kingdom)	Western Sahara	2010	Natural history (shore-based)	Photograph available online at: http://punkbirder.webs.com/sahara2010.htm (accessed on 2 July 2015).
Camphuysen, Kees (Royal Netherlands Institute for Sea Research; NIOZ)	Mauritania	2003	ESAS: boat-based survey	Only record from extensive offshore survey effort between 1986 and 2012 (K. Camphuysen, NIOZ, pers. comm., 24 November 2014).
Campredon, Pierre (International Union for Conservation of Nature; IUCN)	Mauritania	1984–1988	Cetacean: opportunistic	Many of these records were included in the overview by Marchessaux and Campredon (1988) but without specific individual record data on locations, group sizes and dates. This information was provided by P. Campredon (IUCN, pers. comm., 11 January 2015) together with some additional unpublished records.
Collins, Tim (Wildlife Conservation Society, WCS)	Gabon, Congo Republic	2005–2015 (mostly ≥2010)	Cetacean: survey (boat-and shore-based); opportunistic	Most records prior to 2010 have been published (see Table A1).
Cools, Paul	Gambia	2007–2010	N/A: incidental	Opportunistic records provided via Roy Armstrong (University of Cumbria).

Continued

Table A2 Sources of unpublished Atlantic humpback dolphin (*Sousa teuszii*) records used in the analysis—cont'd

Source (and affiliation, where available)	Country	Sousa Record Date Range	Study Type	Notes on Source/Data
Descrienne, Louis-Gerard (l'Office National de la Chasse et de la Faune Sauvage)	Mauritania, Senegal	2010–2013	N/A: incidental	
Gomis, Nicolas (Parc National du Delta du Saloum)	Senegal	2013	N/A: incidental	
Hazevoet, Kees (Instituto de Investigação Científica Tropical; IICT)	Mauritania, Guinea-Bissau	1984–1986	Natural history (boat-based)	
Keith, Lucy (Sea to Shore Alliance)	Guinea Bissau	2013	Manatee	
Leeney, Ruth (Protect Africa's Sawfishes)	Senegal	2014	Cetacean: survey (boat-based); interview	
Nicoll, Martin (World Wildlife Fund, WWF)	Mauritania	1994/95	N/A: incidental	
Spaans, Bernard (Royal Netherlands Institute for Sea Research, NIOZ)	Mauritania	2005	Ornithological	
Ulph, Robert	Western Sahara	2012	Natural history	Provided via Sean Minns.

**Note on study type*. Published studies have been broadly assigned to categories defined as follows: (1) Cetacean—survey: targeted survey work aimed at recording cetaceans (i.e. including systematic field survey effort); (2) Cetacean—opportunistic: study of (or including) cetaceans but without systematic survey effort (e.g. compilations of specimen records, strandings by-catch and/or sightings); (3) Cetacean—interview: study of (or including) cetaceans using interviews/questionnaires of local people to collect information; (4) Ornithological: survey primarily focussed on birds, with cetaceans recorded incidentally (either opportunistically or systematically); (5) Manatee: survey work focussing on the West African manatee (*Trichechus senegalensis*); (6) Natural history: informal study of general wildlife (e.g. University field trips, ecotourism trips) and (7) Incidental: records unrelated to scientific studies.

References to Tables A1 and A2

Altenburg, W., Piersma, T., 1982. Faunistical varia, in: Altenburg, W., Engelmoer, M., Mes, R., Piersma, T. (Eds.), Wintering waders on the Banc d'Arguin, Mauritania. Report of the Netherlands Ornithological Mauritanian Expedition 1980. Stichting Veth tot Steun aan Waddenonderzoek, Leiden, pp. 212–216.

Ayissi, I., Segniagbeto, G.H., Van Waerebeek, K., 2014. Rediscovery of Cameroon dolphin, the Gulf of Guinea population of *Sousa teuszii* (Kükenthal, 1892). ISRN Biodiversity 2014, 819827.

Beaubrun, P.-C., 1990. Un cetacé nouveau pour les côtes sud-marocaines: *Sousa teuszii* (Kükenthal, 1892). Mammalia 54, 162–164.

Busnel, R.G., 1973. Symbiotic relationship between man and dolphins. Transactions of the New York Academy of Sciences 35, 112–131.

Cadenat, J., 1947. Observations de cetaces au Senegal. Notes Africaines 34, 20–23.

Cadenat, J., 1949. Notes sur les cétacés observés sur les côtes du Sénégal de 1941 à 1948. Bulletin de l'Institut Français d'Afrique Noire 11, 1–15.

Cadenat, J., 1956. Un delphinidae encore mal connu de la côte occidentale d'Afrique: *Sotalia teüszii* Kukenthal 1892. Bulletin de l'Institut Français d'Afrique Noire 18A, 555–566.

Cadenat, J., 1957. Observations de cétacés, siréniens, chéloniens et sauriens en 1955–1956. Bulletin de l'Institut Français d'Afrique Noire 19A, 1358–1375.

Cadenat, J., 1959. Rapport sur les petits cétacés ouest-africaines. Résultats des recherches enterprises sur ces animaux jusqu'au mois de mars 1959. Bulletin de l'Institut Français d'Afrique Noire 21A, 1367–1440.

Cadenat, J., Paraiso, F., 1957. Nouvelle observation de *Sotalia* (Cétacé, Delphinidé) sur les côtes du Sénégal. Bulletin de l'Institut Français d'Afrique Noire 19A, 324–332.

Collins, T., Ngouessono, S., Rosenbaum, H.C., 2004. A note on recent surveys for Atlantic humpback dolphins, *Sousa teuszii* (Kukenthal, 1892) in the coastal waters of Gabon. Paper SC/56/SM23 presented to the Scientific Committee of the International Whaling Commission, Sorrento, Italy, 2004.

Collins, T., Boumba, R., Thonio, J., Parnell, R., Vanleeuwe, H., Ngouessono, S., Rosenbaum, H.C., 2010. The Atlantic humpback dolphin (*Sousa teuszii*) in Gabon and Congo: cause for optimism or concern?

Paper SC/62/SM9 presented to the Scientific Committee of the International Whaling Commission, Agadir, 2010.
Duguy, R., 1976. Contribution a l'etude des mammiferes marins de la cote nord-ouest Afrique. Revue des Travaux de l'Institut des Pêches Maritimes, 39, 321–332.
Dupuy, A.R., Maigret, J., 1979. Les mammifères marins des côtes du Sénégal. 3. Observations signalées en 1978. Bulletin de l'Institut Fondamental d'Afrique Noire 41A, 429–439.
Fraser, F.C., 1949. A specimen of *Sotalia tëuszii* Kukenthal from the coast of Senegal. Journal of Mammalogy 30, 274–276.
Fraser, F.C., 1973. Record of a dolphin (*Sousa teuszii*) from the coast of Mauritania. Transactions of the New York Academy of Sciences 35, 132–135.
Fulling, G.L., Foster, J., Fertl, D.C., Fagin, T., 1998. Pilot survey of coastal small cetaceans in the waters of Guinea-Bissau. Unpublished report to Truk Bissau, November 2008.
Kükenthal, W., 1892. *Sotalia tëuszii* n. sp., ein pflanzenfressender (?) Delphin aus Kamerun. Zoologische Jahrbücher, Abteilung für Systematik 6, 442–446.
Leeney, R.H., Weir, C.R., Campredon, P., Regalla, A., Foster, J., 2015. Occurrence of Atlantic humpback (*Sousa teuszii*) and bottlenose (*Tursiops truncatus*) dolphins in the coastal waters of Guinea-Bissau, with an updated cetacean species checklist. Journal of the Marine Biological Association of the United Kingdom, doi: http://dx.doi.org/10.1017/S0025315415000661, 9 pp. Published online: 25 May 2015.
Maigret, J., 1980. Données nouvelles sur l'écologie du *Sousa teuszii* (Cetacea, Delphinidae) de la côte ouest africaine. Bulletin de l'Institut Fondamental d'Afrique Noire 42A, 619–633.
Maigret, J., 1981. Les mammiferes marins des cotes de Mauritanie. 2. Rapport annuel des observations signalees en 1981. Bulletin du Centre National de Recherches Oceanographiques et des Peches de Nouadhibou 10, 81–85.
Maigret, J., Trotignon, J., Duguy, R., 1976. Observations de cetaces sur les cotes de Mauritanie (1971–1975). Conseil International pour l'Exploration de la Mer, Comite des Mammiferes Marins C.M. 1976/N, 4, 1–6 + 1 map.
Marchessaux, D., Campredon, P., 1988. Ecologie des delphinides dans le Parc National du Banc d'Arguin, Republique Islamique de Mauritanie. Rapport Preliminaire, Avril 1988.

Murphy, P.F., Van Waerebeek, K., Jallow, A., 1997. Cetaceans in Gambian coastal waters. Paper SC/49/SM11 presented to the Scientific Committee of the International Whaling Commission, Bournemouth, UK, 1997.

Notarbartolo di Sciara, G., Politi, E., Bayed, A., Beaubrun, P-C., Knowlton, A., 1998. A winter cetacean survey off southern Morocco, with a special emphasis on right whales. Reports of the International Whaling Commission 48, 547–550.

Robineau, D., Vély, M., 1998. Les cétacés de Mauritanie (Afrique du nord-ouest). Particularités et variations spatio-temporelles de répartition: rôle des facteurs océanographiques. Revue d'Ecologie (la Terre et la Vie) 53, 123–152.

Sequeira, M., Reiner, F., 1992. First record of an Atlantic humpback dolphin, *Sousa teuszii* Kukenthal, 1892 (Cetacea, Delphinidae) in Guinea-Bissau. Mammalia 56, 311–313.

Spaans, B., 1990. Dolphins in the coastal area of Guiné Bissau. Lutra 33, 126–131.

van Bree, P.J.H., Duguy, R., 1965. Sur un crâne de *Sotalia teuszii* Kükenthal, 1892 (Cetacea, Delphinidae). Zeitschrift fur Säugetierkunde 30, 311–314.

Van Waerebeek, K., Diallo, M., Djiba, A., Ndiaye, P., Ndiaye, E., 1997. Cetacean research in Senegal 1995–97, an overview. Paper SC/49/SM10 presented to the Scientific Committee of the International Whaling Commission, Bournemouth, UK.

Van Waerebeek, K., Ndiaye E., Djiba, A., Diallo, M., Murphy, P., Jallow, A., Camara, A., Ndiaye, P., Tous, P., 2000. A survey of the conservation status of cetaceans in Senegal, The Gambia and Guinea-Bissau. UNEP/CMS Secretariat, Bonn, Germany. 80 pp.

Van Waerebeek, K., Barnett, L., Camara, A., Cham, A., Diallo, M., Djiba, A., Jallow, A.O., Ndiaye, E., Samba Ould Bilal, A.O., Bamy, I.L., 2003. Conservation of cetaceans in The Gambia and Senegal, 1999–2001, and status of the Atlantic humpback dolphin. WAFCET-2 report. UNEP/CMS Secretariat, Bonn, Germany. 56 pp.

Van Waerebeek, K., Barnett, L., Camara, A., Cham, A., Diallo, M., Djiba, A., Jallow, A.O., Ndiaye, E., Samba Ould Bilal, A.O., Bamy, I.L., 2004. Distribution, status, and biology of the Atlantic humpback dolphin, *Sousa teuszii* (Kükenthal, 1892). Aquatic Mammals 30, 56–83.

Weir, C.R., 2007. Occurrence and distribution of cetaceans off northern Angola, 2004/05. Journal of Cetacean Research and Management 9, 225–239.

Weir, C.R., 2009. Distribution, behaviour and photo-identification of Atlantic humpback dolphins *Sousa teuszii* off Flamingos, Angola. African Journal of Marine Science 31, 319–331.

Weir, C.R., 2015. Photo-identification and habitat use of Atlantic humpback dolphins (*Sousa teuszii*) around the Río Nuñez estuary in Guinea (West Africa). African Journal of Marine Science, doi: 10.2989/1814232X.2015.1069757.

Wolff, W.J. (Ed.), 1998. The end of the east-Atlantic flyway. Waders in Guinea-Bissau. Report on a cooperative research project in the estuaries of Guinea-Bissau, October 1992-May 1993. WIWO Report No. 39. Foundation Working Group for International Waterbird and Wetland Research, Zeist, Utrecht, Netherlands. 94 pp.

Zwart, S.J., Weir, C.R., 2014. Filling in the gaps: first record of *Sousa teuszii* in Benin (Gulf of Guinea: Africa). Marine Biodiversity Records 7, e59.

REFERENCES

Ayissi, I., Segniagbeto, G.H., Van Waerebeek, K., 2014. Rediscovery of Cameroon dolphin, the Gulf of Guinea population of *Sousa teuszii* (Kükenthal, 1892). ISRN Biodivers. 2014, 819827.

Beaubrun, P.-C., 1990. Un cetacé nouveau pour les côtes sud-marocaines: *Sousa teuszii* (Kükenthal, 1892). Mammalia 54, 162–164.

Busnel, R.G., 1973. Symbiotic relationship between man and dolphins. Trans. N.Y. Acad. Sci. 35, 112–131.

Cadenat, J., 1947. Observations de cetaces au Senegal. Notes Afr. 34, 20–23.

Cadenat, J., 1949. Notes sur les cétacés observés sur les côtes du Sénégal de 1941 à 1948. Bull. Inst. Fr. Afr. Noire 11, 1–15.

Cadenat, J., 1956. Un delphinidae encore mal connu de la côte occidentale d'Afrique: *Sotalia teüszii* Kukenthal 1892. Bull. Inst. Fr. Afr. Noire 18A, 555–566.

Collins, T., Boumba, R., Thonio, J., Parnell, R., Vanleeuwe, H., Ngouessono, S., Rosenbaum, H.C., 2010. The Atlantic humpback dolphin (*Sousa teuszii*) in Gabon and Congo: cause for optimism or concern? Paper SC/62/SM9 presented to the Scientific Committee of the International Whaling Commission, Agadir.

Davidson-Arnott, R.G.D., Greenwood, B., 2003. Waves and sediment transport in the nearshore zone. In: Isla, F.I. (Ed.), Coastal Zones and Estuaries. Encyclopedia of Life Support Systems (EOLSS) Publishers, Paris, France, pp. 43–60. Developed under the auspices of the UNESCO.

Dupuy, A.R., 1983. Données complémentaires concernant le statut des mammifères marins du Sénégal. Bull. Inst. Fond. Afr. Noire 45, 380–391.

Dupuy, A.R., Maigret, J., 1979. Les mammifères marins des côtes du Sénégal. 3. Observations signalées en 1978. Bull. Inst. Fond. Afr. Noire 41A, 429–439.

Hamazaki, T., 2002. Spatiotemporal prediction models of cetacean habitats in the mid-western north Atlantic Ocean (from Cape Hatteras, North Carolina, USA to Nova Scotia, Canada). Mar. Mamm. Sci. 18, 920–939.

Karczmarski, L., 2000. Conservation and management of humpback dolphins: the South African perspective. Oryx 34, 207–216.

Kaschner, K., Watson, R., Trites, A.W., Pauly, D., 2006. Mapping world-wide distributions of marine mammal species using a relative environmental suitability (RES) model. Mar. Ecol. Prog. Ser. 316, 285–310.

Kershaw, F., 2008. Mean annual sea surface temperature, for the period 2003–2007. Using data from NASA's Ocean Color database, In: Integrating Highly Migratory Species into High Seas Marine Protected Area Planning: A Global Gap Analysis (Oxford University, Centre for the Environment). UNEP World Conservation Monitoring Centre, Cambridge, UK, p. 113. http://data.unepwcmc.org/datasets/20.

Kükenthal, W., 1892. *Sotalia tëuszii* n. sp., ein pflanzenfressender (?) Delphin aus Kamerun. Zool. Jahrb. Abt. Syst. 6, 442–446.

Leeney, R.H., Weir, C.R., Campredon, P., Regalla, A., Foster, J., 2015. Occurrence of Atlantic humpback (*Sousa teuszii*) and bottlenose (*Tursiops truncatus*) dolphins in the coastal waters of Guinea-Bissau, with an updated cetacean species checklist. J. Mar. Biol. Assoc. U.K. 1–9. http://dx.doi.org/10.1017/S0025315415000661. Published online: 25 May 2015.

MacLeod, C.D., 2009. Global climate change, range changes and potential implications for the conservation of marine cetaceans: a review and synthesis. Endanger. Species Res. 7, 125–136.

Maigret, J., 1980. Données nouvelles sur l'écologie du *Sousa teuszii* (Cetacea, Delphinidae) de la côte ouest africaine. Bull. Inst. Fond. Afr. Noire 42A, 619–633.

Notarbartolo di Sciara, G., Politi, E., Bayed, A., Beaubrun, P.-C., Knowlton, A., 1998. A winter cetacean survey off southern Morocco, with a special emphasis on right whales. Rep. Int. Whaling Commission 48, 547–550.

Spaans, B., 1990. Dolphins in the coastal area of Guiné Bissau. Lutra 33, 126–131.

Spalding, M.D., Fox, H.E., Allen, G.R., Davidson, N., Fernaña, Z.A., Finlayson, M., Halperin, B.S., Jorge, M.A., Lombana, A., Lourie, S.A., Martin, K.D., McManus, E., Molnar, J., Recchia, C.A., Robertson, J., 2007. Marine ecoregions of the world: a bio-regionalization of coastal and shelf areas. Bioscience 57, 573–583.

van Bree, P.J.H., Duguy, R., 1965. Sur un crâne de *Sotalia teuszii* Kükenthal, 1892 (Cetacea, Delphinidae). Z. Säugetierkd. 30, 311–314.

Van Waerebeek, K., Barnett, L., Camara, A., Cham, A., Diallo, M., Djiba, A., Jallow, A.O., Ndiaye, E., Samba Ould Bilal, A.O., Bamy, I.L., 2004. Distribution, status, and biology of the Atlantic humpback dolphin, *Sousa teuszii* (Kükenthal, 1892). Aquat. Mamm. 30, 56–83.

Wang, J.Y., Yang, S.C., Hung, S.K., Jefferson, T.A., 2007. Distribution, abundance and conservation status of the eastern Taiwan Strait population of Indo-Pacific humpback dolphins, *Sousa chinensis*. Mammalia 71, 157–165.

Weir, C.R., 2009. Distribution, behaviour and photo-identification of Atlantic humpback dolphins *Sousa teuszii* off Flamingos, Angola. Afr. J. Mar. Sci. 31, 319–331.

Weir, C.R., 2011. Ecology and conservation of cetaceans in the waters between Angola and the Gulf of Guinea, with focus on the Atlantic humpback dolphin (*Sousa teuszii*). Ph.D. Thesis, University of Aberdeen, UK.

Weir, C.R., 2015. Photo-identification and habitat use of Atlantic humpback dolphins (*Sousa teuszii*) around the Río Nuñez estuary in Guinea (West Africa). Afr. J. Mar. Sci. http://dx.doi.org/10.2989/1814232X.2015.1069757.

Zwart, S.J., Weir, C.R., 2014. Filling in the gaps: first record of *Sousa teuszii* in Benin (Gulf of Guinea: Africa). Mar. Biodivers. Rec. 7, e59.

CHAPTER FIVE

Assessment of the Conservation Status of the Indian Ocean Humpback Dolphin (*Sousa plumbea*) Using the IUCN Red List Criteria

Gill T. Braulik*[,1], Ken Findlay[†], Salvatore Cerchio[‡], Robert Baldwin[§]

*Wildlife Conservation Society—Tanzania Program, Zanzibar, Tanzania
[†]Mammal Research Institute Whale Unit, University of Pretoria, Pretoria, South Africa
[‡]New England Aquarium, Central Wharf, Boston, Massachusetts, USA
[§]Five Oceans Environmental Services LLC, Sultanate of Oman
[1]Corresponding author: e-mail address: GillBraulik@downstream.vg

Contents

1. Introduction 120
2. Taxonomy 121
   2.1 Scientific and Common Names 121
   2.2 Taxonomic Notes 121
3. Geographic Range 122
   3.1 Range Description 122
   3.2 Countries 125
4. Population 125
   4.1 Abundance 125
   4.2 Population Trend 127
5. Habitat and Ecology 128
   5.1 Habitat 128
   5.2 Food and Feeding 129
   5.3 Reproduction 129
   5.4 Migration and Movements 130
6. Use and Trade 131
7. Threats 132
   7.1 Major Threats 132
   7.2 Minor or Local Threats 134
8. Conservation Actions 135
9. IUCN Red List Justification 136
Acknowledgments 137
References 137

Advances in Marine Biology, Volume 72
ISSN 0065-2881
http://dx.doi.org/10.1016/bs.amb.2015.08.004

Abstract

Indian Ocean humpback dolphins (*Sousa plumbea*) are obligate shallow-water dolphins that occur exclusively in the near-shore waters of the Indian Ocean, from South Africa to the Bay of Bengal. They have a narrow habitat preference, restricted distribution and do not appear very abundant across any part of their range. There is no estimate of total species abundance; all populations that have been quantitatively evaluated have been small in size, usually fewer than 200 individuals. Fishing, dredging, land reclamation, construction blasting, port and harbour construction, pollution, boat traffic and other coastal development activities all occur, or are concentrated within, humpback dolphin habitat and threaten their survival. Although data are far from sufficient to make a rigorous quantitative assessment of population trends for this species, the scale of threats is large enough over a significant enough portion of the range to suspect or infer a decline of at least 50% over three generations, which qualifies it for listing on the IUCN Red List as Endangered. The issue primarily responsible is incidental mortality in fisheries, but the loss and degradation of habitat is likely a contributing factor. None of the threats have been adequately addressed in any part of the species' range, even though threat levels are increasing virtually everywhere.

1. INTRODUCTION

Indian Ocean humpback dolphins (*Sousa plumbea*) are obligate shallow-water dolphins that occur exclusively in the near-shore waters of the Indian Ocean from South Africa to the Bay of Bengal (Jefferson and Rosenbaum, 2014). In comparison with most other marine cetaceans, Indian Ocean humpback dolphins have a narrow habitat preference, restricted distribution and do not appear very abundant across any part of their range. Their coastal habitat is the most heavily impacted and exploited part of the ocean by humans, and thus the species is exposed to numerous anthropogenic activities, including fishing and habitat modification, raising serious concerns about their conservation status (IWC, 2002).

Sousa plumbea has only been recognised as a distinct species in its own right since 2014 (Jefferson and Rosenbaum, 2014). Information from many parts of its range is sparse. However, its elevation to a distinct species has increased concern over its status and will likely increase the level of research and conservation attention it receives in the future. In this chapter, we provide a concise review of what is currently known about the status and threats faced by the Indian Ocean humpback dolphin, based on existing published literature and unpublished reports. We then evaluate the conservation status of the species against the IUCN Red List criteria (IUCN, 2001) and make recommendations for its listing on the Red List.

2. TAXONOMY

2.1 Scientific and Common Names

Sousa plumbea (Cuvier, 1829) is commonly referred to as the Indian Ocean humpback dolphin throughout its range. The animal has many different names in the numerous local languages spoken around the Indian Ocean fringe where it occurs, including "kushuku" in some parts of Tanzania, and in Swahili in Tanzania and Kenya "Pomboo nundu" or "Pomboo mweupe". In South Africa, it is known as the "boggelrugdolfyn" in Afrikaans and does not appear to have any specific Zulu or Xhosa terms. In Madagascar, it is typically known by the French name "dauphin à bosse", but there are several names used by Malagasy fishers depending on region and dialect, including "fesodoby", "fesoke manjavany" and "fesobory". (Sutaria et al., 2015) listed the following colloquial names for humpback dolphins in India, depending upon province: "fukariyo", "gada/gad/gaadha reda", "kadal panni", "kadal ongi", "sori vedan", "paru vedan", "thella thoralu" or "goonu". Throughout Arabia, it is generally known as "dukhs". In Pakistan, humpback dolphins are called "malhar" in Sindh Province, and "Goco" or "Gocain" in Balochistan.

2.2 Taxonomic Notes

In 2014, a major taxonomic change in the genus *Sousa* resulted in formal recognition of the Indian Ocean humpback dolphin (*S. plumbea*) as a distinct species (Committee on Taxonomy, 2014; Jefferson and Rosenbaum, 2014). Prior to this change, humpback dolphins from South Africa to Australia were classified as the Indo-Pacific humpback dolphin (*Sousa chinensis*) (Jefferson and Karczmarski, 2001). A molecular genetic assessment using mitochondrial DNA (mtDNA) and nuclear DNA markers found Indian Ocean humpback dolphins to be significantly differentiated from the other *Sousa* species; they have no shared mtDNA control region haplotypes with the other species, and there are two diagnostic sites (Mendez et al., 2013). The skull of *S. plumbea* shows a greater length: breadth ratio (2.2–2.8) than that of *S. chinensis*, and the highest tooth counts (33–39 vs. as low as 27 for the other *Sousa* species) in the genus (Jefferson and Van Waerebeek, 2004). Skulls from South Africa clearly separated from all other *Sousa* spp. skulls in a discriminant function analysis (Mendez et al., 2013).

Figure 1 External morphology of adult Indian Ocean humpback dolphins (*Sousa plumbea*) from (A) Sri Lanka. Photograph: Anouk Ilangakoon. (B) Iran. Photograph: Hamed Moshiri/Plan 4 the Land. (C) Pemba Island, Tanzania. Photograph: Gill Braulik. (D) Mossel Bay, South Africa. Photograph: Renae Logston.

There is considerable variation and population structure within the Indian Ocean humpback dolphin species and it is possible that additional work will reveal several subspecies (Figure 1; Jefferson and Rosenbaum, 2014). There is uncertainty about the taxonomic affinities of the humpback dolphins in eastern India, Sri Lanka, Bangladesh and Myanmar, but there are few samples available from this area to provide clarity (Jefferson and Rosenbaum, 2014). Preliminary genetic analysis indicates that humpback dolphins in Bangladesh are genetically distinct from all other members of the genus (Amaral et al., 2015). Future studies will confirm whether *S. plumbea* actually occurs east of the southern tip of India and will clarify the relationships and taxonomic affinities of humpback dolphins in the Bay of Bengal. However, as the taxonomy is currently not clearly resolved, we tentatively include *Sousa* that occur in the western Bay of Bengal within this *S. plumbea* review.

3. GEOGRAPHIC RANGE

3.1 Range Description

Indian Ocean humpback dolphins occur exclusively in the developing world around approximately 27,000 km of the coastal fringe of the Indian

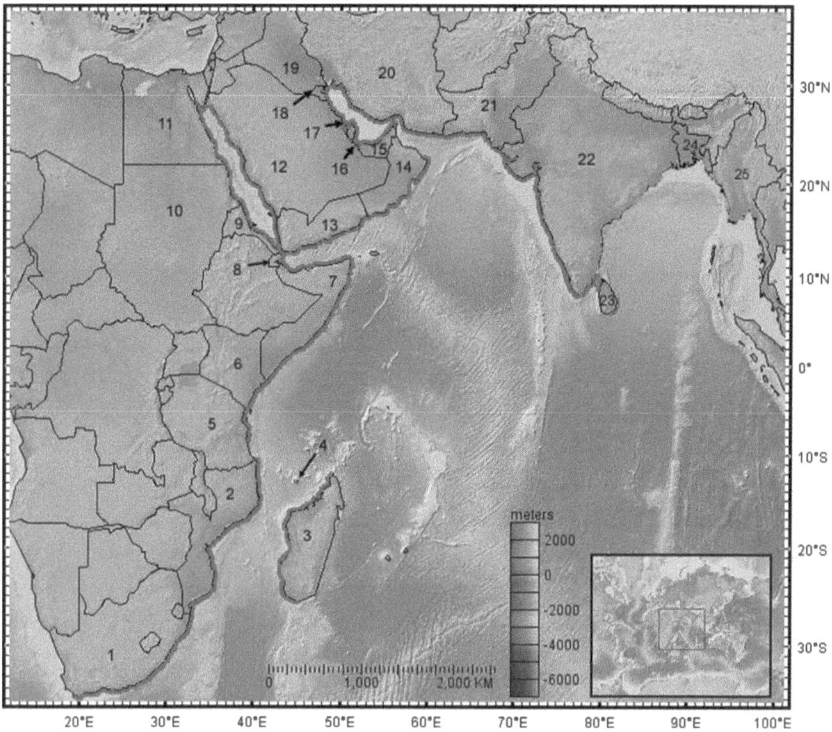

Figure 2 The assumed distribution of the Indian Ocean humpback dolphin (*Sousa plumbea*). Numbers refer to countries specified in Section 3.2.

Ocean (Figure 2). This species is found in a narrow strip of shallow, coastal waters from False Bay, South Africa, in the west, through the coastal waters of southern and eastern Africa, the Middle East and South Asia to at least the tip of India and possibly through the Bay of Bengal. Distribution includes the Red Sea, the Arabian/Persian Gulf, Gulf of Aden, western Madagascar and several offshore islands including the Andamans, Mayotte and Zanzibar. Their distribution covers approximately 65° of latitude (30°N to 35°S) and 60° of longitude (19–80°E). Large portions of the species' range have not been surveyed, and particularly in many parts of north Africa and the Middle East (including Somalia, Yemen, Djibouti, Eritrea, Sudan, Saudi Arabia, Egypt etc.), these animals are known to be present only from a handful of sighting or stranding records (Baldwin et al., 2004). Level of knowledge is greatest in South Africa, the Southwest Indian Ocean and Oman. The species typically occurs less than 3 km from shore and/or in water less than 25 m deep, and populations are usually found in locations with extensive shallows, such as protected bays and estuaries. Their distribution likely reflects the existence of pockets of suitable habitat and possibly local extirpations and range reductions (IWC, 2002). Known areas of *S. plumbea* concentration include Algoa, Richard's and Mossel Bays in South Africa (James et al., 2015; Karczmarski, 2000; Keith et al., 2013); Maputo Bay (Guissamulo and Cockroft, 2004), the Bazaruto Archipelago (Guissamulo and Cockcroft, 1997) and Beira in Mozambique; Nosy Be and the Nosy Iranja/Ampasindava Peninsula in Madagascar (Cerchio et al., 2015); the west coast of Unguja (Stensland et al., 2006) and Pemba Islands (Braulik, unpublished data) in Tanzania; Shimoni in Southern Kenya (Samuel V. Meyler et al., 2011); much of the Arabian Sea coast of Oman; and Goa and Cochin in India (Afsal et al., 2008; Parsons, 1998). They are one of the most common cetaceans in the Arabian/Persian Gulf (Baldwin et al., 2004; Braulik et al., 2010). In India, they are distributed almost continuously along most of the coast and they also occur in northwestern Sri Lanka (Nanayakkara et al., 2014; Sutaria and Jefferson, 2004). There are several sighting records for the northeast Andaman Islands, but none from oceanic archipelagos, such as the Lakshadweep and Maldive Islands (Sutaria and Jefferson, 2004), nor from eastern Madagascar (Cerchio et al., 2015). Humpback dolphins have not been recorded in the Union of Comoros (Kiszka et al., 2010a), however they do exist in very small numbers in the neighboring island of Mayotte (Kiszka et al., 2010b).

Interestingly, the western limit of the species' distribution in South Africa appears to have undergone a recent extension (Findlay et al., 1992). The photographed record of a humpback dolphin from Langebaan Lagoon in 2004/2005 is presumed a vagrant *S. plumbea*, rather than a vagrant *S. teuszii* (Best, 2007). Similarly, the record of a single humpback dolphin on the Mediterranean coast of Israel is a vagrant (Kerem et al., 2001) and these extra-limital records are not depicted on Figure 2.

There is a hiatus in distribution of several hundred kilometres along the Sea of Oman coast of Oman (Baldwin et al., 2004). It is unclear if this results from the influence of recent human population expansion and associated development or has an ecological basis (Baldwin et al., 2004). Recent information indicates that, contrary to earlier assumptions, humpback dolphins do occur along the coasts of eastern Iran and western Pakistan (Kiani and Van Waerebeek, 2015). Survey effort in early 2015 revealed a possible distribution hiatus along several hundred kilometres of the southern Tanzania coast, where the continental shelf is very narrow, and shallow habitat almost non-existent (Braulik, unpublished data). Genetic data suggest that within the western Indian Ocean, there is significant genetic population structure, correlated with large-scale oceanographic regimes distinguishing humpback dolphin populations from Oman, Zanzibar and South Africa/Mozambique (Mendez et al., 2011). More detailed sampling and analysis will likely reveal additional population structure at finer scales. While additional survey effort in unknown parts of this species' range will likely reveal more areas of concentration, it will probably also reveal areas of absence along exposed and deep coastlines, or areas of high human impact, and more discontinuities in distribution than are currently known.

3.2 Countries

Indian Ocean humpback dolphins are believed to occur from the coastal waters of 23 different countries or territories: South Africa (1), Mozambique (2), Madagascar (3), Mayotte (4), Tanzania (5), Kenya (6), Somalia (7), Djibouti (8), Eritrea (9), Sudan (10), Egypt (11), Saudi Arabia (12), Yemen (13), Oman (14), United Arab Emirates (15), Qatar (16), Bahrain (17), Kuwait (18), Iraq (19), Iran (20), Pakistan (21), India (22), and Sri Lanka (23). Humpback dolphins from Bangladesh (24) and Myanmar (25) are of unknown taxonomic status and are included in the *S. chinensis* chapter (Jefferson and Smith, 2016). (Numbers refer to locations in Figure 2)

4. POPULATION

4.1 Abundance

Indian Ocean humpback dolphin abundance has been estimated, using mark-recapture of photo-identified individuals, from several discrete locations in South Africa and the Southwest Indian Ocean (Table 1). There are no absolute abundance estimates from anywhere north or east of Kenya in the species range (Baldwin et al., 2004; Braulik et al., 2010; Gore et al., 2012; Kiani and Van Waerebeek, 2015; Sutaria et al., 2015). All populations that have been quantitatively evaluated have been small in size, always less than 500 individuals and usually fewer than 200 (see Table 1). Richard's Bay in KwaZulu-Natal has received much research focus because it reports the highest incidental catches of humpback dolphins in shark nets in the region (Keith et al., 2002). The population in Richard's Bay was considered to be open, with some residents and other individuals that range along the coast, and over approximately a 15-year period was estimated at between 74 (60–88) (Keith et al., 2002) and 160 (134–229) (Durham, 1994) individuals. No marked animals from Richard's Bay were recorded off Algoa Bay,

Table 1 Indian Ocean Humpback Dolphin (*Sousa plumbea*) Published Abundance Estimates and Confidence Intervals (CI)

Country	Location	Time Period	Estimate (95% CI)	Author
South Africa	KwaZulu-Natal coast (including Richard's Bay)	March 1991 to August 1992	160 (134–229)	Durham (1994)
South Africa	Richard's Bay	1998	74 (60–88)	Keith et al. (2002)
South Africa	Algoa Bay	1991–1994	466 (447–485)	Karczmarski et al. (1999)
South Africa	Mossel Bay	2011–2013	125 (61–260)	James et al. (2015)
South Africa	Plettenberg Bay		112 (75–133)	Jobson (2006)
Mozambique	Maputo Bay	1995–1997	105 (30–151)	Guissamulo and Cockroft (2004)
Mozambique	Bazaruto Archipelago	1992	ca. 60	Guissamulo and Cockcroft (1997)
Tanzania	Zanzibar, Kisimkazi	1999 2001 2002	58 (56–79) 65 (62–102) 63 (57–95)	Stensland et al. (2006)
Kenya	Shimoni archipelago	2006	104 (67–160)	Meyler et al. (2011)

where Karczmarski et al. (1999) estimated the total abundance at 466 individuals. Populations in Plettenberg Bay and Mossel Bay were estimated at about 112 (Jobson, 2006) and 125 individuals (95% CI 61, 260) (James et al., 2015), respectively.

Generally, the sites in the northwest of Madagascar appear to have larger populations of *S. plumbea* than the southwest region (Cerchio et al., 2015). All populations evaluated in Mozambique, Tanzania and Kenya are estimated at approximately 100 individuals or less (see Table 1).

In Oman and much of the Arabian Gulf, this species is among the most commonly recorded coastal cetaceans, and group size can be large, frequently more than 40 individuals, sometimes up to a hundred or more (Baldwin et al., 2004). Pilleri and Pilleri (1979) conducted limited surveys in the Indus Delta of Pakistan and based on these made an "educated guess" that there were approximately 500 individuals in the entire delta, which, if accurate, would make it one of the larger populations of this species. Standardised surveys for *S. plumbea* was recorded in the Gulf of Kachchh and the Goa coast of India, and the *S. plumbea* density was over six times higher in Goa (Sutaria and Jefferson, 2004).

The limited data presented above suggest that *S. plumbea* is not abundant anywhere in its range. There is no estimate of total species abundance, but Jefferson and Rosenbaum (2014) stated that the available estimates for specific populations suggest that the range-wide abundance is probably no higher than the low tens of thousands, which we agree is reasonable.

4.2 Population Trend

Most Indian Ocean humpback dolphin populations are small, comprising no more than a few hundred individuals (Guissamulo and Cockroft, 2004; Meyler et al., 2011; Stensland et al., 2006). Like all cetaceans, they have a low reproductive rate and therefore cannot sustain even a moderate level of anthropogenic mortality (Reilly and Barlow, 1986). Removal of only two or three animals per year is likely to be sufficient to cause a decline in such small populations, and it is clear that mortality rates from fisheries bycatch alone exceed this by several orders of magnitude in most places that *S. plumbea* occurs (see Section 7). In Algoa Bay, there were modelled population growth rates of −3% to +2% per year for that population (Karczmarski, 2000), but there are no quantitative estimates of population dynamics for any of the other small populations recorded across the range. Consequently, trends must be inferred by evaluating the magnitude of mortalities or declines in relative sighting rates. Along the KwaZulu-Natal coast, incidental capture of humpback dolphins in shark nets has been very high; 203 animals were caught between 1980 and 2009 (an average of 6.8 animals per year), of which 61% originated from Richards Bay (Atkins et al., 2013). Between 1984 and 1992, there was a decrease of 50% in incidental sightings of this species by the KwaZulu-Natal Sharks Board, indicating a possible population decrease in this region (Durham, 1994), despite a lack of decline in the bycatch rate suggesting the persistence of the population. Significantly, Atkins et al. (2013) noted that the majority of shark net mortalities were of males.

In the Arabian Gulf, aerial surveys recorded Indian Ocean humpback dolphins as the second most commonly sighted cetacean after the Indo-Pacific bottlenose dolphin *Tursiops aduncus* (Preen, 2004). Between 1986 and 1999, the data indicate a statistically significant decline in abundance of 70% for all cetaceans combined, which included the Indian Ocean humpback dolphin, bottlenose dolphin and Indo-Pacific finless porpoise, *neophocaena phcaenoides* (Preen, 2004).

In Madagascar, there was an observed decline in sighting rates Indian Ocean humpback dolphins in the southwest region (off Anakao), along with a decrease in mean group size between 2004 and 2013, potentially due to hunting, bycatch and habitat degradation (Cerchio et al., 2015). In addition to declining abundance, high levels of anthropogenic mortality in this species are likely to create distribution gaps in areas of high threat.

Although little quantitative data on population trends exist, because of the restricted distribution and small population sizes of this species, in many cases only 4–8 deaths per year is sufficient to cause a 50% decline over three generations necessary to qualify for listing as Endangered. The overwhelming level of threat due to intensive use by humans of their exclusively nearshore habitat and consistently high reported mortality rates throughout the parts of their range where monitoring has occurred, means that the global *S. plumbea* population size is almost certainly decreasing (Friedmann and Daly, 2004; Reeves et al., 2008).

5. HABITAT AND ECOLOGY

5.1 Habitat

Throughout their range, Indian Ocean humpback dolphins occur around a variety of coastal habitats including mangroves, rocky reefs, coastal lagoons and shallow, protected bays. The over-riding habitat preference appears to be for water less than 25 m in depth. In Richards Bay, South Africa, all encounters with humpback dolphins were in water shallower than 20 m. Dolphins used the area within 2 km from shore extensively and were rarely seen beyond 3 km from shore (Atkins et al., 2004). In Algoa Bay, over 80% of sightings were within 400 m of the shore and there was a consistent preference for feeding over shallow rocky reefs throughout the year, despite considerable changes in dolphin abundance (Karczmarski et al., 2000). In Zanzibar, distribution of humpback dolphin groups was restricted to a median distance of 830 m (min–max: 200–1550 m) from the shore and a median water depth of 11 m (min–max: 2–26 m) (Stensland et al., 2006). In Nosy Be, Madagascar, the mean depth of sightings was 8.2 m with 95% of sightings in water less than 20 m deep (Cerchio et al., 2015). In Goa and the Gulf of Kutchchh, India, most groups were sighted in water less than 10 m in depth (Sutaria and Jefferson, 2004), and similarly in Chennai, dolphins occurred in water from between 10 and 25 m deep (Muralidharan, 2013). In Oman and the Arabian Gulf, humpback dolphins generally occur in water depths of less than 20 m. In some areas, they occur along rocky shores with a narrow continental shelf, but in such cases, they occur very close (a few hundred metres at most) from the coast. Humpback dolphins also occur in channels within the mangroves of the massive Indus (Kiani and Van Waerebeek, 2015; Pilleri and Gihr, 1972) and Ganges–Brahmaputra deltas (Smith et al., 2006).

5.2 Food and Feeding

Humpback dolphins feed on a variety of fishes, cephalopods and crustaceans found in coastal areas. In the Eastern Cape of South Africa, common inshore and brackish water fish and an octopus were found in the stomachs of *S. plumbea* (Barros and Cockroft, 1991). In Natal, South Africa, prey items were all fish, 61% were littoral or estuarine species and 25% were demersal species primarily associated with reefs (Ross et al., 1994). Stomach samples collected from eight dead individuals in southern Oman showed a high incidence of cephalopods and crustaceans in the diet (Baldwin et al., 2004). The species generally feeds in shallow near-shore areas, including around reefs and rocky coasts as well as over soft sediments, and in estuaries and mangroves (Baldwin et al., 2004; Kiani and Van Waerebeek, 2015; Ross, 1984; Saayman and Tayler, 1979; Stensland et al., 2006). In Richard's Bay, South Africa, inshore areas were important for feeding, and feeding behaviour decreased with distance offshore and animals moved further offshore to rest (Atkins et al., 2004).

Sousa plumbea have been observed in the Arabian Gulf and the Bazaruto Archipelago in Mozambique, herding fishes onto exposed sand banks and deliberately beaching to seize their prey (Baldwin, 1995; Peddemors and Thompson, 1994).

5.3 Reproduction

There is very little information on the life history of this species, with the vast majority of existing information originating only from South Africa. There is an austral spring or summer calving peak in South Africa, with 70% of births occurring between October and May (Cockcroft, 1989; Karczmarski, 1996; Karczmarski et al., 1999). Mating behaviour was observed in Pakistan only in January and February (Kiani and Van Waerebeek, 2015) and in Maputo, Mozambique, births occurred throughout the year (Guissamulo and Cockroft, 2004). Gestation is believed to last about 10–12 months and length at birth was estimated to be approximately 100 cm (Cockcroft, 1989). In South African waters lactation may last for more than 2 years (Cockcroft, 1989), mother–calf associations remain strong for at least 3–4 years (Karczmarski et al., 1999) and a 3-year calving interval was suggested (Cockcroft, 1989; Karczmarski, 1996; Karczmarski et al., 1999). Age at sexual maturity for South African animals is about 10 years for females and 12–13 years for males (Cockcroft, 1989; Plön et al., 2015).

Sousa plumbea appear to be sexually dimorphic in South Africa, with adult females on average approximately 30 cm shorter than males. Adult males reach around 270 cm, and females 240 cm and the maximum weight recorded was around 260 kg. This species is believed to reach the age of at least 40 years (Cockcroft, 1989). Lengths of 88 specimens from the Arabian Gulf ranged from 93 to 269 cm (Ross et al., 1994). Although it is possible that animals may be larger in the northern Indian Ocean than in South Africa, several records of animals over 3 m in length have been discounted (Jefferson and Rosenbaum, 2014; Jefferson and Van Waerebeek, 2004). Maximum reliable total length for *S. plumbea* is considered to be 279 cm (Jefferson and Rosenbaum, 2014). Generation length of *S. chinensis* and *S. plumbea* combined was modelled to be approximately 25 years (Moore, in press).

5.4 Migration and Movements

The majority of *S. plumbea* populations appear to be composed of long-term resident individuals that show limited movements and a varying proportion of transient animals that range along the shore for tens or hundreds of kilometres (Guissamulo and Cockroft, 2004; Karczmarski et al., 1999; Keith et al., 2002; Parsons, 1998). Humpback dolphins off the southeast tip of Zanzibar showed a very high degree of residency, with 94% of identified individuals re-sighted 2 years later (Stensland et al., 2006). However, Zanzibar is an island surrounded by deep water and dispersal opportunities for this population are therefore fewer than along the coast of mainland Africa. In contrast, the largest known *S. plumbea* population in the Algoa Bay region of South Africa is composed primarily of animals that move along the coast (Karczmarski et al., 1999). Nine photo-identified individuals were recorded to move 140 km between Plettenberg Bay and Mossel Bay in South Africa (James et al., 2015). One dolphin was recorded in Durban in June 1998, and within 8 days had moved around 150 km to Richard's Bay (Keith et al., 2002). Similarly, an animal photo-identified in Kisite in southern Kenya was recently re-sighted in Watamu around 150 km to the north (Kenya Marine Mammal Network, 2012).

The along-shore movement among "populations" in South Africa suggested by photo-identification studies is supported by genetic studies that showed no population structure between sampling sites in South Africa and Mozambique (Mendez et al., 2011). The strong maternal genetic differentiation displayed between humpback dolphins from South Africa/Mozambique, Zanzibar and Oman suggests limited recent gene flow, however some degree of historic migration between these distant areas (Mendez et al., 2011). This suggests that at least along the east coast of the African continent longer movements of humpback dolphins are very rare.

In Algoa Bay, there were two seasonal peaks in abundance, one in summer and one in late winter, possibly due to changes in prey distribution (Karczmarski et al., 1999). In Maputo Bay, Mozambique, there was also an influx of individuals into the study area in summer (Guissamulo and Cockroft, 2004). There is still a great deal to be clarified about the residency, movements and seasonal changes in distribution of this species.

6. USE AND TRADE

Throughout the majority of their range, except perhaps South Africa, accidentally captured humpback dolphins are generally used. Animals are reported to be eaten (termed "marine bushmeat") in numerous countries including Madagascar (Razafindrakoto et al., 2004), Mozambique (Guissamulo, 2008), Tanzania (Amir et al., 2002) and India (Kumarran, 2012). Humpback dolphin meat is sometimes used as bait for sharks (Amir et al., 2002), or for crabs or other fish (Kiani and Van Waerebeek, 2015), and the oil is used as wood preservative for boats (Berggren and Coles, 2009), for cooking, or as medicine (Gore et al., 2012). A dolphin drive hunt in the southwest of Madagascar operates exclusively to capture dolphins for human consumption (Cerchio et al., 2015).

Although they are often utilised, reports of actual sale of animals at market are less common. Humpback dolphins were reported for sale at three different locations on the west coast of India (Goa, Malpe in Karnataka and Trivandrum in Kerala) (Kumarran, 2012; Mohan, 1994), and in Madagascar animals were for sale in the local village (Cerchio et al., 2015). Mohan (1994) reported that one *S. plumbea* was kept in a polythene-lined pool in Calicut for public display.

Non-consumptive uses of humpback dolphins include dolphin watching tourism; however, because of their low abundance and shy behaviour, they are often not the primary target of the activity. Humpback dolphins can be viewed in Kisimkazi, Zanzibar, where the primary target species is Indo-Pacific bottlenose dolphins (Berggren et al., 2007). They are also targeted in Nosy Be Madagascar, and near Salalah, on the Arabian Sea coast of Oman and around the Musandam Peninsula in the Straits of Hormuz, Oman.

7. THREATS

7.1 Major Threats

The preference of Indian Ocean humpback dolphins for shallow waters places them in some of the world's most intensively utilised, fished, shipped, modified and polluted waters. The primary threat to the Indian Ocean humpback dolphin species throughout most, or all, of its range is incidental mortality in fisheries, including in shark control nets in South Africa. There appear to be very few, and possibly no, areas that may be a refuge from this pervasive threat.

With the exception of the KwaZulu-Natal Shark Control Programme, there have been no on-board observer studies from which bycatch estimates could be generated. Although it is impossible to evaluate the magnitude in most areas, incidental fishing mortality is clearly pervasive and certainly unsustainable. In many countries around the Indian Ocean, fisheries are primarily artisanal, boats or canoes are small and frequently oar or sail powered, and therefore fishing concentrates precisely within the preferred near-shore habitat of humpback dolphins. For example, in Pemba in Tanzania approximately 90% of humpback dolphins sighted during recent surveys occurred less than 1 km from shore, and 95% of recorded fishing vessels occurred in precisely the same habitat. Of 27 photo-identified humpback dolphins in the same area, 41% had clear injuries from previous entanglements in fishing nets (Braulik, unpublished data). This clearly demonstrates the degree to which most humpback dolphins have to negotiate and frequently interact with fishing gear and there is every reason to believe that this example from Pemba is the norm throughout the species' range.

Significant incidental mortality of humpback dolphins in coastal gillnets is reported from most countries within their range (IWC, 2002). The best-studied bycatch is in Richard's Bay in South Africa, where a total of 203 humpback dolphins were captured in shark nets in the 30 years between 1980 and 2009 (6.8/year corresponding to 5–10% of the population per annum, which is highly unsustainable (Atkins et al., 2013; Durham, 1994; Keith et al., 2002). In Mozambique, intense coastal fishing effort was considered to be the main threat to this species, and shark fishery gillnets and trawl nets have incidentally killed dolphins throughout the country (Guissamulo, 2008). Incidental catch of humpback dolphins has been documented in gillnets off the south coast of Zanzibar in Tanzania (Amir et al., 2002). There are currently no estimates of the magnitude of the bycatch, but given that even two humpback dolphins taken per year would exceed 2% of the population, it is very likely that bycatch represents a threat to the dolphins in the area (Stensland et al., 2006). In Oman, high incidences of beachcast *S. plumbea* are presumed to result from interactions with fisheries (Collins et al., 2002). The reported mortality rate due to fisheries interactions (both bycatch and directed take) in the southwest region of Madagascar was noted to be almost certainly unsustainable (Cerchio et al., 2015). Although it is impossible to evaluate the magnitude in most areas, in the areas that it has been evaluated incidental fishing mortality of this species appears to be high, unsustainable and resulting in rapid local population declines. There is every reason to believe that interactions with fisheries are equal or possibly even greater elsewhere in the species' range.

There are very few areas within the known range of Indian Ocean humpback dolphins where anthropogenic alteration to habitat has not occurred. Destruction of inshore habitats is likely to be one of the greatest threats for humpback dolphins, particularly in the southern African region and in the Arabian Gulf as well as in many other increasingly developed urban coastal areas (Baldwin et al., 2004; Karczmarski, 2000). Dredging, land reclamation, construction blasting, port and harbour construction, pollution, boat traffic, oil and gas exploration (including seismic surveying) and other coastal development activities all occur, or are concentrated within, humpback dolphin habitat and threaten their survival (IWC, 2002). It was also noted by the IWC (2002) that the continued presence of humpback dolphins in highly degraded habitats does not rule out adverse effects of habitat degradation. This is a pervasive threat that is increasing throughout the species' range and there is no reason to expect this trend to change in the foreseeable future.

In comparison to other marine mammals with wider and more oceanic distributional ranges, the exposure of *S. plumbea* to environmental contaminants is likely to be very high (Jefferson and Karczmarski, 2001). Many of the large urban centres and ports around the Indian Ocean, including Mumbai, Karachi, Dubai, Aden, Mombasa, Dar es Salaam, Maputo and Durban, release a toxic cocktail of agricultural runoff and untreated human and industrial waste into coastal waters, which are inhabited by humpback dolphins. Humpback dolphins in Natal in South Africa had the highest levels of organochlorines in the blubber of any marine mammal in the country. Elsewhere the effects of such pollution on local dolphins have yet to be studied, but may be severe (Gore et al., 2012). For example, in Pakistan very high levels of chemical pollution in creeks of the Indus delta are toxic enough to cause fish kills and are likely to have adverse effects on cetaceans (Kiani and Van Waerebeek, 2015). In Gadani, Pakistan, up to 100 ships per year are scrapped and dismantled, leading to the release of large amounts of heavy metals, asbestos, dioxins and other persistent organic pollutants in coastal waters. These problems are widespread and increasing in many countries; their impact on coastal dolphins has not been evaluated, but pollution is likely contributing to local declines in range and abundance and may have caused extirpations adjacent to major industrial centres.

7.2 Minor or Local Threats

Other threats that appear somewhat less serious (possibly because of lack of information) or only affect a certain portion of the species range include direct killing, boat traffic/harassment and oil spills and exploration (IWC, 2002).

There is little evidence of intense hunting of humpback dolphins except for in Madagascar (IWC, 2002). In the southwest of the country, dolphins were historically taken with harpoons, but are now targeted with gillnets or in drive hunts. Andrianarivelo (2001) estimated a minimum of 61 mortalities of *S. plumbea* between 1985 and 1999 in Anakao related to the directed takes including drive hunts. The hunting tradition is restricted to a single ethnic group and specific villages, but is widespread along a large section of the country and is apparently spreading (Cerchio et al., 2015). Given the relatively small population sizes reported throughout the SWIO region, the reported mortality rate due to hunting is likely unsustainable and contributing to population decline in at least the southwest of Madagascar (Cerchio et al., 2015).

Dolphin hunting used to occur in Menai Bay in Zanzibar; the last hunt occurred in 1996, taking 23 individuals, assumed to be *S. plumbea* and *T. aduncus*. This would represent an annual mortality close to 12% for a combined population estimate of 199 animals for the two dolphin species in the area. This was certainly unsustainable, resulting in a negative impact on the status of the dolphins off the south coast of Zanzibar (Stensland et al., 2006).

Humpback dolphins reportedly were hunted in former years in the Arabian Gulf, and there was some evidence that there was continued hunting of dolphins in Oman from small, motorised boats using harpoons (Baldwin et al., 2004). In Maputo Bay, Mozambique, dolphins were also reported to be hunted for meat in intertidal shallow areas and estuaries (Guissamulo, 2008).

Blast fishing using dynamite or other explosives is an intense threat to humpback dolphins in the few countries in which it occurs frequently: Tanzania (Cagua et al., 2014) and Sri Lanka (Cornelis et al., 2008; Nanayakkara et al., 2014). This illegal activity is concentrated in near-shore areas that are also specifically humpback dolphin habitat.

Humpback dolphins seem to be highly susceptible to disturbance caused by inshore boat traffic in Algoa Bay (Karczmarski et al., 1997) and areas most heavily used by inshore traffic were reported to be seldom visited by humpback dolphins (Karczmarski, 1996). Humpback dolphins do not ride the bow waves of boats and generally actively avoid moving vessels. In Puttalam Lagoon in Sri Lanka, the small area of shallow habitat in which *S. plumbea* is observed is used extensively by high-powered fishing and navy vessels, which present a threat to this small population (Cornelis et al., 2008; Nanayakkara et al., 2014). Increases in the number of small personal water craft, such as jet-skis, to many coastal areas as tourism expands is another potential disturbance within Indian Ocean humpback dolphin habitat and there is also a risk of collisions.

8. CONSERVATION ACTIONS

For many countries and regions throughout their range, Indian Ocean humpback dolphins are extremely poorly known, and although many threats exist, there have been few conservation actions. One exception is KwaZulu-Natal in South Africa, where since 1999 attempts have been made to reduce the large and unsustainable accidental capture of dolphins in shark nets by using acoustic deterrents (pingers) or by making the nets more conspicuous acoustically (e.g. air-filled floats, clangers, etc.) (Cliff and Dudley, 2011), but these have so far been unsuccessful (Atkins et al., 2013). Increasing the mesh size of nets was effective, but compromised the nets ability to repel sharks. Some high catch shark nets were replaced with baited drumlins, which do not catch dolphins (Atkins et al., 2013). Acoustic pingers were also trialled on gillnets in Tanzania.

Dolphins used to be hunted in Menai Bay in the south of Unguja Island, Tanzania; however since 1997, the hunt has been replaced by swim-with-the-dolphin tourism (Berggren et al., 2007). Similarly in Madagascar in the southwest region, there has been a focus on community engagement to mitigate marine mammal hunting and bycatch through the establishment of community-based ecotourism, local protection and enforcement, and social outreach campaigns, which have met with some success (Cerchio et al., 2009, 2014).

9. IUCN RED LIST JUSTIFICATION

In the places where studies have occurred Indian Ocean humpback dolphin populations are small, always less than 500 and often less than 100 individuals in discrete, semi-isolated areas. They have one of the most specific habitat preferences and restricted distributions of any marine megafauna species, and these are both aspects that are proven to reduce the resilience of species to environmental change and anthropogenic threats, thereby increasing their extinction risk (Davidson et al., 2011; Purvis et al., 2000). Indian Ocean humpback dolphins are concentrated in coastal waters less than 2 km from shore and often only a few hundred metres from land. This places them in direct conflict with artisanal fisheries that occur intensively throughout their range and focus in exactly the same near-shore habitat. High and clearly unsustainable mortality rates are reported from several areas and heavy mortality can be inferred from the high degree of scarring (41% of individuals) in Pemba, Tanzania. Although information on population size, threats and mortality is only available from some portions of the species' range, there is every reason to suspect and infer that these will be similar or possibly worse elsewhere.

Mortality of only 4 individuals per year from a population of 100, or 7 from a population of 200 would result in the 50% population decline necessary to qualify this species for Endangered. The available evidence from South Africa and indications from elsewhere in the range suggest that mortality rates are consistently at or above this level. The species' restricted habitat and small populations overlap in both space and time with several ubiquitous and pervasive threats that are increasing in severity, leaving no refuge for this species from anthropogenic mortality. The threats are large enough over a significant enough portion of the range that evidence leads us to suspect and infer a decline of at least 50% over three generations (about 75 years) spanning both the past and the future. The factor primarily responsible for the decline is incidental mortality in coastal artisanal fisheries, which were introduced to the region in the 1960s (Tarbit, 1984), but the loss and degradation of habitat in numerous coastal areas is a contributing and increasing factor. None of the threats have been adequately addressed in any part of the species' range, even though threat levels are increasing virtually everywhere. All evidence suggests that threats and declines will continue and may increase in the future.

The Indian Ocean humpback dolphin therefore qualifies for Endangered under IUCN criteria A4cd, an inferred and suspected population reduction of greater than or equal to 50% over three *S. plumbea* generations (75 years), from approximately 1960 in the past to 2035 in the future, considering that the causes of the suspected/inferred decline in population size—bycatch and decline in habitat quality—have not ceased and are not well understood. This is based on (c) a decline in quality of habitat, and (d) actual or potential levels of exploitation, in this case fisheries bycatch and hunting. As there have been virtually no conservation actions to address these threats, all evidence suggests that they will continue and may escalate in the future, and we strongly suspect that high mortality and continued population declines of this species will occur in the coming years, therefore satisfying criteria A3cd.

ACKNOWLEDGMENTS

We acknowledge the contribution of one reviewer. We also thank the participants of the Workshop to Assess/Reassess IUCN Red Listings for Indo-Pacific Genera of Coastal Small Cetaceans (San Diego, California, 20–21 May 2015) for useful discussions and input that helped with preparation and refinement of this chapter.

REFERENCES

Afsal, V.V., Yousuf, K.M.M., Anoop, B., Anoop, A.K., Kannan, P., Rajagopalan, M., Vivekanandan, E., 2008. A note on cetacean distribution in the Indian EEZ and contiguous seas during 2003–07. J. Cetac. Res. Manag. 10, 209–215.

Amaral, A.R., Smith, B.D., Mansur, R., Brownell, R.L., Rosenbaum, H.C., 2015. Genetic identity of Indo-Pacific humpback dolphins (*Sousa* spp.) in the northern Bay of Bengal, Bangladesh. Document SC/66a/SM/24 Submitted to the International Whaling Commission Scientific Committee.

Amir, O.A., Berggren, P., Jiddawi, N.S., 2002. The incidental catch of dolphins in gillnet fisheries in Zanzibar, Tanzania. Western Indian Ocean J. Mar. Sci. 1, 155–162.

Andrianarivelo, N., 2001. Essai d'évaluation de l'importance de la pêche aux dauphins dans la région d'Anakao (sud-ouest de Madagascar). Unpublished DEA thesis, Institut Halieutique et des Sciences Marines, Université de Toliara, Madagascar.

Atkins, S., Pillay, N., Peddemors, V.M., 2004. Spatial distribution of Indo-Pacific humpback dolphins (*Sousa chinensis*) at Richard's Bay, South Africa: environmental influences and behavioural patterns. Aquat. Mamm. 30, 84–93.

Atkins, S., Cliff, G., Pillay, N., 2013. Humpback dolphin bycatch in the shark nets in KwaZulu-Natal, South Africa. Biol. Conserv. 159, 442–449.

Baldwin, R.M., 1995. Whales and Dolphins of the United Arab Emirates. Emirates Printing Press, Dubai, UAE.

Baldwin, R.M., Collins, M., Van Waerebeek, K., Minton, G., 2004. The Indo-Pacific humpback dolphin of the Arabian region: a status review. Aquat. Mamm. 30, 111–124.
Barros, N.B., Cockroft, V., 1991. Prey of humpback dolphins (*Sousa plumbea*) stranded in eastern Cape Province, South Africa. Aquat. Mamm. 17, 134–136.
Berggren, P., Coles, P., 2009. Whales and Dolphins—A Field Guide to Marine Mammals of East Africa. WIOMSA/East Publishing, Tanzania.
Berggren, P., Amir, O.A., Guissamulo, A., Jiddawi, N.S., Ngazy, Z., Stensland, E., Sarnblad, A., Cockroft, V.G., 2007. Sustainable dolphin tourism in East Africa: MASMA Technical Report. WIOMSA Book Series No. 7, ix+72 pp.
Best, P.B., 2007. Whales and Dolphins of the Southern African Subregion. Cambridge University Press, Cape Town.
Braulik, G.T., Ranjbar, S., Owfi, F., Aminrad, T., Dakhteh, S.M.H., Kamrani, E., Mohsenizadeh, F., 2010. Marine mammal records from Iran. J. Cetac. Res. Manag. 11, 49–63.
Cagua, F., Bassett, C., Kayanda, R., Rubens, J., Machano, H., 2014. Acoustic Monitoring of Blast Fishing. WWF, Dar es Salaam, Tanzania.
Cerchio, S., Andrianarivelo, N., Razafindrakoto, Y., Mendez, M., Rosenbaum, H., 2009. Coastal dolphins hunting in the southwest of Madagascar: status of populations, human impacts and conservation actions. In: Paper SC/61/SM15 presented to the International Whaling Commission Scientific Committee.
Cerchio, S., Andrianarivelo, N., Andrianantenaina, B., Cordi, V., 2014. Ecology, status, fisheries interactions and conservation of coastal Indian Ocean humpback dolphins and Indo-Pacific bottlenose dolphins on the west coast of Madagascar. In: Paper SC/65B/SM21 presented to the International Whaling Commission Scientific Committee.
Cerchio, S., Andrianarivelo, N., Andrianantenaina, B., 2015. Ecology and conservation status of Indian Ocean humpback dolphins (*Sousa plumbea*) in Madagascar. Adv. Mar. Biol., 72, 163–199.
Cliff, G., Dudley, S.F.J., 2011. Reducing the environmental impact of shark-control programs: a case study from KwaZulu-Natal, South Africa. Mar. Freshw. Res. 62, 700–709.
Cockcroft, V.G., 1989. Biology of Indopacific humpback dolphins (*Sousa plumbea*) off Natal, South Africa. In: Paper presented at the Eight Biennial Conference on the Biology of Marine Mammals, Pacific Grove, CA (unpublished). Available at http://www.marinemammalscience.org.
Collins, T., Minton, G., Baldwin, R., Van Waerebeek, K., Hywel Davies, A., Cockroft, V., 2002. A preliminary assessment of the frequency, distribution and causes of mortality of beach cast cetaceans in the Sultanate of Oman, January 1999 to February 2002. In: Paper SC/54/O4 presented to the IWC Scientific Committee, May 2002 (unpublished). 13 pp (available from the Office of this Journal).
Committee on Taxonomy, 2014. List of marine mammal species and subspecies. Society for Marine Mammalogy, www.marinemammalscience.org, consulted on 4th February 2014.
Cornelis, K., Broker, A., Ilangakoon, A., 2008. Occurrence and conservation needs of cetaceans in and around the Bar Reef Marine Sanctuary, Sri Lanka. Oryx 42, 286–291.
Cuvier, G., 1829. Le regne animal distribute d'apres son organisation, pour servir de base a l'histoire naturelle des animaux et d'introduction a l'anatomie comparee. Chez Deterville, Libraire, Paris, France.
Davidson, A.D., Boyer, A.G., Kim, H., Pompa-Mansilla, S., Hamilton, M.J., Costa, D.P., Ceballos, G., Brown, J.H., 2011. Drivers and hotspots of extinction risk in marine mammals. Proc. Natl. Acad. Sci. U.S.A. 109 (9), 3395–3400. www.pnas.org/cgi/doi/10.1073/pnas.1121469109.
Durham, B., 1994. The Distribution and Abundance of the Humpback Dolphin (*Sousa chinensis*) on the Natal Coast, South Africa. University of Natal, Durban, South Africa.

Findlay, K.P., Best, P.B., Ross, G.J.B., Cockcroft, V.G., 1992. The distribution of small odontocete cetaceans off the coasts of South Africa and Namibia. S. Afr. J. Mar. Sci. 12, 237–270.

Friedmann, Y., Daly, B., 2004. Red Data Book of the Mammals of South Africa: A Conservation Assessment. CBSG southern Africa. Conservation Breeding Specialist Group (SSC/IUCN) Endangered Wildlife Trust, Johannesburg, South Africa.

Gore, M.A., Kiani, M.S., Ahmad, E., Hussain, B., Ormond, R.F., Siddiqui, J., Waqas, U., Culloch, R., 2012. Occurrence of whales and dolphins in Pakistan with reference to fishers' knowledge and impacts. J. Cetac. Res. Manag. 12, 235–247.

Guissamulo, A., 2008. Ecological Studies of Bottlenose and Humpback Dolphins in Maputo Bay, Southern Mozambique. Department of Zoology and Entomology. University of Kwazulu-Natal, Durban, South Africa.

Guissamulo, A., Cockcroft, V.G., 1997. Dolphin and dugong occurrence and distribution and fisheries interactions in Maputo and Bazaruto Bays, Mozambique: International Whaling Commission Scientific Committee Report SC/49/SM24. Available at http://iwc.int/sc-documents. 12.

Guissamulo, A., Cockroft, V.G., 2004. Ecology and population estimates of Indo-Pacific humpback dolphins (*Sousa chinensis*) in Maputo Bay, Mozambique. Aquat. Mamm. 30, 94–102.

IUCN, 2001. IUCN Red List Categories and Criteria: Version 3.1. IUCN Species Survival Commission, IUCN, Gland, Switzerland and Cambridge, UK.

IWC, 2002. Annex K. Report of the standings sub-committee on small cetaceans. J. Cetac. Res. Manag. 4 (Suppl.), 325–338.

James, B.S., Bester, M.N., Penry, G.S., Gennari, E., Elwen, S.H., 2015. Abundance and degree of residency of humpback dolphins (*Sousa plumbea*) in Mossel Bay, South Africa. Afr. J. Mar. Sci., in press.

Jefferson, T.A., Karczmarski, L., 2001. Sousa chinensis. Mamm. Species 665, 1–9.

Jefferson, T.A., Rosenbaum, H.C., 2014. Taxonomic revision of the humpback dolphins (Sousa spp.), and description of a new species from Australia. Mar. Mamm. Sci. 30, 1494–1541.

Jefferson, T.A., Smith, B.D., 2016. Re-assessment of the conservation status of the Indo-Pacific humpback dolphin (*Sousa chinensis*) using the IUCN Red List Criteria. Adv. Mar. Biol. 73, (in press).

Jefferson, T., Van Waerebeek, K., 2004. Geographic variation in skull morphology of humpback dolphins (*Sousa* spp.). Aquat. Mamm. 30, 3–17.

Jobson, A.J., 2006. Insights into population size, group dynamics and site fidelity of humpback dolphins (*Sousa chinensis*) in Plettenberg Bay, South Africa. MSc. Thesis, Nelson Mandela Metropolitan University, Port Elizabeth.

Karczmarski, L., 1996. Ecological Studies of Humpback Dolphins *Sousa chinensis* in the Algoa Bay Region, Eastern Cape, South Africa. Ph.D. dissertation, University of Port Elizabeth, Port Elizabeth, South Africa, p. 202.

Karczmarski, L., 2000. Conservation and management of humpback dolphins: the South African perspective. Oryx 34, 207–216.

Karczmarski, L., Thornton, M., Cockroft, V.G., 1997. Description of selected behaviours of humpback dolphins *Sousa chinensis*. Aquat. Mamm. 23, 127–133.

Karczmarski, L., Winter, P.E.D., Cockroft, V.G., McLachlan, A., 1999. Population analyses of Indo-Pacific humpback dolphins *Sousa chinensis* in Algoa Bay, Eastern Cape, South Africa. Mar. Mamm. Sci. 15, 1115–1123.

Karczmarski, L., Cockroft, V., McLachlan, A., 2000. Habitat use and preferences of Indo-Pacific humpback dolphins *Sousa chinensis* in Algoa Bay, South Africa. Mar. Mamm. Sci. 16, 65–79.

Keith, M., Peddemors, V.M., Bester, M.N., Ferguson, J.W.H., 2002. Population characteristics of Indo-Pacific humpback dolphins at Richards Bay, South Africa: implications for incidental capture in shark nets. S. Afr. J. Wildl. Res. 32, 153–162.

Keith, M., Atkins, S., Johnson, A., Karczmarski, L., 2013. Area utilization patterns of humpback dolphins (*Sousa plumbea*) in Richards Bay, KwaZulu-Natal, South Africa. J. Ethol. 31, 261–274.
Kenya Marine Mammal Network, 2012. Newsletter of the Kenya Marine Mammal Network. Issue 1, May 2012.
Kerem, D., Goffman, O., Spanier, E., 2001. Sighting of a single humpback dolphin (*Sousa* sp.) along the Mediterranean coast of Israel. Mar. Mamm. Sci. 17, 170–171.
Kiani, M.S., Van Waerebeek, K., 2015. A review of the status of the Indian Ocean humpback dolphin *Sousa plumbea* in Pakistan. Adv. Mar. Biol. 72, 201–227.
Kiszka, J., Vely, M., Breysse, O., 2010a. Preliminary account of cetacean diversity and humpback whale (*Megaptera novaeangliae*) group characteristics around the Union of the Comoros (Mozambique Channel). Mammalia 74, 51–56.
Kiszka, J., Ersts, P., Ridoux, V., 2010b. Structure of a toothed cetacean community around a tropical island (Mayotte). Afr. J. Mar. Sci. 32, 543–551.
Kumarran, R.P., 2012. Cetaceans and cetacean research in India. J. Cetac. Res. Manag. 12, 159–172.
Mendez, M., Subramanian, A., Collins, T., Minton, G., Baldwin, R., Berggren, P., Särnblad, A., Amir, O.A., Peddemors, V.M., Karczmarski, L., Guissamulo, A., Rosenbaum, H.C., 2011. Molecular ecology meets remote sensing: environmental drivers to population structure of humpback dolphins in the Western Indian Ocean. Heredity, 107, 349–361.
Mendez, M., Jefferson, T.A., Kolokotronis, S.-O., Krützen, M., Parra, G.J., Collins, T., Minton, G., Baldwin, R., Berggren, P., Särnblad, A., Amir, O.A., Peddemors, V.M., Karczmarski, L., Guissamulo, A., Smith, B., Sutaria, D., Amato, G., Rosenbaum, H.C., 2013. Integrating multiple lines of evidence to better understand the evolutionary divergence of humpback dolphins along their entire distribution range: a new dolphin species in Australian waters? Mol. Ecol. 22, 5936–5948.
Meyler, S.V., Felix, H., Crouthers, R., 2011. Abundance and distribution of Indo-Pacific humpback dolphins (*Sousa Chinensis*) in the Shimoni Archipelago, Kenya. Western Indian Ocean J. Mar. Sci. 10 (2), 201–209.
Mohan, L.R.S., 1994. Review of Gillnet Fisheries and Cetacean Bycatches in the Northeastern Indian Ocean. In: Perrin, W.F., Donovan, G.P., Barlow, J. (Eds.), Gillnets and Cetaceans. Report of the International Whaling Commission, Cambridge, UK, pp. 329–346.
Moore, J.E., in press. Intrinsic growth (rmax) and generation time (T) estimates for the cetacean genera Sousa, Orcaella, and Neophocaena, in support of IUCN Red List assessments. NOAA Tech. Memo. NMFS-SWFSC.
Muralidharan, R., 2013. Sightings and behavioral observations of Indo-Pacific humpback dolphins *Sousa chinensis* (Osbeck, 1765) along Chennai Coast, Bay of Bengal. J. Threat. Taxa 5, 5002–5006.
Nanayakkara, R.P., Kusuminda, T., Jefferson, T.A., 2014. Can the Indian Ocean humpback dolphin (*Sousa plumbea*) survive in Sri Lanka? Occurrence of a relict population in Puttalam Lagoon. Aquat. Mamm. 40, 398–406.
Parsons, C.M., 1998. Observations of Indo-Pacific Humpbacked dolphins, *Sousa chinensis*, from Goa, Western India. Mar. Mamm. Sci. 14, 166–170.
Peddemors, V.M., Thompson, G., 1994. Beaching behaviour during shallow water feeding by humpback dolphins *Sousa plumbea*. Aquat. Mamm. 20, 65–67.
Pilleri, G., Gihr, M., 1972. Contribution to the knowledge of the cetaceans of Pakistan with particular reference to the Genera *Neomeris*, *Sousa*, *Delphinus* and *Tursiops* and description of a new Chinese porpoise (*Neomeris asiaorientalis*). Invest. Cetac. 4, 107–162.
Pilleri, G., Pilleri, O., 1979. Observations on the dolphins in the Indus delta (*Sousa plumbea* and *Neophocaena phocaenoides*) in winter 1978–1979. Invest. Cetac. 10, 129–135.

Plön, S., Cockcroft, V.G., Froneman, W.P., 2015. The natural history and conservation of Indian Ocean humpback dolphins (*Sousa plumbea*) in South African waters. Adv. Mar. Biol. 72, 143–162.

Preen, A., 2004. Distribution, abundance and conservation status of dugongs and dolphins in the southern and western Arabian Gulf. Biol. Conserv. 118, 205–218.

Purvis, A., Gittleman, J.L., Cowlishaw, G., Mace, G.M., 2000. Predicting extinction risk in declining species. Proc. R. Soc. Lond. 267, 1947–1952.

Razafindrakoto, Y., Andrianarivelo, N., Rosenbaum, H.C., 2004. Sightings, catches, and other records of Indo-Pacific Humpback dolphins in the coastal waters of Madagascar. Aquat. Mamm. 30, 103–110.

Reeves, R.R., Dalebout, M.L., Jefferson, T.A., Karczmarski, L., Laidre, K., O'Corry-Crowe, G., Rojas-Bracho, L., Secchi, E.R., Slooten, E., Smith, B.D., Wang, J.Y., Zhou, K., 2008. Sousa chinensis. IUCN 2013. IUCN Red List of Threatened Species. Version 2013.2. Downloaded on 04 February 2014, www.iucnredlist.org.

Reilly, S.B., Barlow, J., 1986. Rates of increase in dolphin population size. Fish. Bull. 84, 527–533.

Ross, G.B.G., 1984. The smaller cetaceans of the south east coast of Southern Africa. Ann. Cape. Prov. Mus. 15, 173–410.

Ross, G.J.B., Heinsohn, G.E., Cockcroft, V.G., 1994. Humpback dolphins *Sousa chinensis* (Osbeck, 1765), *Sousa plumbea* (G. Cuvier, 1829) and *Sousa teuszii* (Kükenthal, 1892). In: Ridgway, S.H., Harrison, R. (Eds.), Handbook of Marine Mammals. In: The First Book of Dolphins, vol. 5. Academic Press, New York, pp. 23–42.

Saayman, G.S., Tayler, C.K., 1979. The Socioecology of Humpback dolphins (*Sousa* sp.). In: Winn, H.E., Olla, B.L. (Eds.), Behavior of Marine Animals. Current Perspectives in Research. In: Cetaceans, vol. 3. Plenum Press, New York, London, pp. 165–225.

Smith, B.D., Braulik, G., Strindberg, S., Ahmed, B., Mansur, R., 2006. Abundance of Irrawaddy dolphins (*Orcaella brevirostris*) and Ganges River dolphin (*Platanista gangetica gangetica*) estimated using concurrent counts made by independent teams in waterways of the Sundarbans mangrove forest in Bangladesh. Mar. Mamm. Sci. 22, 527–547.

Stensland, E., Carlen, I., Sarnblad, A., Bignert, A., Berggren, P., 2006. Population size, distribution, and behavior of Indo-Pacific Bottlenose (*Tursiops aduncus*) and humpback (*Sousa chinensis*) dolphins off the South Coast of Zanzibar. Mar. Mamm. Sci. 22, 667–682.

Sutaria, D., Jefferson, T.A., 2004. Records of Indo-Pacific humpback dolphins (*Sousa chinensis*, Osbeck, 1765) along the coasts of India and Sri Lanka: an overview. Aquat. Mamm. 30, 125–136.

Sutaria, D., Panicker, D., Jog, K., Sule, M., Muralidharan, R., Bopardikar, I., 2015. Humpback dolphins (Genus *Sousa*) in India: an overview of status and conservation issues. Adv. Mar. Biol. 72, 229–256.

Tarbit, J., 1984. Inshore Fisheries of the Tanzanian Coast. Overseas Development Administration, London.

CHAPTER SIX

The Natural History and Conservation of Indian Ocean Humpback Dolphins (*Sousa plumbea*) in South African Waters

Stephanie Plön[1,*], Victor G. Cockcroft[†], William P. Froneman[‡]
[*]Coastal and Marine Research Institute, Nelson Mandela Metropolitan University, Port Elizabeth, South Africa
[†]Department of Zoology, Nelson Mandela Metropolitan University, Port Elizabeth, South Africa
[‡]Department of Zoology and Entomology, Rhodes University, Grahamstown, South Africa
[1]Corresponding author: e-mail address: stephanie.plon@nmmu.ac.za

Contents

1. Background: Previous Studies in South African Waters 144
2. Age, Growth and Reproduction 146
3. Diet 151
4. Habitat Utilisation 152
5. Threats 153
 5.1 Bycatch 153
 5.2 Noise Pollution 154
 5.3 Coastal Development/Disturbance in the Coastal Zone 154
6. Health 155
 6.1 Chemical Pollution 155
 6.2 Pathology 156
7. Conclusions 156
References 157

Abstract

Although most knowledge on the biology of *Sousa plumbea* has primarily come from South African waters, a number of research gaps remain on the natural history and status of the species in the region. Research on two populations in South African waters for which some historical data exist may aid in highlighting long-term changes in the biology and natural history of this little known coastal delphinid. Recent studies on the age, growth and reproduction of animals incidentally caught in shark nets in Richards Bay, KwaZulu-Natal, yielded a lower maximum age estimate of 24 (previously 46) growth-layer-groups (GLGs), sexual maturity of 7.5 and 8 GLGs in males and females (previously 12–13 and 10 GLGs, respectively), an ovulation rate of 0.2 and a 5-year calving interval (previously 0.3 and 3-year calving interval) than previously reported. These

Advances in Marine Biology, Volume 72
ISSN 0065-2881
http://dx.doi.org/10.1016/bs.amb.2015.08.005

differences may be due to a difference in the interpretation of GLGs between observers or a predominance of young males being caught in the shark nets. Stomach content analysis revealed a change in the relative proportions of the main prey items over the past 25 years, but no difference in species richness or diversity was found between the sexes. No change in trophic level was recorded between 1972 and 2009. Field studies in Algoa Bay, Eastern Cape, conducted 16 years apart indicated a decline in the mean group size (from 7 to 3 animals), a decline in the maximum group size (from 24 to 13 animals), an increase in solitary individuals (15.4–36%), and a change in behaviour from predominantly foraging (64–18%) to mainly travelling (24–49%). The observed changes are suggestive of a change in food availability, resulting in a range shift or a potential decline in numbers. These studies indicate the importance of long-term studies to monitor population changes and their possible causes. A number of threats, such as shark nets, pollution (noise and chemical), and coastal development and disturbance, to the humpback dolphin populations in South Africa have been identified. Urgent action is required to ensure continued existence of the species in South African waters.

1. BACKGROUND: PREVIOUS STUDIES IN SOUTH AFRICAN WATERS

There has been substantial debate surrounding the taxonomy of the genus *Sousa* (Ross et al., 1994), primarily due to limited material and data. Recent taxonomic revisions have supported the long-standing notion that South African animals belong to the *plumbea* species (Jefferson and Rosenbaum, 2014; Mendez et al., 2013), which is found in shallow waters from Muizenberg, False Bay, South Africa (Best, 2007) to the Bay of Bengal, possibly as far east as the Mergui Archipelago, southeastern Myanmar (Burma), including the Red Sea and Persian (Arabian) Gulf (Jefferson and Rosenbaum, 2014). Off southern Africa, the Indian Ocean humpback dolphin, *Sousa plumbea* occurs in a variety of coastal marine environments (Jefferson and Karczmarski, 2001), with the 25 m isobath representing the critical depth limiting their offshore distribution range (Karczmarski et al., 2000). The population ecology of the species has been studied mainly in Algoa Bay (Karczmarski, 1999; Karczmarski et al., 1999, 2000), Richards Bay (Atkins and Atkins, 2002; Atkins et al., 2004; Keith et al., 2002), Plettenberg Bay/Knysna (Jobson, 2006), and Maputo Bay (Guissamulo, 2008; Guissamulo and Cockcroft, 2004; Figure 1), with their population numbers reported to be low (less than 1000 individuals; see Braulik et al., 2015) across the region.

In the southern African subregion, *S. plumbea* has been recorded from South Africa, Mozambique, Tanzania (including Zanzibar), Kenya, the Comoro Islands and Madagascar (Best, 2007). The most westerly record

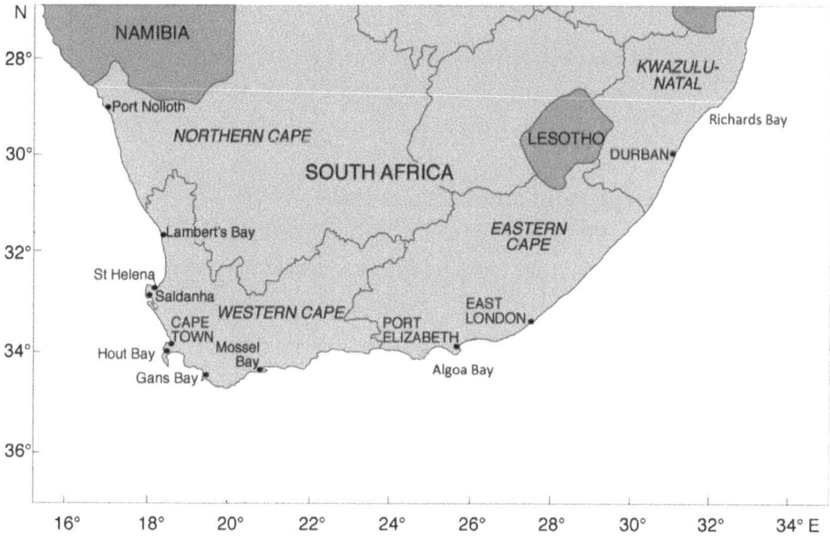

Figure 1 Map of South Africa, highlighting the two study areas discussed in this chapter: Algoa Bay and Richards Bay.

stems from False Bay, Cape Town, with a sighting from Langebaan Lagoon (33° 09′ S, 18° 03′ E) on the west coast being attributed to a vagrant Atlantic humpback dolphin *S. teuszii* (Best, 2007). Contrary to Ross' (1984) previous observations of the species not entering rivers, a few reports of *S. plumbea* have been made about 1 km upriver in the Swartkops and Bushman's River in the Eastern Cape in recent years (S. Plön, personal observations) and in the Knysna Estuary (V.G. Cockcroft, personal observations).

Southern African calves are lighter in colour than adults (Ross et al., 1994). The rostrum is long (6.3–10.1% of body length) (Ross et al., 1994) and the body is robust, but increasingly compressed towards the rear, and there is a pronounced keel on the dorsal and ventral side of the caudal peduncle in animals of all ages (Best, 2007). Morphometrics were given by Best (2007) and the length/weight relationship for 42 animals was: $Y = 4.57 \times 10^{-6} X^{3.183}$; where Y=weight (kg), X=length (cm) (Ross et al., 1994). Neonatal lengths range in South African waters from 97 to 108 cm and maximum recorded body lengths of up to 2.8 m have been recorded for South African waters (Ross et al., 1994). *Sousa plumbea* is the largest of all the *Sousa* species (Jefferson and Rosenbaum, 2014) and animals in excess of 3.0 m have been reported in the Arabian and Indian regions (Parra and Ross, 2009), although the latter are considered unlikely by some authors (Jefferson and Rosenbaum, 2014). Calves are pale grey on the flanks, with a whiter ventral side, but a dark grey on the back and upper surface of the flukes (Best, 2007). A diffuse grey stripe runs from the eye towards the flipper, and the dorsal and ventral keels of the caudal peduncle are off-white (Best, 2007). This pattern remains in juvenile and adult animals, but an overall darkening is observed with age (Best, 2007). In some of the largest adults parts of the dorsal fin and hump become white. Many animals show prominent scars on the dorsal fin and the back behind the hump, some of which have been suggested to have been caused by unsuccessful shark attacks (Cockcroft, 1991). Smaller organ weights in this species indicate that it is a shallow diving, relatively slow moving, coastal delphinid compared to Indo-Pacific bottlenose dolphins, *Tursiops aduncus*, and long-beaked common dolphins, *Delphinus capensis* (Plön et al., 2012). The South African populations in Algoa Bay and Richards Bay (Figure 1) appear to be the most extensively researched, but may not be representative of other populations in the subregion (Best, 2007).

Here, we attempt to synthesise recent research on two populations of humpback dolphins in South African waters for which some historical data exist: Algoa Bay and Richards Bay (see Figure 1). Long-term trends in age, growth and reproduction, as well as diet are examined for the Richards Bay/KwaZulu-Natal population, while group size, site fidelity and behavioural data are examined for Algoa Bay in an attempt to highlight any long-term changes in the biology and natural history of this little known, coastal delphinid.

2. AGE, GROWTH AND REPRODUCTION

Prior studies of the age, growth and reproduction of 118 *S. plumbea* individuals from South African waters between 1980 and 1998 indicated a longevity in excess of 46 years for both males and females, distinct sexual dimorphism, and an asymptotic length and weight of 269 cm and 262 kg for males, 242 cm and 164 kg for females, respectively (Cockcroft, 2002). Assuming one growth-layer-group (GLG) to be equal to one year, females sexually matured at 10 years, males 2-3 years later. The ovulation rate was calculated as 0.3 per year, indicating a 3-year calving interval. Gestation was indicated to last 1 year (as estimated from the relative mass of neonates and females), with size at birth being 115 cm and 14 kg (calculated from a von Bertalanffy growth curve and from two calves with an unhealed umbilicus); nursing occurred up to 2 GLGs (deduced from the presence of milk in the stomachs of calves for which the age was estimated), and by comparison with *T. aduncus* data (Cockcroft, 2002).

Table 1 Growth Parameters for Male ($n=71$) and Female ($n=38$) *S. plumbea* from South African Waters Incidentally Caught in Shark Nets Between 1972 and 2012

Parameter	Males	Females
Shortest animal in the sample (cm)	148	150
Length at birth ([a])	111.6	104.3
Maximum body length recorded (cm)	279	249
Asymptotic length (cm) using Gompertz' growth equation	266.7	241
Maximum age (GLGs)	24	17.7
Maximum weight (kg)	289.4	176

[a]Calculated using Scott's (1949) equation.

While Cockcroft's (2002) study used longitudinal sections of teeth (120 μm thick) etched in 5% formic acid for age determination, a more recent study, using decalcified and haematoxylin- and eosin-stained longitudinal sections of teeth, yielded somewhat different results (Nolte, 2013). In detail, the longevity of males and females from the same population incidentally caught between 1972 and 2012 was estimated at 24 and 17.7 years, respectively ($n=109$) (Table 1; Nolte, 2013). Differences in both length and weight at a similar age were evident between the sexes, in agreement with the previous study, but males were found to become sexually mature at 7.5 years (the lowest age estimate for histologically mature males) and females at 8 years (the lowest age estimate for a female with scars of ovulation present in the ovaries). Size at birth was calculated using Scott's (1949) equation due to a paucity of neonates in the sample, and was estimated at 111.6 cm for males and 104.3 cm for females (Table 1). In addition, a lower ovulation rate of 0.2/year, resulting in a calving interval of 5 years was calculated (Table 2).

To determine whether changes in the age structure of *S. plumbea* from South African waters were due to real changes impacting the populations or due to observer bias, age estimates for 50 specimens incidentally caught in shark nets between 1979 and 1989 (V.G. Cockcroft, unpublished data; conducted for an earlier study (Cockcroft, 2002)) were compared with age estimates obtained in a recent study for 109 animals from the same population incidentally caught between 1972 and 2012 (Nolte, 2013). For both data sets, age estimates were determined from the number of GLGs (*sensu* Perrin and Myrick, 1980) counted in the dentine and cementum of longitudinal sections of teeth. There was a high degree of consistency between

Table 2 Summary of Natural History Parameters for Male and Female *S. plumbea* of Different Maturity Stages from South African Waters (1972–2012)

		Immature		Pubertal[a]		Mature	
	Parameter	**Range**	***n***	**Range**	***n***	**Range**	***n***
Males	Estimated age (GLGs)	2–7.5	27			7.5–23.5	19
	Total body weight (kg)	60–150	27			92–250	19
	Total body length (cm)	167.4–241	27			210.6–269	19
	Combined testis mass (g)	30–546	18			174–1250	16
	Mean testis length (cm)	88.5–107.5	2			200–280	4
	Mean seminiferous tubule diameter (μm)	28.1–79.3	20			64.5–160.3	10
	Index of testis development (g/mm)	0.4–0.5	2			1.8–2.6	4
Females	Estimated age (GLGs)	1.5–7.2	11	5–8.7	3	8–17.3	4
	Total body weight (kg)	47–114	14	79–150	5	140–176	8
	Total body length (cm)	157–222	14	183.5–237	5	220–248	9
	Combined ovary weight (g)	1–6	14	4–13	4	10–22	6
	Ovulation rate/year					0.2	3
	Calving interval					5	3

[a]This maturity stage could only be assessed in female specimens due to the state of the material examined.

the two data sets, with 74% of the age estimates from both data sets being within 1 GLG of one another. However, cemental counts (usually conducted when dentinal counts yielded 12 GLGs or more) vary greatly between the two data sets, which has a significant effect on the maximum age estimated for the species in South African waters (46 in data set 1 vs. 24 in data set 2). Results based on Cockcroft's (1989) findings estimated longevity at an excess of 46 GLGs for both sexes, while Nolte (2013) produced a maximum age estimate of 24 for the same population. The longevity of Indo-Pacific humpback dolphin, *S. chinensis*, in southern Chinese waters has been estimated at 38 GLGs (Jefferson, 2000; Jefferson et al., 2012). The differences in maximum age in *S. plumbea* compared to *S. chinensis* from Hong Kong waters could be attributed to a difference in the interpretation of

GLGs in both dentine and cement. Alternatively, the difference could be attributed to a sampling artefact of a bias in shark net captures, which may have misrepresented the population structure of *S. plumbea* from South African waters (Atkins et al., 2013; Nolte, 2013). Older dolphins reportedly do not frequent inshore netted zones due to differences in foraging behaviour and therefore do not form part of the shark net bycatch (Atkins et al., 2013; Cockcroft, 1994; Cockcroft and Ross, 1990), and this difference in spatial distribution of different age classes could explain the majority of sub-adults reported in the sample (Atkins et al., 2013).

This idea is supported by the majority of dolphins caught in the nets being less than 10 GLGs old (Figure 2), and there were no females between 9 and 12 GLGs in the sample. This supports suggestions by Atkins et al. (2013) that there is a predominance of juvenile males caught in the shark nets, suggesting there is age and sex segregation in the population. Between the ages of 12 and 24 GLGs, a greater number of males than females was caught, and thus the sample of individuals older than 12 GLGs was also skewed towards males (Figure 2). Thus more males than females were present in all age groups in the sample, and certain age groups, such as calves and mature males and females, were underrepresented. This begs the question whether the sample is representative of the population (Perrin and Reilly, 1984).

Examination of the presence of *corpora* in the ovaries of *S. plumbea* females in relation to age (Plön and Bernard, 2007a) yielded an ovulation rate of 0.2 per year, indicating a 5-year calving interval. The ovulation rate of

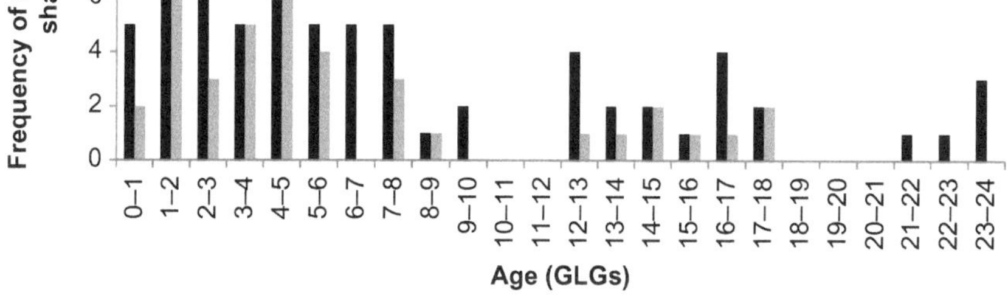

Figure 2 The age distribution of *S. plumbea* caught in shark nets off the coast of KwaZulu-Natal between 1972 and 2012 ($n = 109$).

S. plumbea was previously estimated to be 0.3/year by Cockcroft (1989), resulting in a calving interval of 3.3 years. Jefferson et al. (2012) estimated the mean calving interval for a Chinese population of *S. chinensis* at about 5.2 years, which is closer to the more recent calving interval for South African waters.

Both the examination of female (follicle formation, *corpus* index and births), as well as male (seminiferous tubule diameter, presence/absence of sperm and testis mass) indicators of reproductive seasonality yielded no clear indication of a seasonal cycle in *S. plumbea* (Nolte, 2013). This is in partial agreement with a report by Jefferson et al. (2012) of year-round births for *S. chinensis* in southern Chinese waters, although a strong peak in calving was reported in spring and early summer for those animals.

Sexual dimorphism in odontocetes is commonly displayed in the form of body shape and/or size (Tolley et al., 1995) and the degree of sexual dimorphism generally increases with an increase in body size, with males normally attaining greater body lengths than females (Tolley et al., 1995). Cockcroft (1989) reported clear sexual dimorphism for *S. plumbea* in South Africa, and this is supported by the length/mass at age data for the South African animals, with both the length and mass at a specific age being larger for males than for females (Nolte, 2013).

Testis mass to body mass ratio, sexual dimorphism and group size can be used as indicators of the mating system of a species (Connor et al., 2000; Plön and Bernard, 2007b). Cockcroft (1993) previously reported that the combined testis mass constituted 0.7% of the total body mass in *S. plumbea*. However, while this may be the maximum value (mean of the top 10 testis weights measured), the mean testis mass of sexually mature *S. plumbea* (determined histologically; $n=19$) was recently found to be 0.43% of the total body mass in this species (Plön et al., 2012). Small testes in relation to body mass indicate a monogamous or extremely polygamous mating system (see Plön and Bernard, 2007b). The small testis size in association with sexual dimorphism and small group size reported in *S. plumbea*, agrees with the findings of Cockcroft (1993) and the roving male mating strategy proposed by Plön and Bernard (2007b). This strategy involves males actively searching for receptive females to maximise their reproductive opportunities, rather than monopolising and fighting over a number of females (i.e. a harem) (Plön and Bernard, 2007b). Small group size in *S. plumbea* thus allows for larger males to dominate smaller males, and subsequently deny them access to females (Plön and Bernard, 2007b). Durham (1994) found that females in northern KwaZulu-Natal waters form relatively stable, resident groups, while males tend not to be resident, and move between several female groups; this was supported with molecular data by Smith-Goodwin (1997), indicating female philopatry.

3. DIET

Studies on the diet of *S. plumbea* in southern African waters have been restricted to individuals incidentally caught in shark nets along the KwaZulu-Natal coastline. Humpback dolphins feed predominantly on reef-associated, estuarine and demersal fish (Barros and Cockcroft, 1991; Ross et al., 1994). Cockcroft and Ross (1983), examining 17 humpback dolphin stomachs from the KwaZulu-Natal coastline, found that glassnosed anchovy (*Thryssa vitrirostris*) made up the largest numerical percentage of the total prey species consumed, followed by ribbon fish (*Trichiurus lepturus*), olive grunter (*Pomadasys olivaceum*) and longtooth kob (*Otolithes ruber*). More recent stomach content analysis of 22 *S. plumbea* (13 males, 9 females) incidentally caught between 2005 and 2009, indicated that the animals feed on 59 different prey species. The numerically dominant prey species were broadly similar to the previous study, although bearded croaker (*Johnius amblycephalus*) was also present in the diet (Venter, 2009). An Analysis of Variance (ANOVA) revealed no significant differences in species richness ($F(_{1,21})=2.7645$; $P=0.11123$) or diversity ($F(_{1,21})=0.24843$; $P=0.62336$) in the diet between male and female dolphins (Venter, 2009). The lack of any difference was most likely due to the dolphins being generalist feeders as well as the narrow niche they occupy along the coast. This is supported by recent stable isotope analyses, showing that humpback dolphins had the most enriched $\delta\ ^{13}C$ and $\delta\ ^{15}N$ values of four inshore dolphin species (*T. aduncus, S. plumbea, D. capensis*, and the striped dolphin, *coeruleoalba*) examined off the KwaZulu-Natal coastline (Browning et al., 2014).

Although relative proportions of the five main prey items have changed, there has been little other change in the diet from this area over the last 25 years. This result is supported by the isotope analysis of teeth from 21 individuals (10 males, 11 females) between 1972 and 2009, which indicated no significant differences in the carbon ($\delta\ ^{13}C$) and nitrogen ($\delta\ ^{15}N$) signatures between decades ($p>0.05$) (Weltz, 2010). However, the overall trend of $\delta\ ^{15}N$ values showed a decrease in isotopic values over time and the $\delta\ ^{15}N$ ratios between 1970 and 2008, were significantly different from one another ($P=0.024$), but changed less than 3–4‰ from 1970 to 2008, suggesting that no shift in trophic level feeding had occurred (Weltz, 2010). Similarly the overall trend for $\delta^{13}C$ values showed a depletion from 1970 to 2008; these were non-significant between any of the years ($p > 0.05$ in all cases) (Venter, 2009; Weltz, 2010).

Moreover, the mid-lateral and mid-ventral blubber thickness demonstrated no significant change between the 1980s and 2000s ($p > 0.05$) (S. Plön, unpublished data). These data support results from stomach content analysis that the diet of humpback dolphins along the south-east coast of southern Africa has not changed over the past four decades, and thus there has been no impact on body condition of the animals. Diet data for Eastern Cape waters are restricted to two stranded specimens examined by Barros and Cockcroft (1991).

4. HABITAT UTILISATION

A recent 1-year study (2010–2011) in Algoa Bay, Eastern Cape (Koper et al., in press) aimed to link up with work conducted 16 years previously by Karczmarski (1991–1994). The results indicated a similar pattern of annual and diurnal occurrence within the bay, thought to be related to prey availability (Karczmarski et al., 1999). However, the mean group size recorded was three, which is smaller than that observed elsewhere (Baldwin et al., 2004; Guissamulo, 2008; Parsons, 1998), and half of the previous mean of about six individuals recorded for Algoa Bay (Karczmarski, 1999; Karczmarski et al., 1999). In addition, maximum group sizes recorded had decreased by almost half, from 24 individuals previously, to 13 in the recent study. A separate boat-based study between 2008 and 2011 confirmed a mean group size of 3.2 individuals (median group size: 2; range: 1–23). The decrease in mean group size is primarily reflected by an increase in the percentage of solitary individuals from 15.4% to 36% over the past 16 years. Associated changes in behaviour were also recorded; while previously the majority of animals were observed foraging in the bay (64%), this has declined to less than 1/5 (18%). Furthermore, the occurrence of travelling (*sensu* Karczmarski et al., 1999) has almost doubled from 24% previously to 49% more recently, while other behavioural states, such as resting and socialising remained similar.

Under poor environmental conditions, during which food patches are small, the costs associated with increased competition among group members can be greater than the advantages associated with cooperation, leading to smaller group sizes (Krebs and Davies, 2007). This assumption is supported by a change in dominant behaviour from "foraging" in 1991–1994 to "travelling" in 2010–2011 within Algoa Bay. This change is likely to indicate a home range expansion as a consequence of an increased amount of time spent moving between different foraging patches, which is needed when local prey availability is too low for sufficient food supply. Dicken et al. (2012) demonstrated that the catch-per-unit-effort and mean size of line fish along the Eastern Cape coastline of South Africa, including Algoa Bay, has significantly decreased over the past three decades. The observed decrease was thought to be the result of the overexploitation of stocks by line fishers and, more recently, as a result of habitat change associated with the construction of a harbour in Algoa Bay (Dicken et al., 2012). Such behavioural changes are of concern, as they can cause shifts in habitat use (Lusseau, 2005), temporary displacement (Bejder et al., 2006a) or a reduced energy consumption with an increased energy expenditure (Bishop, 1999; Frid and Dill, 2002) and therefore in the long-term potentially negatively affect individual survival rates or population size (Bejder et al., 2006b).

From the above studies, it appears that significant changes in habitat utilisation and social structure of humpback dolphins in Algoa Bay have been observed over the past three decades. Possible explanations are a range shift, with animals moving out of the study area, or a potential decline in numbers. Either scenario indicates population changes in excess of those that have been recognised in long-term datasets on the natural history or diet of the species. This is cause for concern, because population sizes could decline to dangerously low levels before any management action is taken (Thompson et al., 2000). The absence of long-term studies on *S. plumbea* from other regions within their distributional range means that we are unable to determine whether the patterns observed in Algoa Bay are cause for concern or reflect subtle changes in their behaviour in response to both environmental and anthropogenic forcing.

Below we highlight a number of imminent threats, all of which demand immediate attention to ensure the long-term survival of the species in South African waters.

5. THREATS

5.1 Bycatch

Many of the threats to humpback dolphins are difficult to quantify, with the exception of shark nets in KwaZulu-Natal waters, which are of concern due to the low population size (Atkins et al., 2013; Peddemors et al., 1991). A known total of 203 animals were caught between 1980 and 2009

(Atkins et al., 2013). Catch rates were skewed towards males between 2.2 and 2.5 m in length, and juveniles formed the majority of the bycatch (56%) (Atkins et al., 2013). However, mortality appeared low for dependent calves and intermediate for adults (Atkins et al., 2013). The composition of the catch was thought to either reflect the spatial distribution of humpback dolphin age/sex classes, particularly the coincidence of shark nets with a feeding area, or be indicative of a propensity for risk-taking that is higher in male juveniles than other sex/age classes (Atkins et al., 2013). A number of bycatch-reduction techniques have been proposed, including pingers, metal-oxide infused nets to enhance acoustic detectability, time/area closures, reduction of fishing effort and gear changes (Atkins et al., 2013). Efforts to reduce bycatch of humpback dolphins in shark nets are urgently required to ensure continued existence of this important *Sousa* population in South African waters. Bycatch in other types of nets and fishing operations in South Africa is not well understood.

5.2 Noise Pollution

Anthropogenic sound in the ocean and its potential effects on marine life has increasingly been recognised as a potentially major disturbance to cetaceans (see Koper and Plön, 2012). Unfortunately, to date no research has been conducted on Indian Ocean humpback dolphins in South African waters. As a result, there is currently a virtual absence of data on the auditory abilities of *Sousa* spp., with the exception of an audiogram from a single *S. chinensis* in captivity (Li et al., 2012). Measurements carried out on this approximately 13-year-old male showed that the animal had a hearing sensitivity between 20 and 120 kHz. This indicates a hearing ability similar to those of many other odontocete species, although caution should be employed when extending the results from this one individual to the genus as a whole. In order to assess potential impacts from anthropogenic sources in the coastal zone and their effects on the behaviour of the animals, it is important to learn more about individual variation by measuring the hearing capabilities of more than one individual, including animals of different ages and under various environmental conditions (Li et al., 2012).

5.3 Coastal Development/Disturbance in the Coastal Zone

Coastal development, tourism and disturbance in the coastal zone are all of concern to the welfare of inshore coastal cetaceans (Allen et al., 2012; Beijder et al., 2012). Previous research has indicated that boat traffic disturbs humpback dolphin behaviour and masks their vocalisations (Ng and Leung, 2003; Sims et al., 2011; Van Parijs and Corkeron, 2001a,b). Both feeding and social vocalisations (specifically burst pulses) are masked by small vessel noise, and disturbance can occur over 1.5 km (Van Parijs and Corkeron, 2001a,b). In addition, fast-moving vessels elicit strong disturbance reactions (Ng and Leung, 2003) and underwater noise levels are largely dependent on boat speed (Koper and Plön, 2012). Recent work in Algoa Bay indicated that the dolphins displayed a number of negative responses towards high speed vessels and personal water craft as expected, but also towards surf skis, and to a lesser degree towards swimmers and surfers (Koper et al., in press). These negative reactions included changes in direction of movement, increased dive durations and group splits, corresponding to previously observed reactions in other dolphins (Koper et al., in press). In addition, the overall increase in travelling and milling behaviour with a decrease in foraging behaviour for the humpback dolphins in Algoa Bay corresponds to documented behavioural responses of a number of small cetacean species to nearby vessels (see Ng and Leung, 2003). These results suggest that humpback dolphins in Algoa Bay are prone to disturbance from human activities, raising concern on the potential negative effects of anthropogenic impacts on the energy budget and thus on the long-term survival and fecundity (Gill et al., 2001) of humpback dolphins in this area. However, it also highlights that tourism activities, such as surf skiing, swimming and surfing, which are usually perceived to be of no impact (as they are quiet activities), may also have an adverse effect on these coastal dolphins. Urgent management and conservation actions should include the establishment of dolphin protection zones or sanctuaries to the exclusion of any human activity.

6. HEALTH

6.1 Chemical Pollution

Early work on organochlorines in cetaceans from the east and south-east coast of South Africa indicated higher levels of PCBs and DDT in inshore waters versus pelagic waters, and lower levels from Eastern Cape waters than from KwaZulu-Natal waters (Cockcroft, 1999; Cockcroft et al., 1991). Amongst the coastal dolphin species investigated, humpback dolphins were characterised by the highest organochlorine concentrations of any marine mammal in South African waters (Gardner et al., 1983 in: Cockcroft, 1999). Although the use of DDT and other persistent organic pollutants (POPs) has been banned in most countries, these chemicals are still in use in South Africa for malaria and insect pest control. Recent investigations on 11 POP compounds in blubber samples from animals either bycaught or stranded between 2005 and 2009, indicated that *S. plumbea* was the most contaminated dolphin by total POPs (21,100 ng g^{-1} lw) (D. Gui et al., unpublished data). Overall DDT was the major pollutant, representing more than 60% of the total concentration combined of all POPs analysed, followed by PCBs (30%) (D. Gui et al., unpublished data).

Cockcroft et al. (1991) highlighted the importance of continued monitoring and assessment of potential impacts from organic pollutants, yet little progress has been made in the 25 years since that study was published. Although shown to have a negative impact on reproduction in other marine mammals (e.g. polar bears, *Ursus maritimus*, and pinnipeds; Reijnders, 1980), it is unknown what impacts these compounds may have on humpback dolphin reproductive success, particularly in the absence of long-term population studies, which could provide vital information on calving interval and calf mortality (Tezanos-Pinto et al., 2014). Given the relatively small population sizes in this species (see Braulik et al., 2015) in South African waters, such impacts would urgently need to be quantified and most likely reversed. In view of the continued recreational fishing in the coastal zone for subsistence purposes (Mann et al., 2000), detailed information of pollutant levels of coastal reef and estuarine fish will also have implications for human health and reproduction.

6.2 Pathology

Recent systematic health investigations of dolphins incidentally caught in shark nets off KwaZulu-Natal have indicated that, although most lesions were mild to moderate, humpback dolphins appear to show a greater amount of histopathological lesions than *T. aduncus* (S. Plön, personal observations). The majority of lesions are parasite related (Lane et al., 2014), but appear to have become more numerous in recent years (S. Plön, personal observations). These preliminary observations remain to be quantified, but may be indicative of the host–parasite interface changing in response to the degradation of the local marine environment. Such observations will need to be investigated in conjunction with information on pollutant load to clarify the exact impacts on dolphin population health (Lane et al., 2014).

7. CONCLUSIONS

Coastal dolphins are amongst the most threatened dolphins due to their proximity to anthropogenic impacts (Allen et al., 2012; Beijder et al., 2012; Thompson et al., 2000). Due to their longevity requiring long-term studies, however, populations could decline to dangerously low levels before management action is taken (Thompson et al., 2000). Calls for improved management and conservation strategies for coastal delphinids have been numerous internationally (e.g., Allen et al., 2012; Beijder et al., 2012; Thompson et al., 2000), and have included urgent calls for action regarding the local *Sousa* populations in South African waters (Atkins et al., 2013; Karczmarski et al., 1998; Keith et al., 2013; Ross et al., 1994), yet virtually no progress has been made. A recent evaluation of the conservation status of *S. plumbea* in South African waters painted a bleak picture, with a possible decline in population sizes amongst a glaring lack of baseline data, particularly on population structure and connectivity (Plön et al., 2015). Unfortunately, efforts to determine trends in population sizes in South African waters have presented somewhat of a "shot-gun" approach and highlighted the need for long-term studies over a larger geographic extent to take seasonal immigration and emigration rates into account. Recent studies on humpback whales (*Megaptera novaeangliae*) have indicated "that assessments of depleted animal populations that do not consider the social behaviour of a species are missing a potentially vital component of the picture" (Clapham and Zerbini, 2015).

Research on long-lived mammals requires long-term data sets to develop effective management and conservation plans. The first studies on *Sousa* in South African waters (Saayman and Tayler, 1979; Saayman et al., 1972), as well as more recent studies, have shown that due to their coastal nature, humpback dolphins may in some locations be easily studied from elevated observation points along the shore (Koper et al., in press). Such studies are low in cost, but can yield important long-term data on site fidelity, patterns of occurrence, group size and local movements. However, they are insufficient to answer all questions currently pertinent to the humpback dolphin populations in South African waters (Plön et al., 2015). Data on social structure, longevity, calving interval and calf survival, resource utilisation, pollutant loads, as well as population structure and connectivity, are all sorely needed to ensure the long-term survival of the species in local waters.

REFERENCES

Allen, S.J., Cagnazzi, D.D., Hodgson, A.J., Loneragan, N.R., Bejder, L., 2012. Tropical inshore dolphins of north-western Australia: unknown populations in a rapidly changing region. Pac. Conserv. Biol. 18, 56–63.

Atkins, S., Atkins, B., 2002. Abundance and site fidelity of Indo-Pacific humpback dolphins (*Sousa chinensis*) at Richards Bay, South Africa: International Whaling Commission.

Scientific Committee Report SC/54/SM25. 54th annual meeting of the International Whaling Commission, Shimonoseki, Japan, May 2002.

Atkins, S., Pillay, N., Peddemors, V.M., 2004. Spatial distribution of Indo-Pacific humpback dolphins (*Sousa chinensis*) at Richards Bay, south Africa: environmental influences and behavioural patterns. Aquat. Mamm. 30 (1), 84–93.

Atkins, S., Cliff, G., Pillay, N., 2013. Humpback dolphin bycatch in the shark nets in KwaZulu-Natal, South Africa. Biol. Conserv. 159, 442–449.

Baldwin, R.M., Collins, M., Van Waerebeek, K., Minton, G., 2004. The Indo-Pacific humpback dolphin of the Arabian region: a status review. Aquat. Mamm. 30, 111–124.

Barros, N.B., Cockcroft, V.G., 1991. Prey of humpback dolphins (*Sousa plumbea*) stranded in eastern Cape Province, South Africa. Aquat. Mamm. 17, 134–136.

Beijder, L., Hodgson, A., Loneragan, N., Allen, S., 2012. Coastal dolphins in north-western Australia: the need for re-evaluation of species listings and short-comings in the environmental impact assessment process. Pac. Conserv. Biol. 18, 22–25.

Bejder, L., Samuels, A., Whitehead, H., Gales, N., Mann, J., Connor, R., Heithaus, M., Watson-Capps, J., Flaherty, C., Krützen, M., 2006a. Decline in relative abundance of bottlenose dolphins exposed to long-term disturbance. Conserv. Biol. 20, 1791–1798.

Bejder, L., Samuels, A., Whitehead, H., Gales, N., 2006b. Interpreting short-term behavioural responses to disturbance within a longitudinal perspective. Anim. Behav. 72, 1149–1158.

Best, P.B., 2007. Whales and Dolphins of the Southern African Subregion. Cambridge University Press, Cape Town. 338 pp.

Bishop, C.M., 1999. The maximum oxygen consumption and aerobic scope of birds and mammals: getting to the heart of the matter. Proc. Zool. Soc. Lond B 266, 2275–2281.

Braulik, G., Cerchio, S., Findlay, K., Baldwin, R., 2015. Assessment of the conservation status of the Indian Ocean humpback dolphin *Sousa plumbea* using the IUCN Red List Criteria. Adv. Mar. Biol. 72, 119–141.

Browning, N.E., Cockcroft, V.G., Worthy, G.A.J., 2014. Resource partitioning among South African delphinids. J. Exp. Mar. Biol. Ecol. 457, 15–21.

Clapham, P.J., Zerbini, A.N., 2015. Are social aggregation and temporary immigration driving high rates of increase in some Southern Hemisphere humpback whale populations? Mar. Biol. 162, 625–634. http://dx.doi.org/10.1007/s00227-015-2610-3.

Cockcroft, V.G., 1989. Biology of Indopacific humpback dolphins (*Sousa plumbea*) off Natal, South Africa. In: 8th Biennial Conference on the Biology of Marine Mammals, Pacific Grove, California, USA, 7–11 December. Abstract only.

Cockcroft, V.G., 1991. Incidence of shark bites on Indian Ocean hump-backed dolphins (*Sousa plumbea*) off Natal, South Africa. Cetaceans and cetacean research in the Indian Ocean sanctuary. UNEP Mar. Mamm. Tech. Rep, 278–282.

Cockcroft, V.G., 1993. Size: the only male reproductive strategy? In: 10th Biennial Conference on the Biology of Marine Mammals, Galveston, Texas, USA, 11-15 November.

Cockcroft, V.G., 1994. Is there common cause for dolphin capture in gillnets? A review of dolphin catches in shark nets off Natal, South Africa. Report of the International Whaling Commission (Special Issue 15), 541–547.

Cockcroft, V.G., 1999. Organochlorine levels in cetaceans from South Africa: a review. J. Cet. Res. Manage. (Special Issue 1), 169–176.

Cockcroft, V.G., 2002. Natural history of Indopacific humpback dolphins (*Sousa chinensis*) off Southern Africa: International Whaling Commission. Report SC/54/SM22, Shimonoseki, Japan. 1 pp.

Cockcroft, V.G., Ross, G.J.B., 1983. Feeding of three inshore delphinid species in Natal waters. In: 5th International Oceanographic Symposium, Rhodes University, Grahamstown, South Africa, 24-28 January 1983.

Cockcroft, V.G., Ross, G.J.B., 1990. Age, growth and reproduction of bottlenose dolphins *Tursiops truncatus* from the east coast of Southern Africa. Fish. Bull. US 88 (2), 289–302.

Cockcroft, V.G., Ross, G.J.B., Connell, A.D., Gardner, B.D., Butler, A.C., 1991. Occurrence of organochlorines in stranded cetaceans and seals from the east coast of Southern Africa. Cetaceans and cetacean research in the Indian Ocean Sanctuary. UNEP Mar. Mamm. Tech. Rep. 272–276.

Connor, R.C., Read, A., Wrangham, R., 2000. Male reproductive strategies and social bonds. In: Mann, J., Connor, R.C., Tyack, P.L., Whitehead, H. (Eds.), Cetacean Societies: Field Studies of Whales and Dolphins. University of Chicago Press, Chicago, pp. 247–269.

Dicken, M.L., Smale, M.J., Booth, A.J., 2012. Long-term catch and effort trends in 11 Eastern Cape Angling Week competitions. Afr. J. Mar. Sci. 34, 259–268.

Durham, B., 1994. The Distribution and Abundance of Humpback Dolphins (*Sousa chinensis*) Along the Natal coast, South Africa. MSc thesis, University of Natal, Durban, South Africa. 95 pp.

Frid, A., Dill, L.M., 2002. Human-caused disturbance stimuli as a form of predation risk. Conserv. Ecol. 6, 11. Available at www.consecoLorg/vol6/iss1/art11.

Gardner, B.D., Cornell, A.D., Eagle, G.A., Molden, A.G.S., Oliff, W.D., Orren, M.J., Watling, R.J., 1983. South African marine pollution survey report 1976–1979. S. Afr. Sci. Program. Rep. 73, 87–91.

Gill, J.A., Norris, K., Sutherland, W.J., 2001. Why behavioural responses may not reflect the population consequences of human disturbance. Biol. Conserv. 97, 265–268.

Gui, D., Karczmarski, L., Yu, R.-Q., Plön, S., Chen, L., Tu, Q., Cliff, G., Wu, Y. n.d. Profiling and spatial variation analysis of persistent organic pollutants in South African delphinids. Environ. Sci. Technol. (unpublished data).

Guissamulo, A.T., 2008. Ecological Studies of Bottlenose and Humpback Dolphins in Maputo Bay, Southern Mozambique. PhD thesis, University of KwaZulu-Natal, Pietermaritzburg, South Africa. 290 pp.

Guissamulo, A.T., Cockcroft, V.G., 2004. Ecology and population estimates of the Indo-Pacific humpback dolphins (*Sousa chinensis*) in Maputo Bay, Mozambique. Aquat. Mamm. 30, 94–102.

Jefferson, T.A., 2000. Population biology of the Indo-Pacific hump-backed dolphin in Hong Kong waters. Wildl. Monogr. 144, 1–65. 104.

Jefferson, T.A., Hung, S.K., Robertson, K.M., Archer, F.I., 2012. Life history of the Indo-Pacific humpback dolphin in the Pearl River Estuary, southern China. Mar. Mamm. Sci. 28 (1), 84–104.

Jefferson, T.A., Karczmarski, L., 2001. Sousa chinensis. Mamm. Species 655, 1–9.

Jefferson, T.A., Rosenbaum, H.C., 2014. Taxonomic revision of the humpback dolphins (*Sousa* spp.), and description of a new species from Australia. Mar. Mamm. Sci. 30 (4), 1494–1541.

Jobson, A.J., 2006. Insight Into Population Size, Group Dynamics and Site Fidelity of Humpback Dolphins (*Sousa chinensis*) in Plettenberg Bay, South Africa. M.Sc. thesis. Nelson Mandela Metropolitan University, Port Elizabeth, South Africa. 182 pp.

Karczmarski, L., 1999. Group dynamics of humpback dolphins (*Sousa chinensis*) in the Algoa Bay region, South Africa. J. Zool. Lond. 249, 283–293.

Karczmarski, L., Cockcroft, V.G., McLachlan, A., Winter, P.E.D., 1998. Recommendations for the conservation and management of humpback dolphins *Sousa chinensis* in the Algoa Bay region, South Africa. Koedoe 41 (2), 121–129.

Karczmarski, L., Cockcroft, V.G., McLachlan, A., 1999. Group size and seasonal pattern of occurrence of humpback dolphins *Sousa chinensis* in Algoa Bay, South Africa. S. Afr. J. Mar. Sci. 21, 89–97.

Karczmarski, L., Cockcroft, V.G., McLachlan, A., 2000. Habitat use and preferences of Indo-Pacific humpback dolphins *Sousa chinensis* in Algoa Bay, South Africa. Mar. Mamm. Sci. 16 (1), 65–79.

Keith, M., Peddemors, V.M., Bester, M.N., Ferguson, J.W.H., 2002. Population characteristics of Indo-Pacific humpback dolphins at Richards Bay, South Africa: implications for incidental capture in shark nets. S. Afr. J. Wildl. Res. 32 (2), 153–162.

Keith, M., Atkins, S., Johnson, A.E., Karczmarski, L., 2013. Area utilization patterns of humpback dolphins (*Sousa plumbea*) in Richards Bay, KwaZulu-Natal, South Africa. J. Ethol. 31, 261–274.

Koper, R.P., Plön, S., 2012. The Potential Impacts of Anthropogenic Noise on Marine Animals and Recommendations for Research in South Africa. EWT Research & Technical Paper No. 1, Endangered Wildlife Trust, Johannesburg, South Africa. 103 pp.

Koper, R.P., Karczmarski, L., du Preez, D., Plön, S. In press. Sixteen years later: occurrence, group size, and habitat use of humpback dolphins (*Sousa plumbea*) in Algoa Bay, South Africa. Mar. Mamm. Sci.

Krebs, J.R., Davies, N.B., 2007. Living in groups. In: Krebs, J.R., Davies, N.B. (Eds.), An Introduction to Behavioural Ecology. Blackwell Science, Oxford, United Kingdom, pp. 120–146.

Lane, E.P., de Wet, M., Thompson, P., Siebert, U., Wohlsein, P., Plön, S., 2014. A systematic health assessment of Indian Ocean bottlenose (*Tursiops aduncus*) and Indo-Pacific humpback (*Sousa plumbea*) dolphins, incidentally caught in shark nets off the KwaZulu-Natal coast, South Africa. PLoS One 9 (9), e0107038.

Li, S., Wang, D., Wang, K., Taylor, E., Cros, E., Shi, W., Wang, Z., Fang, L., Chen, Y., Kong, F., 2012. Evoked-potential audiogram of an Indo-Pacific humpback dolphin (*Sousa chinensis*). J. Exp. Biol. 215, 3055–3063.

Lusseau, D., 2005. Residency pattern of bottlenose dolphins *Tursiops* spp. in Milford Sound, New Zealand, is related to boat traffic. Mar. Ecol. Prog. Ser. 295, 265–272.

Mann, B.Q., Radebe, P.V., Beckley, L.E., 2000. Southern African marine line fish status reports. Oceanogr. Res. Inst. 7, 1–260.

Mendez, M., Jefferson, T.J., Kolokotronis, S.-O., Krützen, M., Parra, G.J., Collins, T., Minton, G., Baldwin, R., Berggren, P., Särnblad, A., Amir, O.A., Peddemors, V.M., Karczmarski, L., Guissamulo, A., Smith, B., Sutaria, D., Amato, G., Rosenbaum, H.C., 2013. Integrating multiple lines of evidence to better understand the evolutionary divergence of humpback dolphins along their entire distribution range: a new dolphin species in Australian waters? Mol. Ecol. 22, 5936–5948. http://dx.doi.org/10.1111/mec.12535.

Ng, S.L., Leung, S., 2003. Behavioral response of Indo-Pacific humpback dolphin (*Sousa chinensis*) to vessel traffic. Mar. Environ. Res. 56, 555–567.

Nolte, Z., 2013. The Natural History of the Humpback Dolphin, *Sousa chinensis*, in KwaZulu-Natal: Age, Growth and Reproduction. MSc thesis, Department of Zoology & Entomology, Rhodes University, Grahamstown, South Africa. 150 pp.

Parra, G.J., Ross, G.J.B., 2009. Humpback dolphins *S. chinensis* and *S. teuszii*. In: Perrin, W.F., Würsig, B., Thewissen, J.G.M. (Eds.), Encyclopedia of Marine Mammals. Academic Press, San Diego, pp. 576–582.

Parsons, E.C.M., 1998. The behaviour of Hong Kong's resident cetaceans: the Indo-Pacific hump-backed dolphin and the finless porpoise. Aquat. Mamm. 24, 91–110.

Peddemors, V.M., Cockcroft, V.G., Wilson, R.B., 1991. Incidental dolphin mortality in the Natal shark nets: a preliminary report on prevention measures. Cetaceans and cetacean research in the Indian Ocean sanctuary. UNEP Mar. Mamm. Tech. Rep. pp. 129–137.

Perrin, W.F., Myrick, A.C., 1980. Report of the workshop. Rep. Int. Whal. Commn. (Special Issue 3), 1–50.

Perrin, W.F., Reilly, S.B., 1984. Reproductive parameters of dolphins and small whales of the Family Delphinidae. Rep. Int. Whal. Commn. (Special Issue 6), 97–133.

Plön, S., Bernard, R.T.F., 2007a. Anatomy with particular reference to the female. In: Miller, D.L. (Ed.), Reproductive Biology and Phylogeny of Cetacea—Whales, Dolphins and Porpoises. Science Publishers, Enfield, New Hampshire, pp. 147–169.

Plön, S., Bernard, R.T.F., 2007b. Testis, spermatogenisis, and testicular cycles. In: Miller, D.L. (Ed.), Reproductive Biology and Phylogeny of Cetacea—Whales, Dolphins and Porpoises. Science Publishers, Enfield, New Hampshire, pp. 215–244.

Plön, S., Albrecht, K.H., Cliff, G., Froneman, P.W., 2012. Organ weights of three dolphin species (*Sousa chinensis*, *Tursiops aduncus* and *Delphinus capensis*) from South Africa: implications for ecological adaptation? J. Cet. Res. Manage. 12 (2), 265–276.

Plön, S., Cockcroft, V.G., Pistorius, P., Conry, D., Atkins, S., 2015. A conservation assessment of *Sousa plumbea*. In: Child, M.F., Raimondo, D. (Eds.), The Red List of Mammals of South Africa, Swaziland and Lesotho. South African National Biodiversity Institute, Pretoria.

Reijnders, R.J.H., 1980. Organochloride and heavy metal residues in harbour seals from the Wadden Sea and their possible effects on reproduction. Neth. J. Sea Res. 14, 3–65.

Ross, G.J.B., 1984. The smaller cetaceans of the south east coast of Southern Africa. Ann. Cape Prov. Mus. (Nat. Hist.) 15 (2), 173–410.

Ross, G.J.B., Heinsohn, G.E., Cockcroft, V.G., 1994. Humpback dolphins *Sousa chinensis* (Osbeck, 1765), *Sousa plumbea* (G. Cuvier, 1829) and *Sousa teuszii* (Kükenthal, 1892). In: Ridgway, S.H., Harrrison, R.J. (Eds.), Handbook of Marine Mammals 5: The First Book of Dolphins. Academic Press, New York, pp. 23–42.

Saayman, G.S., Tayler, C.K., 1979. The socioecology of humpback dolphins (*Sousa* sp.). In: Winn, H.E., Olla, B.L. (Eds.), Behaviour of Marine Mammals. Plenum Press, New York, pp. 165–226.

Saayman, G.S., Bower, D., Tayler, C.K., 1972. Observations on inshore and pelagic dolphins on the south-eastern Cape coast of South Africa. Koedoe 15, 1–24.

Scott, E.O.G., 1949. Neonatal length as a function of adult length in Cetacea. Pap. Proc. R. Soc. Tasmania, 75–93.

Sims, P.Q., Vaughn, R., Hung, S.K., Würsig, B., 2011. Sounds of Indo-Pacific humpback dolphins (*Sousa chinensis*) in West Hong Kong: a preliminary description. J. Acoust. Soc. Am. 131 (1), EL48–EL53.

Smith-Goodwin, J.A., 1997. A Molecular Genetic Assessment of the Population Structure and Variation in Two Inshore Dolphin Genera on the East Coast of South Africa. PhD thesis, Rhodes University, Grahamstown, South Africa. 267 pp.

Tezanos-Pinto, G., Constantine, R., Mourão, F., Berghan, J., Baker, C.S., 2014. High calf mortality in bottlenose dolphins in the Bay of Islands, New Zealand—a local unit in decline. Mar. Mamm. Sci. 31 (2), 540–559.

Thompson, P.M., Wilson, B., Grellier, K., Hammond, P.S., 2000. Combining power analysis and population viability analysis to compare traditional and precautionary approaches to conservation of coastal cetaceans. Conserv. Biol. 14 (5), 1253–1263.

Tolley, K.A., Read, A.J., Wells, R.S., Urian, K.W., Scott, M.D., Irvine, A.B., Hohn, A.A., 1995. Sexual dimorphism in wild and captive bottlenose dolphins (*Tursiops truncatus*) from Sarasota, Florida. J. Mamm. 76 (4), 1190–1198.

Van Parijs, S.M., Corkeron, P.J., 2001a. Boat traffic affects the acoustic behaviour of Pacific humpback dolphins, Sousa chinensis. J. Mar. Biol. Assoc. UK 81, 533–538.

Van Parijs, S.M., Corkeron, P.J., 2001b. Vocalizations and behaviour of Pacific humpback dolphins, *Sousa chinensis*. Ethology 107, 701–716.

Venter, K., 2009. The Diet of Humpback Dolphins (*Sousa chinensis*) From the Anti-Shark Nets Off KwaZulu-Natal. BSc Hons. thesis, Department of Zoology & Entomology, Rhodes University, Grahamstown, South Africa. 28 pp.

Weltz, K., 2010. Historical Isotopic Signatures From Teeth of Humpback Dolphins *Sousa chinensis*. BSc Hons. thesis, Department of Zoology & Entomology, Rhodes University, Grahamstown, South Africa. 27 pp.

CHAPTER SEVEN

Ecology and Conservation Status of Indian Ocean Humpback Dolphins (*Sousa plumbea*) in Madagascar

Salvatore Cerchio*,†,1, Norbert Andrianarivelo‡, Boris Andrianantenaina‡

*New England Aquarium, Central Wharf, Boston, Massachusetts, USA
†Woods Hole Oceanographic Institution, Woods Hole, Massachusetts, USA
‡Institut Halieutique et des Sciences Marines, Universite de Toliara, Toliara, Madagascar
[1]Corresponding author: e-mail address: scerchio@neaq.org

Contents

1. Introduction 164
   1.1 Current Knowledge on *S. plumbea* in Madagascar 164
2. Dolphin Surveys on the West Coast of Madagascar 167
   2.1 General Methodology 168
   2.2 Effort Summary 169
   2.3 Sightings and Encounter Rates 169
3. Assessing By-catch and Hunting Through Interview Surveys 179
   3.1 Results 180
4. General Discussion 187
Acknowledgements 196
References 196

Abstract

The Indian Ocean humpback dolphin (*Sousa plumbea*) has been studied in several range states in the Southwest Indian Ocean, however little information exists on populations in Madagascar. Here, we review available literature and describe a study on *S. plumbea* conducted between 2004 and 2013 on the west coast of Madagascar, involving boat-based field surveys in the southwest and northwest regions, and interview surveys with local fishers from villages along most of the west coast. Field surveys in the southwest region of Anakao/St. Augustine Bay revealed low encounter rates and mean group size, and markedly declining trends in both from 1999 to 2013. Conversely, in the northwest region around Nosy Be and Nosy Iranja, encounter rates were higher, as were mean group sizes, suggesting an apparently more abundant and less impacted population. Interview surveys revealed by-catch of coastal dolphins along the entire west coast, including *S. plumbea*, as well as other species. Directed hunting, including drive hunts

Advances in Marine Biology, Volume 72
ISSN 0065-2881
http://dx.doi.org/10.1016/bs.amb.2015.09.004

of groups of dolphins, was reported primarily in the southern regions, in the range of the Vezo Malagasy ethnicity; however, there was evidence of hunting starting in one area in the northwest, where hunting dolphins is normally considered taboo for the predominant Sakalava ethnicity. Thus, the conservation status of *S. plumbea* in Madagascar appears to be spatially heterogeneous, with some areas where the local population is apparently more impacted than others. Conservation measures are recommended to mitigate further decline in the southwest of Madagascar, while protecting habitat and ensuring resilience in the northwest.

1. INTRODUCTION

The Indian Ocean humpback dolphin (*Sousa plumbea* G. Cuvier, 1829) has recently been re-established as a distinct species, separated from *Sousa chinensis* after a protracted series of studies and reviews debating the phylogeny and taxonomy of the genus (Cockcroft et al., 1997; Frère et al., 2008; Jefferson and Karczmarski, 2001; Jefferson and Rosenbaum, 2014; Jefferson and Van Waerebeek, 2004; Mendez et al., 2013; Rice, 1998). The most recent and comprehensive assessments (Jefferson and Rosenbaum, 2014; Mendez et al., 2013) support a range extent for *S. plumbea* encompassing the coastal waters of the western Indian Ocean from approximately E20° to E95° longitude, or from western South Africa to Myanmar. In this chapter, we briefly review currently available information on the species in Madagascar, and then report general results of (i) extensive field survey work from 2004 to 2013 in locations on the southwest and northwest coasts of Madagascar, and (ii) socio-ecological interview surveys from 2010 to 2013 with local fishers in villages spread along the entire west coast of Madagascar.

1.1 Current Knowledge on *S. plumbea* in Madagascar

Within the Southwest Indian Ocean (SWIO), populations of *S. plumbea* have been documented and studied to varying extents in South Africa (Algoa Bay: Karczmarski, 1999; Karczmarski et al., 1998, 1999a,b; Richards Bay/KwaZulu Natal: Atkins and Atkins, 2002; Atkins et al., 2004; Keith et al., 2002, 2013), Mozambique (Maputo Bay: Guissamulo, 2007; Guissamulo and Cockcroft, 2004; Peddemors and Thompson, 1994), Tanzania (Zanzibar: Amir et al., 2002, 2005b; Stensland et al., 2006) and Madagascar (Andrianarivelo, 2001; Cerchio et al., 2009a; Razafindrakoto et al., 2004). Significant population structure in the mtDNA control region has recently been reported along the East Africa coast, with little/no evidence for genetic structure between South Africa and Mozambique samples, but strong genetic differentiation between South Africa/Mozambique and Tanzania, and between Tanzania and Oman samples (Mendez et al., 2011). Two Madagascar samples included in the analysis were generally associated with the Tanzania samples in a haplotype network, suggesting a closer population connection with Tanzania than with South Africa/Mozambique; however, the sample size was too small to include in statistical analyses.

Information on *S. plumbea* in Madagascar is highly limited and currently consists of a single published review of existing records (Razafindrakoto et al., 2004), and several unpublished reports and documents many of which focused on broader topics (Andrianarivelo, 2001; Cerchio et al., 2009a; Cockcroft and Young, 1998; Cockcroft, 1993; Collins et al., 2009; Laran et al., 2012; Razafindrakoto et al., 2008; Robineau and Rose, 1984; Van Canneyt et al., 2010). There is documentation of *S. plumbea* in several locations along the west coast of Madagascar: in the far northwest off Nosy Be and Nosy Mitsio (Cerchio et al., 2009a; Cockcroft and Young, 1998; Robineau and Rose, 1984); on the lower northwest coast in the Loza Lagoon system, approximately 150 km north of Mahajunga, during the response to a mass stranding of melon-headed whales (*Peponocephala electra*) (Collins et al., 2009), and off Mahajanga (Cockcroft and Young, 1998; Razafindrakoto et al., 2004); during aerial surveys along extensive stretches of coast around Madagascar, in the northwest between Mahajanga and Nosy Be, in the mid-west south of Maintirano and in the southwest north of Morombe and near Andavadoake (Laran et al., 2012; Van Canneyt et al., 2010); and most extensively in the southwest off Anakao (Andrianarivelo, 2001; Cerchio et al., 2009a; Razafindrakoto et al., 2004). There are no documented observations of *S. plumbea* along the northeast coast of Madagascar, despite extensive cetacean surveys in near-shore waters of Antongil Bay from 1996 to 2006 (Cerchio et al., 2009b; Ersts and Rosenbaum, 2003), regular whale-watching tours around Ile Saint Marie from 1999 to 2008 (Vely et al., 2009), shorter expeditions along the Masoala Peninsula's east coast (M. Vely, Megaptera NGO, personal communication, 23 March 2015) and west coast (S. Cerchio, personal observation, July 2005), and aerial surveys from the Masoala peninsula to Tamatave (Laran et al., 2012). On the southeast coast, Razafindrakoto et al. (2004) reported the absence of sightings from Fort Dauphin. However, there are some anecdotal reports from Fort Dauphin, and for the deep-south coast near Cap St.

Marie in July/August of 2000 (M. Nicoll, personal communication, 24 March 2015). There are no reports known to us along the large stretch of east coast from Fort Dauphin to Ile St. Marie, or north of the Masoala Peninsula, but there is also a complete lack of survey effort in these regions. Therefore, the range is suspected to be restricted largely to the west coast, and likely around the south coast to Fort Dauphin.

By-catch and hunting, along with consumption and sale of dolphin meat has been reported from most sources assessing fisheries interactions in Madagascar (Cockcroft and Young, 1998; Razafindrakoto et al., 2004, 2008; Robineau and Rose, 1984). Socio-ecological interview surveys in the southwest in 1999 indicated extensive hunting of coastal dolphins by Anakao fishermen of Vezo ethnicity (Andrianarivelo, 2001). Dolphin species most often taken were the spinner dolphin, *Stenella longirostris*, the Indo-Pacific bottlenose dolphin, *Tursiops aduncus* and *S. plumbea*. Vezo fishers hunted coastal dolphins for local consumption and sale of meat, and the most serious drive hunts are associated with the period between August and December, when fishermen aggregate to fish for bonefish, *Albula vulpes*, and Indian anchovy, *Stolephorus indicus*. Fishers thus have the opportunity to cooperate for corralling and hunting dolphins in a drive hunt. Andrianarivelo (2001) estimated that over 6000 individual dolphins, predominantly *Stenella* spp. and *Tursiops* spp., but also *S. plumbea*, were taken between 1985 and 1999 from a single village, Anakao, with a substantial increase in catches evident during 1995–1999 accounting for 57% of all catches (Cerchio et al., 2009a). The trend suggested an increase in intensity of hunting and in the impact on populations; this was likely associated with a change in hunting technique (from harpoons to nets) in the mid to late 1980s, and depletion of other food resources (e.g. decrease in fish populations), and subsequent increased adoption of cetacean hunting and consumption throughout the 1990s.

Based on the interview surveys, Andrianarivelo (2001) estimated a minimum of 61 mortalities of *S. plumbea* between 1985 and 1999 in Anakao, 26 reported as directed hunting, and 35 as 'strandings', which the author interpreted as likely related to drive hunts. As with the overall mortality numbers, the majority occurred between 1995 and 1999, including 58% of reported hunts and 49% of 'strandings'. Interestingly, there were no reports of *S. plumbea* by-catch and few by-catch reports of other species. During 2005 a renewed interview effort, including other surrounding villages, revealed a reported 56 by-catch events between 2000 and 2005, accounting for 79 taken individuals, 12 of which were *S. plumbea* (Razafindrakoto et al., 2008). All *S. plumbea* by-catches occurred in a large-mesh gillnet called 'Jarifa' used primarily to catch sharks. Considering that only a fraction of fishers in this region were interviewed in these studies, the numbers reported for hunting and by-catch likely underestimate the total mortality. There is no reliable abundance estimate or trend for the population, but given the relatively small population sizes reported throughout the SWIO region, the reported mortality rate due to fisheries interactions in the Anakao region is almost certainly unsustainable.

The remainder of this chapter reports on research conducted between 2004 and 2013, investigating the diversity, distribution and status of cetacean populations on the west coast of Madagascar, which focused on coastal dolphin species *S. plumbea* and *T. aduncus*. Small boat-based field surveys were conducted in two regions, the southwest around Anakao and the northwest around Nosy Be, and socio-ecological interviews were conducted in villages spread widely along the majority of the west coast.

2. DOLPHIN SURVEYS ON THE WEST COAST OF MADAGASCAR

Between 2004 and 2013, coastal dolphins including *S. plumbea* and other cetacean populations were assessed in two major regions of Madagascar (Figure 1). On the southwest coast during 2004–2009 the site was the Anakao/Saint Augustine Bay region, where previous socio-ecological interview work indicated extensive dolphin hunting as well as incidental by-catch in artisanal fisheries (Andrianarivelo, 2001; Razafindrakoto et al., 2004, 2008).

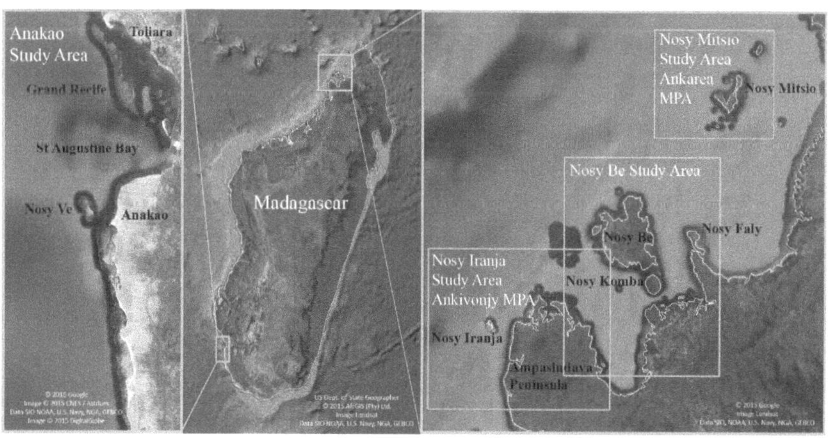

Figure 1 Dolphin survey study sites on Madagascar from 2004 to 2013, showing details of the southwest (Anakao) and northwest (Nosy Be, Nosy Mitsio and Nosy Iranja) study areas.

On the northwest coast during 2007–2013 the site included Nosy Be, Nosy Mitsio and Nosy Iranja/Ampasindava Peninsula, regions where there were anecdotal reports of high encounter rates of *S. plumbea* around Nosy Be (J. Kiszka, University of La Rochelle, personal communication, early 2007), and conservation concerns related to two recently created marine protected areas (MPAs), the Ankarea MPA covering 173,690 ha, and the Ankivonjy MPA covering 196,659 ha.

2.1 General Methodology

Boat surveys were conducted from an eight metre outboard boat with two to four observers. Previous work in Madagascar (Andrianarivelo, 2001; Razafindrakoto et al., 2004) and elsewhere in the SWIO (Atkins et al., 2004; Karczmarski et al., 2000; Keith et al., 2013; Stensland et al., 2006) has indicated that *S. plumbea* is almost exclusively distributed in shallow water (<25 m depth), often very close to coasts and fringing reefs. Initial effort in 2004 and 2005 in the southwest, employed standardized line-transect surveys with parallel transect lines that were orthogonal to the bathymetric gradient. However, weather-related time constraints and sparse distribution yielded low effort in shallow water and low encounter rates for coastal dolphins, and consequently these transects were highly inefficient. Therefore, near-shore transects parallel to the coast and fringing reefs were employed to effectively cover shallow-water coastal habitat and maximize encounters with coastal dolphins. For deep water surveys past the shelf break in the northwest, saw-tooth and parallel transects were conducted orthogonal to the bathymetric gradient in order to ensure equal-coverage of different depth regimes. Daily boat tracks and precise records of effort were logged so as to calculate standardized sighting per unit effort (SPUE) measures. Three types of search effort were logged: (1) coastal waters (generally shallower than 20 m), (2) open shelf waters (generally, but not always, greater than 20 m and less than 100 m), and (3) deep offshore waters (generally greater than 100 m; exclusively off Nosy Iranja). When groups of dolphins were encountered, every possible effort was made to obtain individual identification photographs of right and left sides of dorsal fins from each individual. To infer depth of sightings, bathymetric data were downloaded from the 2014 GEBCO dataset (www.gebco.net), and sighting depths in the northwest were improved from georeferenced digital scans of marine navigational charts. Due to gross inaccuracy of available data for the southwest, depth inference was not attempted there.

2.2 Effort Summary

Effort is reported in terms of the actual daily boat tracks (Figures 2–4), kilometres of track line, total boat-hours, and hours searching in three categories of habitat (Table 1). In any given year/region, a concerted attempt was made to spread effort throughout the region within the safe working range of our boat, focusing on different areas on consecutive days. In the Anakao region, surveys were conducted in six years between 2004 and 2013, primarily in the austral winter/spring months of July to October (Tables 1 and 2), and encompassed all coastal waters north and south of Anakao (Figure 2). The year 2004 was a pilot season by a different team, before effort recording protocol was established and without the collection of GPS track data, so 2004 is not represented in track data or SPUE calculations. In the Nosy Be region, surveys were conducted each year from 2007 to 2012, with all effort prior to 2012 in the austral winter/spring months of July to October, the 2012 effort primarily in December (see Tables 1 and 2), and encompassed all coastal waters around Nosy Be, adjacent small islands, and the Madagascar mainland (Figure 3). The year 2007 was a short pilot season of 1 week, and the 2010 season was also a single-week short visit. During July and November 2011, surveys were conducted in the Nosy Mitsio island group (Ankarea MPA), and were focused in all coastal waters around the main island and all surrounding small islands, as well as offshore around near-by shallow banks. In the Ankivonjy MPA, surveys were conducted during 2012 and 2013 in coastal waters around Nosy Iranja and along the coast of the Ampasindava peninsula, on the shallow shelf between Nosy Iranja and the coast, and offshore past the shelf break (see Table 1; Figure 3).

2.3 Sightings and Encounter Rates

Species diversity ranged among the sites from 15 cetacean species sighted off Anakao and 14 species sighted off Nosy Iranja, to six species in the Nosy Be region and three species off Nosy Mitsio (S. Cerchio, unpublished data). In each site, Indian Ocean humpback dolphins and Indo-Pacific bottlenose dolphins were sighted with varying frequency. Although here we focus on *S. plumbea*, we also note the presence of *Tursiops* spp., since they broadly co-occur in the same habitat and were regularly sighted in mixed species groups. In the north study sites, *T. aduncus* and probably the common bottlenose dolphin, *T. truncatus* were observed, both at times in association with *S. plumbea*. *Tursiops truncatus* was clearly more robust, moderately larger in length, and had an apparent lack of spots on the belly.

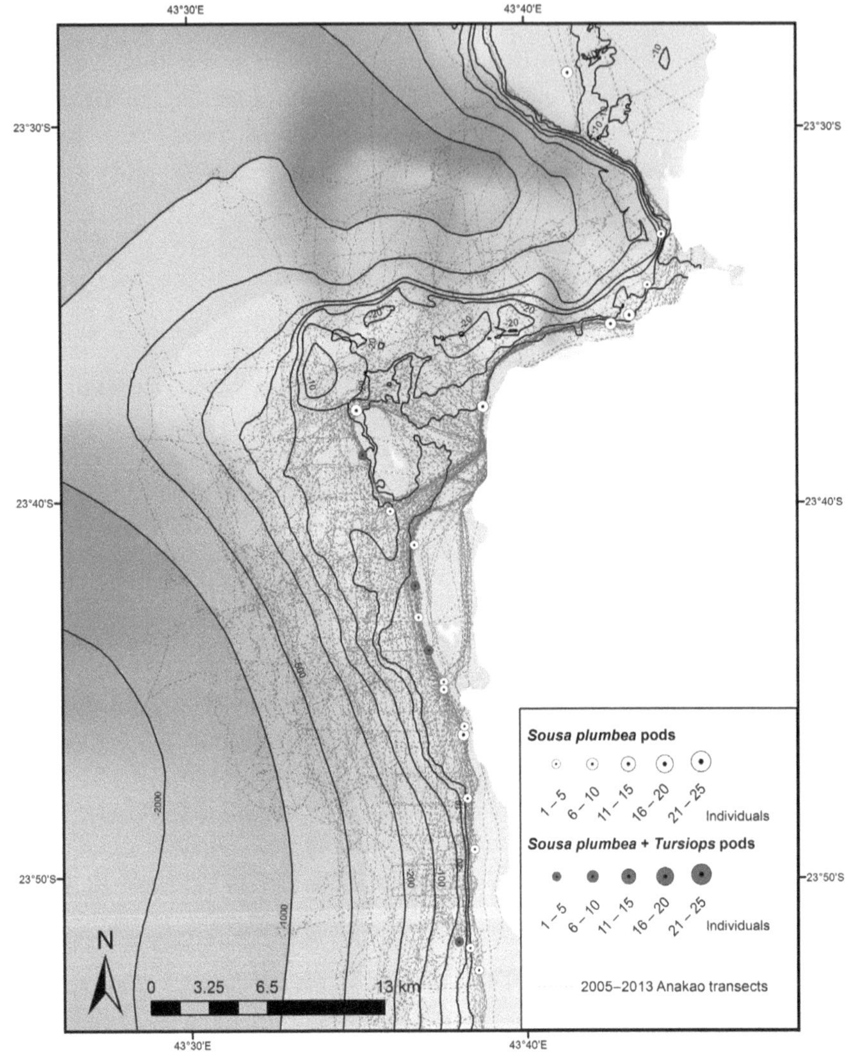

Figure 2 Effort and coastal dolphin sightings in the Anakao study area, all years 2004–2013. Bathymetric data were downloaded from the 2014 GEBCO dataset (www.gebco.net) in the form of a 1 arc second raster format and contour lines were extruded from the underlying raster using the Contour tool. However, inaccuracy of the data along the coastline led us to amend the GEBCO dataset with soundings taken in the field defining fringing reefs before re-interpolating a more accurate bathymetric raster using the Topo to Raster tool in ArcGIS v.10.2.

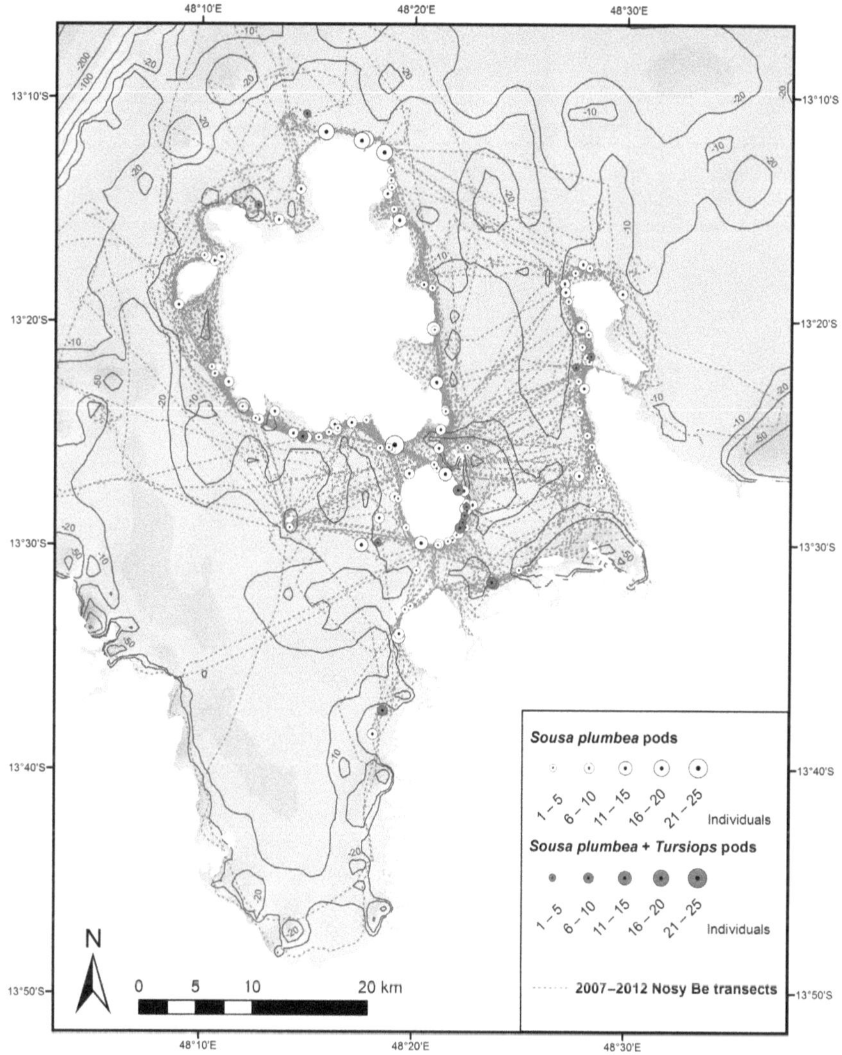

Figure 3 Effort and coastal dolphin sightings in the Nosy Be study area, all years 2007–2012. Bathymetric data were downloaded from the 2014 GEBCO dataset (www.gebco.net) in the form of a 1 arc second raster format and contour lines were extruded from the underlying raster using the Contour tool.

2.3.1 Anakao Region

Groups of *S. plumbea* were encountered on 23 occasions during the 6 study years off Anakao, and were moderately more common than *T. aduncus* (Table 3). Group size of *S. plumbea* was generally small, ranging from

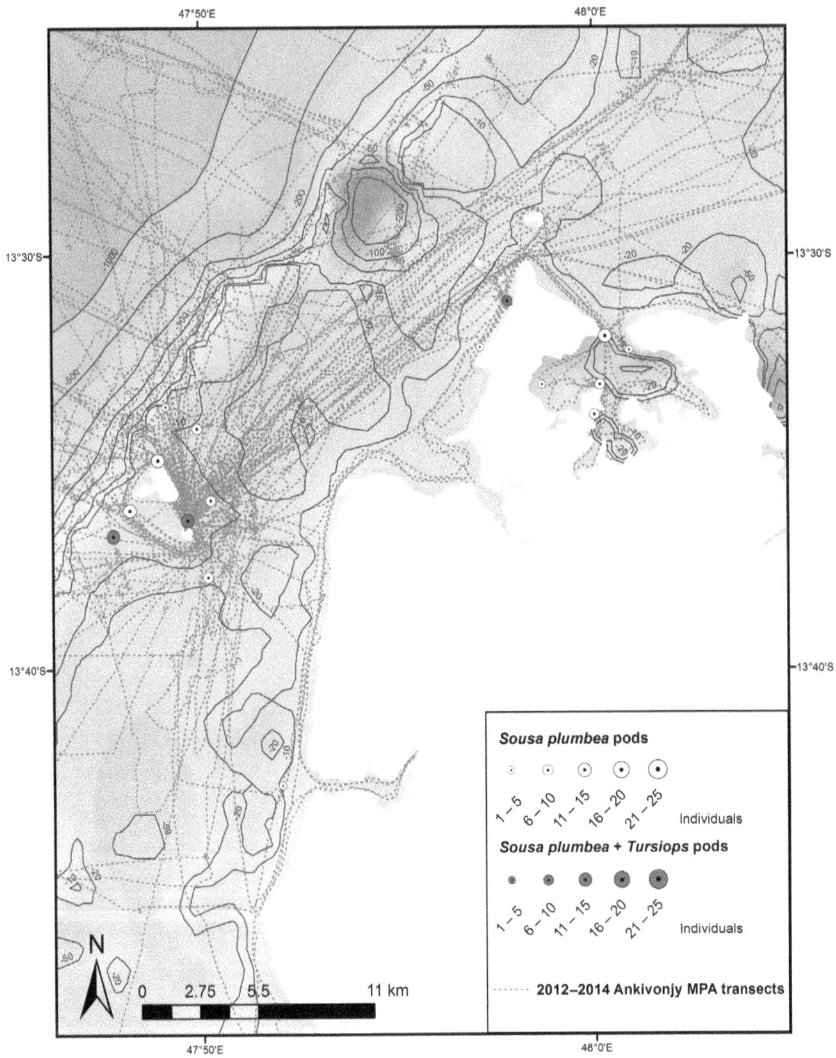

Figure 4 Effort and coastal dolphin sightings in the Nosy Iranja/Ankivonjy MPA study area. Bathymetric data were downloaded from the 2014 GEBCO dataset (www.gebco.net) in the form of a 1 arc second raster format, and bathymetric contour lines were extruded from the underlying raster using the Contour tool.

one to nine individuals, with a mean of 3.6 (SD 2.7); mean group size for *T. aduncus* was somewhat larger at 6.1 (SD 3.5) individuals (Table 4). Mixed species groups of *S. plumbea* and *T. aduncus* were encountered on five occasions.

Table 1 Distribution of Dolphin Survey Effort By Year Off Anakao, Nosy Be, Nosy Mitsio, and Nosy Iranja

Southwest				**Anakao**			
Effort Type	**2004**	**2005**	**2006**	**2007**	**2009**	**2013**	**Total**
Boat days	23	25	35	19	19	11	132
Total km		1442	2243	1399	1300	1084	7467
Total hours		105	187	142	133	95	663
Search effort hours							
Coastal		20	34	13	16	35	117
Open shelf + deep offshore		27	44	33	24	13	141
Total		47	78	46	40	48	258

Northwest				**Nosy Be**				**Nosy Mitsio**	**Nosy Iranja**		**Grand Total**
Effort Type	**2007**	**2008**	**2009**	**2010**	**2011**	**2012**	**Total**	**2011**	**2012**	**2013**	
Boat days	6	21	17	5	24	15	88	17	17	24	146
Total km	517	1772	1502	340	2092	1384	7607	1336	1561	1954	12,458
Total hours	50	160	135	31	185	118	679	113	132	181	1105
Search effort hours											
Coastal	18	67	47	13	65	40	251	41	19	20	331
Open shelf	11	17	18	2	23	32	103	35	4	41	183
Deep offshore									40	31	70
Total	29	84	65	15	88	72	354	76	63	92	584

Effort is expressed as total working boat days, kilometres of track line, and hours on water. Time spent actively searching for cetaceans is expressed in three broad habitat categories: along the coast or fringing reefs (Coastal), on the open water shelf (Open shelf), or in deep water past the shelf break (Deep offshore); in Anakao no distinction was made between the latter two categories due to the lack of a distinct shelf break on available digital charts.

Distribution of *S. plumbea* was exclusively along the coasts and fringing reefs of the study area (see Figure 2), typically within 100–200 m of the break zone. During 2004, a concerted effort of standardized track lines (not depicted in Figure 2, due to lack of GPS track data) was made to search

Table 2 Distribution of Dolphin Survey Effort by Month off Anakao and Nosy Be

	Anakao							Nosy Be						
Month	**2004**	**2005**	**2006**	**2007**	**2009**	**2013**	**Total**	**2007**	**2008**	**2009**	**2010**	**2011**	**2012**	**Total**
June				22.5			22.5							
July				119.6	12.8		132.4					74.8		74.8
August			9.5		120.5	95.4	225.4			16.6		109.9		126.5
September			120.3				120.3	6.6	160.1	118.6				285.3
October		80.1	57.5				137.7	43.2			31.3			74.6
November		25.2					25.2						12.5	12.5
December													105.8	105.8

Effort is expressed in total hours on water.

Table 3 Encounters with Groups of Coastal Dolphins Off Anakao, Nosy Be, Nosy Mitsio, and Nosy Iranja

Southwest	**Anakao**						
Species	**2004**	**2005**	**2006**	**2007**	**2009**	**2013**	**Total**
Sousa plumbea	6	5	4	4	3	1	23
Tursiops aduncus	4	3	2	2	1	1	13

Northwest	**Nosy Be**						**Nosy Mitsio**	**Nosy Iranja**		
Species	**2007**	**2008**	**2009**	**2010**	**2011**	**2012**	**2011**	**2012**	**2013**	**Total**
Sousa plumbea	9	35	18	5	40	19	2	10	6	144
Tursiops aduncus	2	8	4	1	11	1	0	3	6	36
Tursiops sp.	0	4	4	1	2	3	5	0	1	20

Values represent the number of groups encountered while surveying; mixed species groups (see Table 5) are counted twice, once for each species.

Table 4 Group Size of Coastal Dolphins Encountered Off Anakao, Nosy Be, Nosy Mitsio, and Nosy Iranja

Southwest	**Anakao**						
Species	**2004**	**2005**	**2006**	**2007**	**2009**	**2013**	**Overall**
Sousa plumbea	5.5	4.4	2.8	1.5	3.0	1.0	3.6
Tursiops aduncus	10.0	4.3	2.5	4.5	4.0	8.0	6.1

Northwest	**Nosy Be**							**Nosy Mitsio**	**Nosy Iranja**		
Species	**2007**	**2008**	**2009**	**2010**	**2011**	**2012**	**Overall**	**2011**	**2012**	**2013**	**Overall**
Sousa plumbea	4.9	4.7	5.7	5.0	6.7	7.3	5.9	1.0	6.5	8.2	7.1
Tursiops aduncus	2.5	5.8	6.8	7.0	4.6	4.0	5.2		7.0	8.2	7.8

Values represent means of 'best' estimates of group size across all encountered pods of each species, and include single individuals as a group size = 1. During sightings, group size is recorded as minimum, best and maximum estimates, all of which being equivalent when size is confidently determined.

the shallow lagoons within the fringing reef stretching from Anakao to the south, and within the Grand Récif de Toliara, north of St. Augustine Bay (see Figure 1), with no sightings of *S. plumbea*. Only a single incidental lagoon sighting was made inside the Grand Récif during a transit to Toliara in 2005 (see Figure 2). Besides the lack of sightings within lagoons, there were no obvious indications of habitat avoidance or preference, or spatial clustering of encounters, with sightings spread fairly evenly south of Anakao and along the coast of St. Augustine Bay.

Encounter rates for *S. plumbea* and *T. aduncus* were calculated as SPUE for both groups and individuals, using the effort spent searching in coastal waters (no sightings occurred during other search effort categories) for each year individually and across all years (Table 5). Encounter rate for *S. plumbea* was generally low with an overall mean of 0.42 individuals sighted per hour of search time (ind/h). The lowest encounter rate was observed in 2013 (0.03 ind/h), when the team returned to Anakao four years after the main study period and encountered only one individual, despite 34.7 h of coastal search effort representing the most in any single year. Groups of *T. aduncus* were less frequently encountered (0.08 gp/h overall), but had a similar individual encounter rate (0.33 ind/h) due to the larger mean group size as compared to *S. plumbea*.

Table 5 Group Encounter Rates (Top Rows) and Individual Encounter Rates (Bottom Rows) for Coastal Dolphin Species By Year

Southwest				**Anakao**			
Species	**2004**	**2005**	**2006**	**2007**	**2009**	**2013**	**Overall**
Sousa plumbea	n.a.	0.25	0.12	0.31	0.19	0.03	0.14
	n.a.	1.11	0.32	0.46	0.57	0.03	0.42
Tursiops aduncus	n.a.	0.15	0.06	0.15	0.06	0.03	0.08
	n.a.	0.66	0.15	0.70	0.25	0.23	0.33

Northwest				**Nosy Be**				**Nosy Mitsio**		**Nosy Iranja**	
Species	**2007**	**2008**	**2009**	**2010**	**2011**	**2012**	**Overall**	**2011**	**2012**	**2013**	**Overall**
Sousa plumbea	0.49	0.52	0.38	0.38	0.61	0.48	0.50	0.05	0.52	0.30	0.41
	2.40	2.44	2.18	1.90	4.10	3.46	2.95	0.05	3.38	2.43	2.89
Tursiops aduncus	0.11	0.12	0.08	0.08	0.17	0.03	0.11		0.13	0.10	0.11
	0.27	0.69	0.57	0.53	0.78	0.10	0.56		1.09	2.43	1.78

To calculate encounter rates for *S. plumbea* and *T. aduncus*, the total number of groups or individuals encountered was divided by the total effort searching to derive number of groups or individuals sighted per hour search effort.

2.3.2 Nosy Be Region

Sousa plumbea was the most commonly sighted cetacean around Nosy Be by nearly fivefold, with 126 encounters of groups over the six years, as compared to 27 groups of *T. aduncus* (Table 3). There were many re-sightings of individuals across encounters (mark–recapture using photographic identification will be useful for future estimates of population abundance for *S. plumbea*). Average group size for *S. plumbea* was 5.9 (SD 4.2) individuals across the entire six years (Table 4), with a maximum group size of 22 individuals encountered in 2011, and a total of 22 groups with greater than 10 individuals across all years. Average group size for *T. aduncus* was similar (see Table 4), however, with lower maximum size of 10 individuals, observed on two occasions. Groups of *S. plumbea* were sighted in association with both forms of *Tursiops* spp., but much more commonly with *T. aduncus* (11 of 14 encounters of mixed species groups). During these encounters, individuals of the two species were clearly associating, typically forming subgroups of one or two members of each species, as opposed to subgroups of entirely the same species segregated from each other. In some instances, it appeared that multiple *Tursiops* were pursuing or chasing a single *S. plumbea*, but on at least one occasion the reverse was occurring.

Distribution of *S. plumbea* around the Nosy Be region was heterogeneous, with several areas appearing to be favoured by dolphins (see Figure 3). The pattern of geographic distribution was consistent across the four main sampling years. There were several cases noted of individuals being re-sighted across encounters and years within the same general area, suggesting there might be site fidelity within the high sighting probability areas and restricted individual ranges. Heterogeneous distribution and site fidelity can be tested by spatially modelling encounter probability and assessing photographic recapture within areas of high sighting probability. The mean depth of sightings was 8.2 m (SD 5.15 m), with a maximum depth of 25 m and 95% of sightings <20 m.

Sousa plumbea in the northwest had the highest encounter rates for coastal dolphin species across the entire study, and around Nosy Be ranged from a low in 2009 at 1.90 ind/h, to a high in 2011 at 4.10 ind/h (Table 5). Considering variation across the six survey months (July to December), the months of July and August had the highest encounter rates (Table 6). However, year and month are conflated, since all July effort and the majority of August effort occurred in 2011 (see Table 2), so it is difficult to distinguish between seasonal and yearly effects without further sampling. Moreover, it is yet to be determined if this variation in SPUE is real or a product of variation

Table 6 Individual Encounter Rates for Dolphins by Month for Nosy Be

Species	July	Aug	Sept	Oct	Nov	Dec
Sousa plumbea	4.64	3.72	2.19	2.41	n.a.	2.98
Tursiops aduncus	1.01	0.52	0.65	0.42	n.a.	0.36

Encounter rates are calculated as in Table 5, with all encounters and effort being summed across all years for each month. November was excluded due to having too little coastal search effort (3.4 h).

in spatial coverage from year to year, coupled with a heterogeneous distribution pattern. The spatial assessment of effort and sighting probability will be used to assess this. The SPUE for *S. plumbea* was four- to sevenfold greater than for *T. aduncus*, the next most frequently encountered species, for all years and months (see Tables 5 and 6).

2.3.3 Nosy Mitsio Region

Numbers of sightings of coastal dolphins were low in both July and November expeditions, with only one sighting of *S. plumbea* during each month, and three and two sightings, respectively, of *Tursiops* sp. (Table 3). The identity of *Tursiops* species sighted is not certain, but was likely to have been *T. truncatus*, as opposed to *T. aduncus* based on the size, behaviour and apparent lack of belly spots. The encounter rate for *S. plumbea* (0.05 ind/h) was dramatically less than that documented in the Nosy Be region (Table 5).

2.3.4 Nosy Iranja Region

Along the shallow coastal waters of Nosy Iranja, and the extensive coastline of Ampasindava, we sighted 16 groups of *S. plumbea*, nine groups of *T. aduncus* (associated in a mixed species group on two occasions) and 1 group of undetermined *Tursiops* sp. (Table 3; Figure 4). Based upon observations of photographic recaptures, it appeared that at least two different social groups of *S. plumbea* were encountered on multiple occasions, one repeatedly seen around the shallow waters of Nosy Iranja and another around the north of the Ampasindava peninsula (Figure 4). Both were found with relatively large group sizes, at times in excess of 10 individuals, and mean group size was somewhat larger than documented around Nosy Be (Table 4). As in the Nosy Be region, *S. plumbea* was sighted only in the shallow, near-shore waters; however, *T. aduncus* was sighted frequently on the shallow shelf in open water inshore of Nosy Iranja, particularly in 2013. For this reason, encounter rates for *S. plumbea* were calculated as in Nosy Be using only coastal search effort, but for *T. aduncus* using coastal and shelf search effort combined. The encounter rate for *S. plumbea* in the Nosy Iranja region was similar to that documented in the Nosy Be region, both of which were dramatically higher than that documented on Nosy Mitsio (Table 5). The encounter rate for *T. aduncus* was also very similar for Nosy Iranja and Nosy Be, and consistently lower than documented for *S. plumbea* in both regions (Table 5). The mean depth of sightings for *S. plumbea* was 10.5 m (SD 7.09 m) with a maximum depth of 23 m and 82% of sightings <20 m.

3. ASSESSING BY-CATCH AND HUNTING THROUGH INTERVIEW SURVEYS

A programme of rapid assessment socio-ecological interview surveys has been conducted since 1999 to assess local communities' perceptions of and interactions with coastal marine mammals along the west coast of Madagascar. These interviews provided an initial understanding of artisanal fisheries interactions, focusing on coastal dolphin hunting and by-catch; early efforts led to discoveries of, and conservation work on, unsustainable hunting in the southwest of Madagascar, in the Anakao region (Andrianarivelo, 2001; Razafindrakoto et al., 2008). Following the 1999 interviews, a targeted interview campaign was conducted in villages along a ca. 1400 km stretch of the west coast of Madagascar from Toliara to Nosy Be, between 2008 and 2013. This included distinct efforts: in the Nosy Be region in 2008–2009; in the extended northwest region (the Ankarea and Ankivonjy MPAs, and Baie d'Ambaro on the Madagascar mainland to the east of Nosy Be) during 2011–2013; and during a long-range sailing vessel expedition in 2010, conducting interviews in villages on the route between Toliara and Mahajanga. The general objectives were to assess fisheries interactions in terms of the geographic scope and magnitude of directed hunting and by-catch. Additionally, we collected local knowledge on cetacean species encountered by fishers, complementing results from boat-based surveys.

Since hunting of marine mammals is illegal in Madagascar, the collection of data regarding hunting and by-catch of marine mammals is very sensitive (Andrianarivelo, 2001; Razafindrakoto et al., 2008). Therefore, interview protocols were aimed at creating a relaxed atmosphere and gaining the confidence of interviewees, and involved an informal, standardized set of questions that were delivered by an exclusively Malagasy team through casual conversation. Reports on marine mammals were categorized into one of four different event types: hunting, by-catch, stranding, or live sighting.

For hunting and by-catch events, the type of fishing gear was recorded along with if, and how, the marine mammal meat was used (i.e. consumed, sold). The number of individuals hunted or by-caught in each event was recorded, along with the timing of the event by the year of occurrence. More detailed description of interview methods can be found in Cerchio et al. (2014).

3.1 Results

Surveyed villages in close proximity were grouped into separate 'locations' (Figure 5; Table 7), and among the 15 locations surveyed, a total of 78 villages were visited and 1066 fishers interviewed during 339 interview sessions (see Table 7). Estimated population size, summed for each location among only the villages surveyed varied widely, from 14,170 in Nosy Be to 60 in the Barren Islands; thus, the estimated percentage of fishers interviewed varied similarly from a low of 1% to high of 40% (see Table 7). Estimated age of interviewees ranged from 15 to 80 (children under 15 were not considered in the interviews), with a modal age of 31–40 years representing 31% of all interviewees (Figure 6). Interviewees were predominantly men (89%), and all but one woman interviewed was within a focus group, reflecting the role bias toward men being the fishers in this traditional society (although women routinely participate in collecting invertebrates from exposed reefs at low tide).

Although focusing on *S. plumbea* for this chapter, we report here interview results for *Tursiops* spp. as well, given the similarity and overlap of habitat of *S. plumbea* and *T. aduncus*, and the likelihood that the two species are subjected to similar pressures and threats from artisanal fisheries. In only two cases fishers reported large offshore common bottlenose dolphins, *Tursiops truncatus*, and we assume all reports of *Tursiops* to represent coastal animals, predominantly *T. aduncus*. However, due to the observations of *T. turncatus* in the coastal waters of Nosy Be, here we refer to *Tursiops* spp.

There were a total of 194 reports of *S. plumbea* and 471 reports of *Tursiops* spp. across all locations, years, and categories of report (Table 8). The most frequently reported category was live sighting; however, there were substantial reports of hunting and by-catch, being predominantly attributed to affect *Tursiops* spp. (see Table 8). *Sousa plumbea* was reported geographically widespread from the northwest to southwest (Figure 7), with the highest incidence of reports in the northwest locations of Nosy Be, Baie d'Ambaro and Nosy Faly, having 92%, 92% and 82% of interview sessions, respectively, reporting the species. Among the northwest locations, Nosy Mitsio and

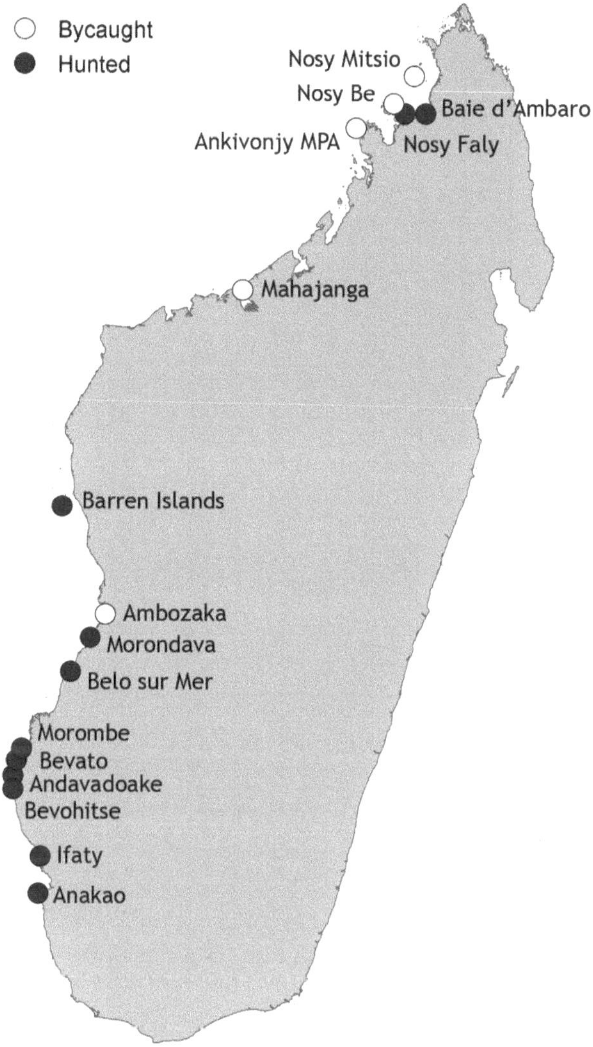

Figure 5 Position of locations where interviews were conducted along the west coast of Madagascar. Anakao, in the southwest, was surveyed previously in 1999 (Andrianarivelo, 2001) and 2004 (Razafindrakoto et al., 2008), and all sites north of Anakao were surveyed from 2010 to 2013 as part of the reported study. Marker colour indicates whether coastal dolphins (either *Sousa plumbea* or *Tursiops aduncus*) were reported by fishers as either hunted or by-caught, but not hunted. In all locations, at least one species was reported as either hunted or by-caught. See Figure 7 and Table 9 for specific reports for each species.

Table 7 Description of All West Coast Locations Were Sampled and Interviews Were Conducted Since 2008

	Location Stats.				Interview Stats.				
Location	**Villages**	**Est. Pop.**	**Est. # Fishers**	**%Pop. Fishers**	**Tot. # Interv.**	**Single Interv.**	**Focus Groups**	**Tot. Individ.**	**%Fisher Interv.**
Nosy Mitsio	18	547	142	26%	41	31	10	58	40%
Nosy Be/Komba	8	14,170	686	5%	24	14	10	143	20%
Baie d'Ambaro	11	6727	1883	28%	60	27	33	218	12%
Nosy Faly/GT*	5	2979	1115	37%	22	6	16	173	16%
Ankivonjy MPA	9	2983	270	9%	56	35	21	97	36%
Mahajanga	4	2300	1390	60%	13	4	9	27	2%
Barren Islands	3	60	60	100%	5	0	5	20	33%
Ambozaka	1	2040	1800	88%	4	1	3	13	1%
Morondava	2	8980	1700	19%	4	0	4	12	1%
Belo sur Mer	3	5187	3440	66%	30	17	13	107	3%
Morombe	2	9144	2220	24%	7	2	5	21	1%
Bevato	1	437	300	69%	9	3	6	37	12%
Andavadoake	4	892	691	77%	28	14	14	57	8%
Bevohitse	3	475	251	53%	24	14	10	57	23%
Ifaty	4	11,540	3730	32%	12	7	5	29	1%
Total	78				339	175	164	1069	

Villages indicate the number of separate villages that were visited at each location; the estimated population (Est. Pop.), and estimated number and percentage of fishers (Est. # Fishers and %Pop. Fishers) are the summations for all indicated villages (as estimated by an elder or mayor in each village). Interviews are defined as each separate session (Tot. # Interv.) irrespective of number of individuals; Single Interv. indicate number of single person sessions, and Focus Groups indicate the number of sessions with focus groups of two or more people; Tot. Individ. is the summation of all people present in all sessions, and %Fisher Interv. is the percentage of the estimated number of fishers represented by Tot. Individ.

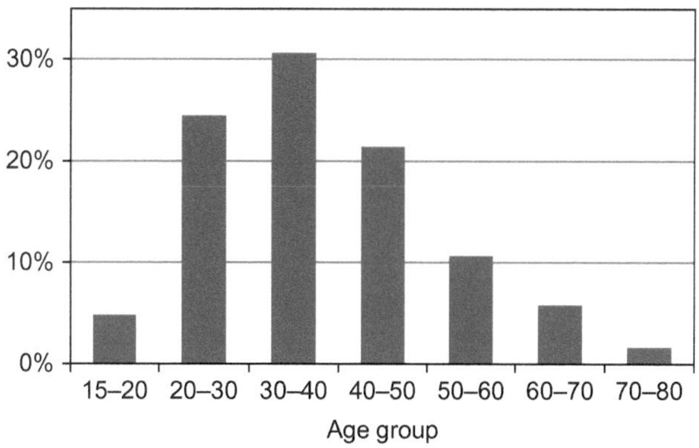

Figure 6 Age breakdown of all interviewed fishers from all locations combined.

Table 8 Summation of All Reports of Coastal Dolphins in All Interviews for All Locations, Irrespective of Time Period Being Reported (Ranging from 1950 to 2010)

Species	Live	Hunt	By-catch	Stranded	Total
Sousa plumbea	173	3	16	2	194
Tursiops spp.	251	93	103	24	471
Total	424	96	119	26	665

Values represent numbers of events reported, not summation of individual dolphins reported, nor number of interview sessions in which events were reported, such that a single interview session may have resulted in multiple reports of events, and a single event may involve multiple dolphin individuals. Reports are grouped by whether dolphins were sighted live, hunted, by-caught or stranded.

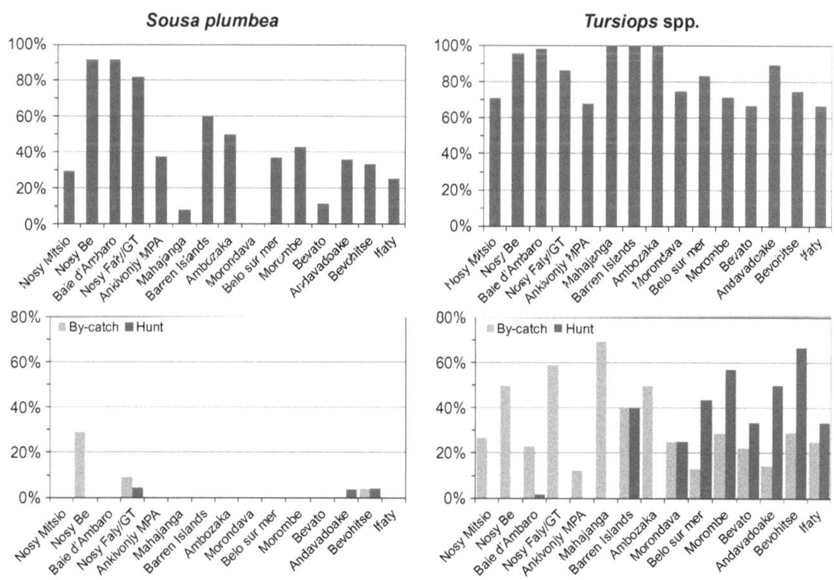

Figure 7 Percentage of interview sessions in which *Sousa plumbea* and *Tursiops* spp. were reported, irrespective of type of sighting (top row), and percentage in which hunting and by-catch were reported (bottom row). Locations are arranged left to right from north to south.

Ankivonjy MPA had a lower incidence of reports, represented in only 29% and 38% of interview sessions, respectively. These findings are generally congruent with the encounter rates of *S. plumbea* found in the northwest field research surveys, with the exception of the Ankivonjy MPA. There were no reports of *S. plumbea* from Morondova, and all southwest locations had a generally lower incidence with reports ranging from 11% in Bevato to 43% in Morombe. *Tursiops* spp. were more commonly reported than *S. plumbea* and more uniformly reported across locations from north to south (Figure 7).

Both hunting and by-catch were reported extensively, with relatively few for reports *S. plumbea* as compared to *Tursiops* spp. (Tables 8 and 9; Figure 7). Hunting reports were predominantly made in southwest locations from the Barren Islands to Ifaty, and overwhelmingly involved *Tursiops* spp. There were only two reports of *S. plumbea* hunted in the southwest: one report from Andavakoake of six individuals taken in 2008 (technique/gear not reported), and one report from Bevohitse of 10 individuals taken in 2010 (by drive hunt). Although hunting was rarely reported in the northwest, a single event for *S. plumbea* was reported from the Nosy Faly/Grande Terre (the main land of Madagascar, hereafter referred to as Nosy Faly/GT) village of Ampasimbary, of two individuals taken in 2009 using nets. Notably, by-catch of *S. plumbea* was reported in 30% of interviews from Nosy Be, involving 13 different events for 31 individuals killed, of which eight events were reported to have occurred off Nosy Faly. In addition, two by-catch events of *S. plumbea* reported from Nosy Faly/GT (using an unspecified net type) were from the same village as the reported hunt. Finally, a single by-catch event (of *Tursiops truncatus*) was observed during our 2009 research surveys, also off the same village near Nosy Faly involving gillnets (described below).

Reports of hunting for *Tursiops* spp. were made in all eight locations in the southwest represented in 25–67% of interviews, with the highest incidence in Bevohitse (see Figure 7). A total of 1414 individual dolphins were killed in 92 reported events, with 56% of individual mortalities reported since the year 2000 (Table 9). By-catch of *Tursiops* spp. was reported in all locations, and reported in 13–69% of interviews, with the highest incidence in Mahajanga (which notably had no reports of hunting). A total of 366 individuals were reported killed in by-catch events, with 90% reported since 2000 (see Table 9). A single case of by-catch was directly documented during the boat surveys in the Nosy Faly/GT area during 2009 involving a relatively large group of *T. truncatus* inshore, and is worth noting here. A group of 10–15 dolphins were encountered already associated with a small group of traditional fishers that had set gillnets, on which

Table 9 Summation of *S. plumbea* and *Tursiops* spp. Individual Mortality Reports for all Interviews

	All Reports				Reports Since 2000			
	Sousa plumbea		*Tursiops* spp.		*Sousa plumbea*		*Tursiops* spp.	
Location	**Hunt**	**By-catch**	**Hunt**	**By-catch**	**Hunt**	**By-catch**	**Hunt**	**By-catch**
Nosy Mitsio	0	0	0	29	0	0	0	23
Nosy Be	0	31	0	12	0	24	0	3
Baie d'Ambaro	0	0	5	116	0	0	0	112
Nosy Faly/GT*	2	3	0	126	2	3	0	118
Ankivonjy MPA	0	0	0	21	0	0	0	15
Mahajanga	0	0	0	24	0	0	0	24
Barren Islands	0	0	7	2	0	0	7	2
Ambozaka	0	0	0	6	0	0	0	6
Morondava	0	0	25	2	0	0	25	2
Belo sur mer	0	0	272	6	0	0	112	6
Morombe	0	0	89	3	0	0	53	3
Bevato	0	0	24	2	0	0	24	2
Andavadoake	6	0	279	7	6	0	71	6
Bevohitse	10	1	655	8	10	1	455	7
Ifaty	0	0	58	2	0	0	50	2
Total	18	35	1414	366	18	28	796	330

Results are presented irrespective of time period being reported (ranging from 1950 to 2010), and for only reports after the year 2000. Values represent actual numbers of individual dolphins reported and are grouped by whether they were hunted or by-caught.
*GT="Grande Terre", referring to nearby main land of Madagascar.

the dolphins were apparently depredating. An immature dolphin became entangled, and the fishers hauled it into their canoe and allowed us to briefly examine it and take pictures. The fishers claimed that the animal had died in the water (drowned), and that this was a rare event and they do not normally catch dolphins. However, it appeared that they had the severed tail of a second dolphin in the canoe, and were acting very wary and uneasy at our approach and questions. The fishers said that it is not taboo (*Fady* in Malagasy) to eat dolphin, contrary to reports from the majority of Sakalava villages in the northwest, and that they would take the by-catch with them back to their village to sell and/or consume. The impression of our Malagasy team discussing this event with the fishers, was that the fishers may have been deliberately setting gillnets with the hope of entangling dolphins in this specific area (where we had repeatedly seen fishers setting nets, and encountered *Tursiops* spp. and *S. plumbea* on many occasions; near the Nosy Faly/GT village of Ampasimbary).

A total of 74 reports of hunting and 88 reports of by-catch were accompanied by information on the technique and gear used during the event (Figure 8). The majority of directed hunting events were done by drive hunts, accounting for 83% of *S. plumbea* and 90% of *Tursiops* spp. hunt mortalities. The reported *S. plumbea* drive hunt was a single recent event in 2010 in the southwest village of Bevohitse, with a take of 10 individuals. Drive hunts for *Tursiops* spp. were more frequently reported, accounting for 977 reported mortalities, and restricted to the southwest locations of Bevohitse (29 reports), Andavadoake (13 reports), Belo sur Mer (six reports) and Bevato and Morombe (one report each). Drive hunts were reported as far back as the 1970s, however they occurred predominantly in

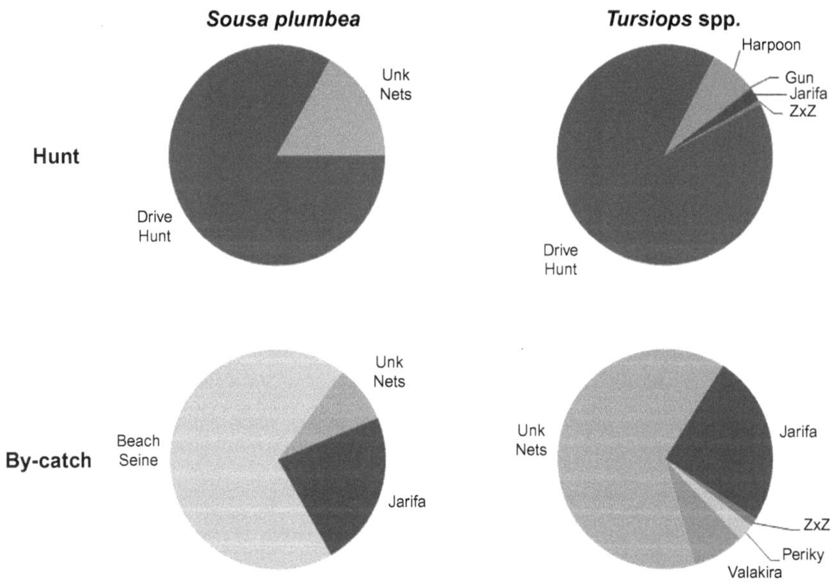

Figure 8 Fishing gear/technique to be reported used during hunting and by-catch, of *Sousa plumbea* and *Tusiops* spp. representing percentages of individuals caught. 'Jarifa' and 'ZxZ' are large-mesh gillnet (12–25 and 8–10 cm, respectively), both used primarily for large pelagic fish and shark; 'Periky' is a small-mesh gillnet (20–35 mm) used to catch sardines; and 'Valakira' is a barrage fence style trap, 150–300 m long and V-shaped, used to catch shrimp.

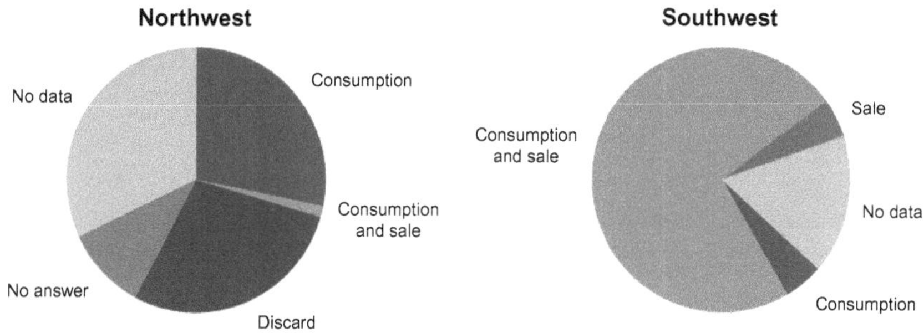

Figure 9 Reported use of by-catch for 16 *Sousa plumbea* and 103 *Tursiops* spp. events combined, separated by Northwest (all locations north of Mahajanga; 78 events) and Southwest (Ifaty north to and including Mahajanga; 41 events).

the 2000s (39 reports for 508 mortalities). A fewer number of reports, from 19 interviews for 74 mortalities, indicated the use of harpoon, dating back to the 1950s. By-catch of *S. plumbea* was reported primarily in beach seine nets (all reported occurred off Nosy Faly, for 24 individuals), with the remainder made with 'Jarifa', a very large-mesh (12–25 cm) gillnet, used primarily for large pelagic fishes and sharks (Gough et al., 2009). The majority of *Tursiops* spp. by-catch was reported to have occurred in an unspecified net type, and was accounted in 25 reports for 155 mortalities. The net type was likely a large-mesh gillnet similar to the widely used Jarifa, which was reported in 38 interviews for 62 mortalities. In all cases where information was collected on the use of dolphin meat from hunts (83 of 93 reports), interviewees reported both direct consumption and local sale in their home village or near-by villages. Use of by-catch on the other hand, showed more variation and varied markedly between the northwest and southwest locations (including Mahajanga with all more southerly locations; Figure 9). 'Direct consumption and local sale' accounted for 73% of by-catch reports from the southwest, whereas only 1% from the northwest; 'local consumption with no sale' accounted for only 5% of reports in the southwest, but 28% in the northwest. By-catch was reported as 'discarded without consumption or sale' in 28% of northwest events, but never in the southwest. There was no indication of use of by-catch or hunted animals as fishing bait.

4. GENERAL DISCUSSION

The field surveys and interview surveys reported here provide the most extensive data collected to date on *S. plumbea* in Madagascar. The survey effort and interviews, combined with previous reports, suggests a range extent in Madagascar that covers primarily the western coasts (Figure 10). Moreover, the spread of field effort between two distinct regions, coupled with interview surveys that revealed patterns of artisanal hunting and by-catch, suggests important implications for the status and conservation of the species in Madagascar. Generally, the northwest region appears to have larger, potentially healthier populations of *S. plumbea* than the southwest region, and interviews indicated that this difference is at least correlated with the level of hunting and artisanal fisheries interactions with coastal dolphins that are reported by local fishers.

The distinction between the conservation status of *S. plumbea* in southwest and northwest Madagascar is most apparent when directly comparing the SPUE encounter rates of individuals and mean group sizes (Figure 11). The highest encounter rate of individuals in Nosy Be, occurred in 2011, is fourfold greater than the highest rate that occurred for Anakao, in 2005 (see Figure 11; Table 5); overall the Nosy Be rate is sevenfold greater (see Table 5). This marked difference is clearly a combination of encounter rate of groups (see Table 5) and mean group size (see Table 4). The population around Anakao is characterized by smaller groups that are more sparsely distributed throughout the shallow coastal habitat (reflected in group encounter rate), relative to the Nosy Be region population.

In the southwest site of Anakao, there are alarming negative trends suggested by our data from 2004/2005 to 2013 for both encounter rate and group size of *S. plumbea* (see Figure 11). This is evident when considering only the primary field effort years of 2004 to 2009, and more dramatic when considering the 2013 data, in which only one *S. plumbea* individual was sighted, despite comparatively high effort. However the 2013 study period in Anakao was shorter in total duration than all other years, with only 11 days of boat time over a 14-day span, and this may have negatively biassed the 2013 data. Still considering reports that existed prior to our study, the evident negative trend becomes even more dramatic. Razafindrakoto et al. (2004) reported five groups encountered in 1999 for 65 individuals during 98.41 h total effort, which resulted in a crude encounter rate of 0.66 ind/h and a mean group size of 13 animals (SD 7.61, range 5–25). Comparable crude encounter rates for our data, using total hours on the water and total hours searching irrespective of depth zone, yield encounter rates of 0.12 and 0.32 ind/h, respectively, which is approximately half the rate reported for 1999. The mean group size reported for 1999 is more than twofold greater than the largest mean group size in 2004 (see Figure 11).

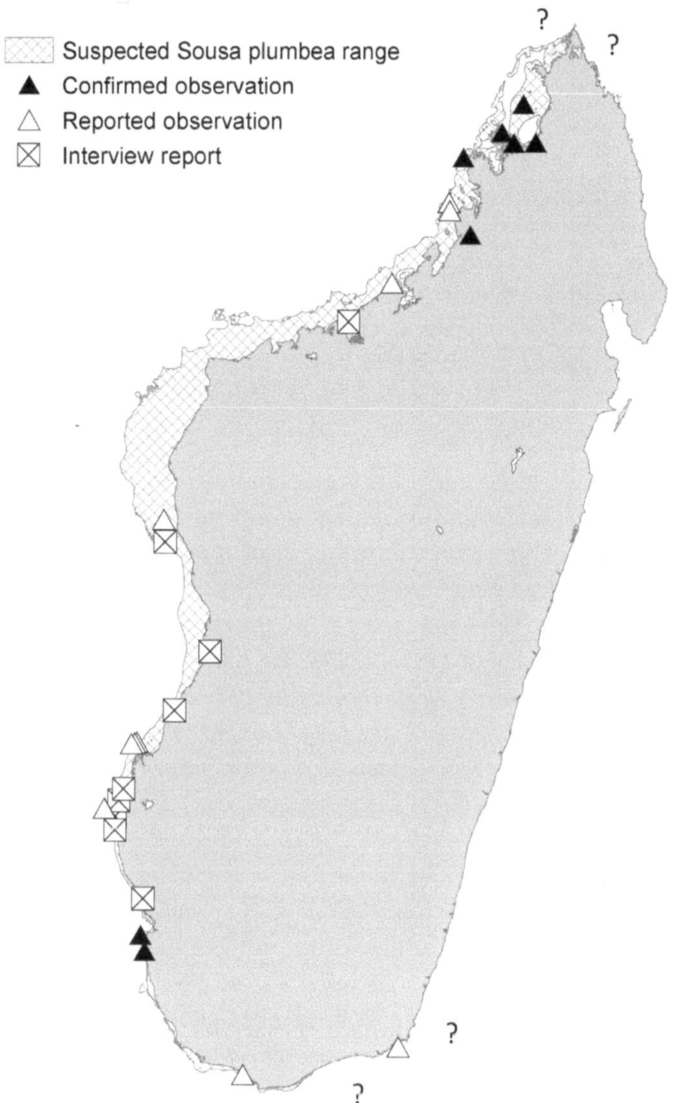

Figure 10 Suspected range of *Sousa plumbea* in Madagascar based upon existing knowledge, including known locations of sightings. The inferred distribution of the species is represented as a continuous band of coastal waters within the 30 m isobath where it is suspected to occur, with no allowance for discontinuities where the species may be absent. 'Confirmed observations' refer to locations where direct sightings or photographic evidence was collected or confirmed by the authors; 'Reported Observations' refer to visual sightings reported to the authors by trusted sources (west coast sighting positions from REMMOA survey 2010, Agence des aires marines protégées/Observatoire PELAGIS, Laran et al., 2012; Van Canneyt et al., 2010; south coast sighting positions from M. Nicoll, World Wildlife Fund, personal communication, 24 March 2015); 'Interview reports' refer to locations reported in interviews with local fishers conducted by the authors.

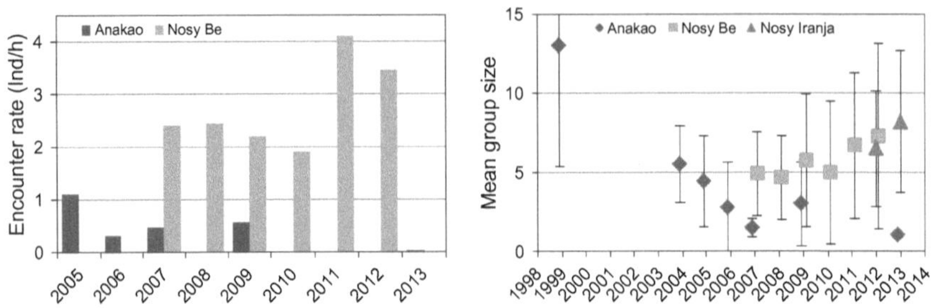

Figure 11 Comparison of individual encounter rate (left) and group size (right) for *Sousa plumbea* in the southwest (Anakao) and the northwest (Nosy Be and Nosy Iranja) across the study period. The 1999 value for mean group size in Anakao is from Razafindrakoto et al. (2004).

The largest group reported in 1999 was 25 individuals, compared to 9 individuals in 2004 representing the largest group sighting between 2004 and 2013. Although it is difficult to estimate a rate of population decline from these data, this change represents a roughly 14–18% per annum decline in mean group size between 1999 and 2013. It is conceivable that the 1999 values reported by Razafindrakoto et al. (2004) are inflated relative to 2004–2013 data reported here, since it was reported by different researchers (with the exception of N. Andrianarivelo) that was relatively inexperienced at the time. However, even if the 1999 estimates of group size were inflated twofold (which we consider an unlikely extreme bias), the trend in encounter rate and group size would still indicate a decline. Furthermore, irrespective of the comparability of the 1999 data, the trend and the observations in the final year 2013, suggest that the population of *S. plumbea* around Toliara/St. Augustine Bay/Anakao may be near extirpation. However, given the brevity of the 2013 field effort, this conclusion should be considered with some caution, and further monitoring for an extended period should be conducted to more rigorously establish the current population status in this region.

Northwest field survey results suggest that coastal dolphin species utilize shallow water areas as important habitat around the Nosy Be region, Nosy Iranja/Ampasindava Peninsula (Ankivonjy MPA) and to a lesser extent the Nosy Mitsio island group (Ankarea MPA). The encounter rate and apparent abundance of *S. plumbea* were similar between the coastal/shallow water areas of Nosy Be region and the Ankivonjy MPA, both of which were dramatically greater than that for the Nosy Mitsio island group (see Table 5) or any other region assessed by our team, including the southwest (Anakao) and other western shallow water habitat surveyed for other projects (Morombe, Belo sur Mer and Barren Islands; S. Cerchio, unpublished data).

The Nosy Be region and Ankivonjy MPA also appear to be important for *T. aduncus*. Factors influencing lower encounter rates of coastal dolphins around Mitsio are not yet understood. Group size of *S. plumbea* in the Nosy Be and Nosy Iranja regions was similar to that reported in South Africa for Algoa Bay (Karczmarski et al., 1999a, six individuals, SD 2.72). The reported South Africa individual encounter rates of 3.78 and 3.33 ind/h for Algoa Bay and St. Francis Bay, respectively (Karczmarski et al., 1999b), were also similar to that found in Nosy Be (2.95 ind/h) and Nosy Iranja (2.89 ind/h). This suggests some broad similarities between these populations. Group sizes reported by Laran et al. (2012) during aerial surveys along the west coast of Madagascar are broadly congruent with the observations reported here, both in value and declining trend from north to south: in the northwest, a mean of 5.0 individuals for three sighted groups; in the central west a single group of 4.0 individuals; and in the southwest a mean of 2.2 individuals for five sighted groups.

There were no strong trends in yearly or seasonal variation in sightings in the northwest, in part due to limitations of the dataset. There was an apparent increase in encounter rate in Nosy Be during 2011 and 2012 at approximately twofold from 2007 to 2009, and a slight increase in group size (Figure 11). However, it is difficult to determine whether this might reflect an increase in numbers due to yearly, seasonal, or spatial effort variation. Irrespective, there is no indication of decline in population numbers, as detected in Anakao, and there is potentially a suggestion of population increase in this area. Given the restricted seasonal coverage of the Nosy Be data (with no summer or autumn data), it was not possible to assess yearly seasonal variation in group size or encounter rate. However, the nominal peak in encounter rates during July/August, compared to September through December (see Table 6), was not congruent with Karczmarski et al. (1999a), who reported a low in encounter rate during May/June/July in Algoa Bay. Therefore, if there is seasonal variation in Nosy Be related to movement of individuals in and out of the area, it is likely influenced by different processes than those operating in Algoa Bay. It is not yet possible to report whether the population around Nosy Be more closely resembles the transient nature of the Algoa Bay population (Karczmarski, 1999), or the highly resident population of southern Zanzibar (Stensland et al., 2006). Analysis of photographic capture–recapture data may be useful address this question. Currently, casual observations of recaptures within and across years (S. Cerchio, unpublished data), suggests at least some level of residency, as reported for Maputo, Mozambique (Guissamulo and Cockcroft, 2004) and Richards Bay, South Africa (Keith et al., 2002).

The distribution of all sightings in both the northwest and southwest regions, concurs with the understanding that *S. plumbea* is an obligate shallow water and coastal species. There is some preliminary indication of preference for certain areas within the study range in the Nosy Be region (see Figure 3), which would be similar to results of other studies (Atkins et al., 2004; Karczmarski et al., 2000; Keith et al., 2013; Stensland et al., 2006). Karczmarski et al. (2000) reported no seasonal variation in habitat preference despite seasonal shifts in group size and influx of individuals, suggesting that these dolphins are dependent on specific habitat type within the already limited shallow water coastal zone.

The geographic spread of reports of *S. plumbea* from interview surveys with fishers, provides further evidence that the species ranges along the entire west coast, likely in a continuous distribution. Reports from interviews were most common in the northwest and displayed a decreasing trend from north to south (see Figure 7), broadly congruent with inferences on population status from the field surveys. There were relatively few reports of hunting and by-catch of *S. plumbea* throughout the range. However, this was offset by high levels of both hunting and by-catch reported for *Tursiops* spp. (see Table 6; Figure 7). Given the apparent takes of *Tursiops* spp., it is evident from the reports that both by-catch and/or hunting pressure on coastal dolphins are substantial along the entire west coast of Madagascar. The discrepancy between reports for *S. plumbea* versus *Tursiops* spp. could have several explanations. It may reflect a general difference in population abundance between the species, which seems most plausible for the southern part of the range where *S. plumbea* appears to be at lower abundance; however, this seems unlikely in the northwest, where field surveys indicate *S. plumbea* to be markedly more common. It could also be due to misidentification by fishers between the two species, yet, in most cases interviewees gave the impression of being well aware of the differences between the species and familiar with both (particularly among the southwest Vezo that specialize in dolphin hunting). It may also represent a difference in the behaviour of the two species, which seems plausible from field observations suggesting that *Tursiops* spp. tend toward greater curiosity and gregariousness, which may make them more vulnerable to both hunt and by-catch, whereas *S. plumbea* tend to display more wariness of humans and boats. Irrespective, the interview data evidence suggests that there is substantial pressure from fisheries interactions on coastal dolphins in general, and that *S. plumbea* are taken in both hunting and by-catch events, albeit in smaller numbers than reported for *Tursiops* spp. in Madagascar.

Regarding takes of *S. plumbea*, by-catch was reported predominantly from Nosy Be, where the species is documented as most common from both field surveys and interviews. Two-thirds of these events were reported to have occurred in the eastern channel between Nosy Be and Nosy Faly, closer to the latter island; this locale appears to be a hotspot for both by-catch and suspected hunting in the region. By-catch of *S. plumbea* in this area was reported to occur entirely in beach seine nets, unlike *Tursiops* spp. by-catch, which was never reported in this net type. This may be a function of the more shallow/coastal habitat of *S. plumbea*, or an idiosyncrasy of the specific locale. Hunting for *S. plumbea* was reported rarely, but in both the southwest and northwest primarily in locations that reported the highest incidences of dolphin hunting overall, Bevohitse in the southwest and Nosy Faly/GT in the northwest. All reported hunts and most by-catch events for *S. plumbea* were reported to have occurred in the 2000s, therefore recently relative to the sum of interview reports as a whole. Due to the non-randomized sampling for interviews and likely heterogeneity of fishing pressure, we have not attempted to extrapolate the interview reports to estimate the total number of individuals taken for entire villages or regionally. However, we emphasize that the numbers reported in interview surveys likely represent only a fraction of *S. plumbea* mortalities region-wide, given the small proportion of active fishers interviewed (see Table 7). For example, considering the southwest village of Bevohitse, the 10 *S. plumbea* hunted and one by-caught (see Table 9), would extrapolate to 48 *S. plumbea* mortalities over a decade period from this single village, considering only 23% of fishers were interviewed (see Table 7). Although this may be speculative, given the small population sizes reported for *S. plumbea* in other parts of its range and the generally slow reproductive rates of delphinids, the mortality reported in our interview surveys is almost certainly unsustainable.

Hunting of coastal dolphins was clearly more prevalent in surveyed areas from the Barren Islands south, with a sharp division between Mahajanga and the Barren Islands (see Figure 7). This is also the general boundary between the two major maritime-culture ethnic groups on the west coast, the Vezo in the south and the Sakalava in the north. The Vezo are most prevalent in the southwest regions and are semi-nomadic making migrations to the north at least as far as the Barren Islands to fish in other regions (Gough et al., 2009). All villages encountered and interviewed on the Barren Islands were in fact Vezo from Anakao or Andavadoake regions that had setup temporary fishing camps. Vezo are known to be dolphin hunters, historically using harpoons, and have developed the drive hunt tradition in the southwest, apparently in

Anakao, that has been culturally transmitted to other villages along the coast, such as Bevohitse (Andrianarivelo, 2001; Cerchio et al., 2009a). Throughout the southwest hunting communities, Vezo fishers report eating dolphin meat without hesitation, despite the common after effect of inducing gastrointestinal discomfort or diarrhoea, as reported during interviews. This is apparent in the reported use of by-catch and hunted dolphins, with the majority of southwest fishers stating that meat was consumed as well as sold locally, which indicates a local demand and market beyond simple subsistence for fishers' families. Conversely, Sakalava fishers in villages north of Mahajunga, predominantly reported that killing and eating dolphins was *Fady*, or taboo. Discarding of by-catch was reported only in the northwest, and when not discarded, fishers reported consumption alone almost exclusively, with only a single report of sale of meat. Within the northwest, the one notable exception to the predominant trend of no hunting and discarding of by-catch, was a small group of villages near Nosy Faly, to the east of Nosy Be. Interview surveys indicated a relatively high prevalence of by-catch and at least one reported hunting event in this area. This region had the only report of sale of dolphin meat and nearly 75% of reports of consumption of by-catch. Furthermore, this was the locale of the by-catch event that we observed during boat field surveys. The concentration of these events in the same specific area (despite the widespread distribution of dolphins and the regional coverage of the interview effort in the northwest), and the use of nets for apparent hunting (consequently blurring the distinction between hunting and by-catch), suggests that a few specific villages in this area are recently developing a hunting tradition. Moreover, it appears this tradition may be originating from incidental by-catch, stepping through a progression to 'non-targeted-deliberate' acquisition, and then to 'targeted' acquisition (as described in Robards and Reeves, 2011), with fishers deliberately setting their nets in an area they know to be frequented by dolphins, and targeting dolphins that exhibit the behaviour of depredating their nets.

The data we have presented here, indicate that the status of *S. plumbea* populations is heterogeneous along the west coast of Madagascar. By-catch and directed takes, as described in the interview results and documented directly during our surveys, are clearly important conservation concerns, as has been documented globally (Reeves et al., 2013; Robards and Reeves, 2011). The presence of an active hunting tradition in the southwest of Madagascar, particularly using drive hunt techniques to capture large numbers of dolphins in a single event, is unique in the SWIO, to the best of our knowledge. The impact of the hunting tradition and incidental by-catch on dolphin populations is likely substantial, given the numbers of animals reported in interview surveys, and the marked difference in apparent abundance of *S. plumbea* between the northwest and southwest populations. Conservation efforts in the southwest region (specifically in Anakao and the Andavadoake/Bevohitse/Bevato communities) have focused on community engagement to mitigate marine mammal hunting and by-catch. Establishment of community-based ecotourism, local traditional governance structures (*Dina*; see Rakotoson and Tanner, 2006), and social outreach campaigns, have met with some success (Cerchio et al., 2009a, 2014; Razafindrakoto et al., 2008). Indicators include the creation of local conservation associations and detailed *Dina* in four separate southwest Communes (political entity of a group of villages), and increase of whale-watching tourism in Anakao from six local operators serving 143 tourists in 2010, to 25 local operators serving 1322 tourists in 2013, grossing over US$23,000.

Although hunting and by-catch are likely to be the greatest threats, other factors likely contribute to the decline of *S. plumbea* populations in the southwest, including habitat degradation and prey depletion (Brenier et al., 2011; Laroche and Ramananarivo, 1995). The Grand Recif de Toliara barrier reef complex is particularly degraded, as the result of overfishing, destructive fishing practices, pollution and sedimentation, and recovery of the system is considered by some to be unlikely (Andréfouët et al., 2013; Harris et al., 2010). Given that reefs and associated habitats may be critical for coastal dolphins (as inferred by Amir et al., 2005a; Karczmarski et al., 2000), the cumulative impacts present a bleak outlook for the conservation of these populations. Conversely, the northwest region appears to support more abundant populations of *S. plumbea*. This may be attributed directly to the cultural characteristics of the Sakalava fishers, who do not have a hunting tradition or routinely eat dolphin meat, unlike the Vezo of the southwest. However, the indication that 'deliberate' by-catch and a directed hunting tradition on dolphins are developing in one locale within the surveyed northwest region, is of notable concern. Conservation efforts should be directed to work with local communities in the northwest to stop the development and spread of hunting and mitigate existing by-catch. Together with continued field research to define priority habitat and population status, the information presented here should be used to support the development of management policy at the community and government levels, within and beyond the regional MPAs, and should inform conservation measures and policy actions throughout the region.

ACKNOWLEDGEMENTS

The authors wish to acknowledge many people and organizations that contributed to this project. The authors were employed by the Wildlife Conservation Society during the duration of the work for the study. Within WCS, we thank in particular Howard Rosenbaum and Yvette Razafindrakoto who played key roles during the early years of the project. In addition, we are grateful for the support of the WCS Madagascar Program, including: Alison Clausen, Christopher Holmes, Lisa Gaylord, Helen Crowley, Ambroise Brenier, Herilala Randriamahazo, Bemahafaly Randriamanantsoa, Francisco Ramananjatovo, Santisy Andriamiravo, Luccianie Raonison and Sylviane Raharivelo for institutional support; and Tahina Rasoloarijao, Lovasoa Andriambola and Dina Andrianaivoravelona for assistance in the field. From the WCS headquarters, we thank Josh Ginsberg, Grace Seo, Devon Litherland, Sarah Pacyna, Robert Rose and Melinda Rekdahl. Christopher Spagnoli contributed substantively for map creation. We thank the staff of IHSM in Toliara, in particular, Daniel Ramampiherika and Thierry Lavitra, and CNRO in Nosy Be for their support of and involvement in our work. For logistical support in the Nosy Be region, we thank Julien and Christian Tombozafy, Nosy Komba Aventure, the Zahir Hotel and Nosy Iranja Lodge; in the Anakao region, we thank Thierry Bournonville and the staff of Lalandaka Lodge, Michel Agou and Madam Diamondra from Le Prince Hotel and the Ministry of Fisheries PACP project. Funding for this work was provided by the International Whaling Commission Small Cetacean Conservation Fund, the U.S. Marine Mammal Commission, the Kate Sanderson Bequest to the IUCN Species Survival Commission (SSC), the Flora Family Foundation, the Eppley Foundation for Research and an anonymous donor.

REFERENCES

Amir, O.A., Berggren, P., Jiddawi, N.S., 2002. The incidental catch of dolphins in gillnet fisheries in Zanzibar, Tanzania. Western Indian Ocean J. Mar. Sci. 1, 155–162.

Amir, O.A., Berggren, P., Ndaro, S.G., Jiddawi, N.S., 2005a. Feeding ecology of the Indo-Pacific bottlenose dolphin (*Tursiops aduncus*) incidentally caught in the gillnet fisheries off Zanzibar, Tanzania. Estuar. Coast. Shelf Sci. 63, 429–437.

Amir, O.A., Jiddawi, N.S., Berggren, P., 2005b. The occurrence and distribution of dolphins in Zanzibar, Tanzania, with comments on the differences between two species of Tursiops. Western Indian Ocean J. Mar. Sci. 4, 85–94.

Andréfouët, S., Guillaume, M.M., Delval, A., Rasoamanendrika, F.M.A., Blanchot, J., Bruggemann, J.H., 2013. Fifty years of changes in reef flat habitats of the Grand Récif of Toliara (SW Madagascar) and the impact of gleaning. Coral Reefs 32, 757–768.

Andrianarivelo, N., 2001. Essai d'évaluation de l'importance de la pêche aux dauphins dans la région d'Anakao (sud-ouest de Madagascar). DEA Thesis. Institut Halieutique et des Sciences Marines, Université de Toliara, Madagascar.

Atkins, S., Atkins, B.L., 2002. Abundance and site fidelity of Indo-Pacific hump-backed dolphins (*Sousa chinensis*) at Richard's Bay, South Africa. In: Paper SC/54/SM25 Presented to the International Whaling Commission Scientific Committee.

Atkins, S., Pillay, N., Peddemors, V.M., 2004. Spatial distribution of Indo-Pacific humpback dolphins (*Sousa chinensis*) at Richard's Bay, South Africa: environmental influences and behavioural patterns. Aquat. Mamm. 30, 84–93.

Brenier, A., Ferraris, J., Mahafina, J., 2011. Participatory assessment of the Toliara Bay reef fishery, southwest Madagascar. Madag. Conserv. Dev. 6, 60–67.

Cerchio, S., Andrianarivelo, N., Razafindrakoto, Y., Mendez, M., Rosenbaum, H., 2009a. Coastal dolphin hunting in the southwest of Madagascar: Status of populations, human impacts and conservation actions. In: International Whaling Commission Scientific Report SC/61/SM15.

Cerchio, S., Ersts, P., Pomilla, C., Loo, J., Razafindrakoto, Y., Leslie, M., Andrianarivelo, N., Minton, G., Dushane, J., Murray, A., Collins, T., Rosenbaum, H., 2009b. Updated estimates of abundance for humpback whale breeding stock C3 off Madagascar. In: International Whaling Commission Scientific Report SC/61/SH7.

Cerchio, S., Andrianarivelo, N., Andrianantenaina, B., Cordi, V., 2014. Ecology, status, fisheries interactions and conservation of coastal Indian Ocean humpback dolphins and Indo-Pacific bottlenose dolphins on the west coast of Madagascar. In: International Whaling Commission Scientific Report SC/61/SM21.

Cockcroft, V.G., 1993. A preliminary assessment of the status of Madagascar coastal resources, with special reference to marine mammals and reptiles. South African Department of Foreign Affairs and African Wildlife. Final Report for United Nations Environment Programme, 16 pp.

Cockcroft, V.G., Young, D.D., 1998. An investigation of the status of coastal marine resources along the West Coast of Madagascar. Mission report to WWF, unpublished, 58 pp.

Cockcroft, V.G., Leatherwood, S., Goodwin, J., Porter, L.J., 1997. The phylogeny of humpback dolphins genus *Sousa*: insights through mtDNA analyses. In: International Whaling Commission Scientific Report SC/49/SM25.

Collins, T., Moore, K., Cerchio, S., McAloose, D., Harry, C.T., Calle, P., Razafindrakoto, Y., Randriamanantsoa, B., McClave, C., Rosenbaum, H., 2009. Madagascar Melon-Headed Whale (*Peponocephala electra*) Mass Stranding Event of May/June 2008 Final Report—Part I: Site Description and Mass Stranding Timeline. Wildlife Conservation Society and International Fund for Animal Welfare. Available from: https://iwc.int/2008-mass-stranding-in-madagascar.

Ersts, P.J., Rosenbaum, H.C., 2003. Habitat preference reflects social organization of humpback whales (*Megaptera novaeangliae*) on a wintering ground. J. Zool. (Lond.) 260, 337–345.

Frère, C.H., Hale, P., Porter, L., Cockcroft, V.G., Dalebout, M.L., 2008. Phylogenetic analysis of mtDNA sequences suggests revision of *Sousa* taxonomy is needed. Mar. Freshw. Res. 59, 259–268.

Gough, C., Thomas, T., Humber, F., Harris, A., Cripps, G., Peabody, S., 2009. Vezo fishing: an introduction to the methods used by fishers in Andavadoaka Southwest Madagascar. Blue Ventures Conservation Report.

Guissamulo, A.T., 2007. Ecological studies of bottlenose and humpback dolphins of Maputo Bay, Southern Mozambique. Ph.D. Thesis, University of Natal.

Guissamulo, A., Cockcroft, V.G., 2004. Ecology and population estimates of Indo-Pacific humpback dolphins (*Sousa chinensis*) in Maputo Bay, Mozambique. Aquat. Mamm. 30, 94–102.

Harris, A., Manahira, G., Sheppard, A., Gough, C., Sheppard, C.R.C., 2010. Demise of Madagascar's once great barrier reef—change in coral reef condition over 40 years. Atoll Res. Bull. 574, 1–16.

Jefferson, T.A., Karczmarski, L., 2001. *Sousa chinensis*. Mamm. Species 655, 1–9.

Jefferson, T.A., Rosenbaum, H.C., 2014. Taxonomic revision of the humpback dolphins (*Sousa* spp.), and description of a new species from Australia. Mar. Mamm. Sci. 30, 1494–1541. http://dx.doi.org/10.1111/mms.12152.

Jefferson, T.A., Van Waerebeek, K., 2004. Geographic variation in skull morphology of humpback dolphins (*Sousa* spp.). Aquat. Mamm. 30, 3–17.

Karczmarski, L., 1999. Group dynamics of humpback dolphins (*Sousa chinensis*) in the Algoa Bay region, South Africa. J. Zool. (Lond.) 249, 283–293.

Karczmarski, L., Cockcroft, V.G., Maclachlan, A., Winter, P.E.D., 1998. Recommendations for the conservation and management of humpback dolphins *Sousa chinensis* in the Algoa Bay region, South African perspective. Koedoe 41, 121–129.
Karczmarski, L., Cockcroft, V.G., Mclachlan, A., 1999a. Group size and seasonal pattern of occurrence of humpback dolphins *Sousa chinensis* in Algoa Bay, South Africa. S. Afr. J. Mar. Sci. 21, 89–97.
Karczmarski, L., Winter, P.E.D., Cockcroft, V.G., Mclachlan, A., 1999b. Population analyses of Indo-Pacific humpback dolphins *Sousa chinensis* in Algoa Bay, Eastern Cape, South Africa. Mar. Mamm. Sci. 15, 1115–1123.
Karczmarski, L., Cockcroft, V.G., McLachlan, A., 2000. Habitat use and preferences of Indo-Pacific humpback dolphins *Sousa chinensis* in Algoa Bay, South Africa. Mar. Mamm. Sci. 16, 65–79.
Keith, M., Peddemors, V.M., Bester, M.N., Ferguson, J.W.H., 2002. Population characteristics of Indo-Pacific humpback dolphins at Richards Bay, South Africa: implications for incidental capture in shark nets. S. Afr. J. Wildl. Res. 32, 153–162.
Keith, M., Atkins, S., Johnson, A.E., Karczmarski, L., 2013. Area utilization patterns of humpback dolphins (*Sousa plumbea*) in Richards Bay, KwaZulu-Natal, South Africa. J. Ethol. 31, 261–274.
Laran, S., Van Canneyt, O., Dorémus, G., Mannocci, L., Ridoux, V., Watremez, P., 2012. Distribution et abondance de la mégafaune marine dans le sud sud-ouest de l'océan Indien tropical. REMMOA—Océan Indien—Rapport final pour L'Agence des Aires Marines Protégées, 170 pp (in French).
Laroche, J., Ramananarivo, N., 1995. A preliminary survey of the artisanal fishery on coral reefs of the Tulear region (southwest Madagascar). Coral Reefs 14, 193–200.
Mendez, M., Subramaniam, A., Collins, T., Minton, G., Baldwin, R., Berggren, P., Särnblad, A., Amir, O.A., Peddemors, V.M., Karczmarski, L., Guissamulo, A., Rosenbaum, H.C., 2011. Molecular ecology meets remote sensing: environmental drivers to population structure of humpback dolphins in the Western Indian Ocean. Heredity 107, 349–361.
Mendez, M., Jefferson, T.A., Kolokotronis, S.O., Krützen, M., Parra, G.J., Collins, T., Minton, G., Baldwin, R., Berggren, P., Särnblad, A., Amir, O.A., Peddemors, V.M., Karczmarski, L., Guissamulo, A., Smith, B., Sutaria, D., Amato, G., Rosenbaum, H.C., 2013. Integrating multiple lines of evidence to better understand the evolutionary divergence of humpback dolphins along their entire distribution range: a new dolphin species in Australian waters? Mol. Ecol. 22, 5936–5948.
Peddemors, V.M., Thompson, G., 1994. Beaching behaviour during shallow water feeding by humpback dolphins *Sousa plumbea*. Aquat. Mamm. 20, 65–67.
Rakotoson, L.R., Tanner, K., 2006. Community-based governance of coastal zone and marine resources in Madagascar. Ocean Coast. Manag. 49, 855–872.
Razafindrakoto, Y., Andrianarivelo, N., Rosenbaum, H.C., 2004. Sightings, catches, and other records of Indo-Pacific humpback dolphins in the coastal waters of Madagascar. Aquat. Mamm. 30, 103–110.
Razafindrakoto, Y., Andrianarivelo, N., Cerchio, S., Rasoamananto, I., Rosenbaum, H.C., 2008. Preliminary assessment of cetacean incidental mortality in artisanal fisheries in Anakao, Southwestern region of Madagascar. Western Indian Ocean J. Mar. Sci. 7, 175–184.
Reeves, R.R., McClellan, K., Werner, T.B., 2013. Marine mammal bycatch in gillnet and other entangling net fisheries, 1990 to 2011. Endanger. Species Res. 20, 71–97.
Rice, D.W., 1998. Marine Mammals of the World. Systematics and Distribution. Society for Marine Mammalogy, Lawrence, KS.
Robards, M.D., Reeves, R.R., 2011. The global extent and character of marine mammal consumption by humans: 1970–2009. Biol. Conserv. 144, 2770–2786.

Robineau, D., Rose, J.-M., 1984. Les cétacés de Djibouti Bilan de connaissances actuelles sur la faune cétologique de la Mer Rouge et du Golfe d'Aden. Bull. Mus. Hist. Natl. Paris 6, 219–249. (In French).

Stensland, E., Carlén, I., Särnblad, A., Bignert, A., Berggren, P., 2006. Population size, distribution, and behavior of indo-pacific bottlenose (*Tursiops aduncus*) and humpback (*Sousa chinensis*) dolphins off the south coast of Zanzibar. Mar. Mamm. Sci. 22, 667–682.

Van Canneyt, O., Dorémus, G., Laran, S., Ridoux, V., Watremez, P., 2010. Distribution et abondance de la mégafaune marine dans le sud sud-ouest de l'océan Indien tropical. Campagne REMMOA—Océan Indien—Rapport préliminaire. Université La Rochelle (in French).

Vely, M., Bastid, J.J., Perri, M., Mayer, F.X., Rabearisoa, N., Ibrahima, S.M., Bertrand, N., 2009. Humpback whale monitoring at Isle Sainte-Marie, Madagascar, and in the Comoros Archipelago (Western Indian Ocean), from 1999 to 2008, based on whale-watching. In: Indian Ocean Cetacean Symposium, 18–20 July, 2009.

CHAPTER EIGHT

A Review of the Status of the Indian Ocean Humpback Dolphin (*Sousa plumbea*) in Pakistan

Muhammad Shoaib Kiani*, Koen Van Waerebeek†,[1]
*Institute of Marine Science, University of Karachi, Karachi, Pakistan
†Peruvian Centre for Cetacean Research (CEPEC), Lima, Peru
[1]Corresponding author: e-mail address: corewam@yandex.com

Contents

1. Introduction 202
2. Distribution in Pakistan 203
   2.1 General 203
   2.2 Indus River Delta 205
   2.3 Balochistan Coast 207
3. Stock Structure 208
4. Abundance 211
5. Habitat 212
6. Biology and Behaviour 213
   6.1 General Behaviour 213
   6.2 Feeding 214
   6.3 Sexual Behaviour 214
   6.4 Group Size and Composition 215
   6.5 Residency 217
   6.6 Interaction with Boats 217
7. Parasites and Pathology 217
8. Anthropogenic Threats and Conservation 218
   8.1 General 218
   8.2 Fisheries Interactions 218
   8.3 Water Pollution 221
   8.4 Other Threats and Legislation 222
9. Conclusions and Recommendations 223
Acknowledgements 224
References 224

Abstract

Limited historical and new information on Indian Ocean humpback dolphins, *Sousa plumbea*, in Pakistan are reviewed. Although present along most of the coast, *S. plumbea* concentrates in the mangrove-lined creek system of the Indus Delta

Advances in Marine Biology, Volume 72
ISSN 0065-2881
http://dx.doi.org/10.1016/bs.amb.2015.09.002

(Sindh), Miani Hor (Sonmiani Bay), Kalmat Lagoon, Gwadar and the Dasht River estuary (Gwater Bay, Jiwani). Other areas of distribution comprise the Karachi coast, Kund Malir, Ormara and Pasni. In the Indus Delta, 46 small-boat surveys conducted monthly (minus July and October) in 2005–2009, documented 112 sightings (439 individuals) in major creeks, smaller channels and nearshore waters. Group sizes ranged from 1–35 animals (mean = 3.92 ± 4.60). Groups of 1–10 animals composed 91% of total (27.9% single animals). An encounter rate of 0.07–0.17 dolphins km^{-1} lacked a significant trend across survey years. A discovery curve remained steep after 87 dolphins were photo-identified, suggesting the population is vastly larger. In Sonmiani Bay, Balochistan, during 9 survey days in 2011–2012, group sizes ranged from 1–68 animals (mean = 11.9 ± 13.59; $n = 36$), totalling 428 dolphins. Incidental entanglements, primarily in gillnets, pollution (especially around Karachi), overfishing and the ship breaking industry in Gaddani, pose major threats. Incidental catches occur along the entire Pakistani coast. Of 106 stranded cetaceans, 24.5% were *S. plumbea*. Directed takes in Balochistan, driven by demand for bait in shark fisheries, have reportedly declined following dwindling shark stocks. Habitat degradation threats include depletion of prey and increased maritime traffic. Domestic sewage and solid waste pollution are predominant on the Balochistan coast, especially at Miani Hor, Kund Malir, Ormara, Kalmat Lagoon, Pasni, Gwadar and Jiwani. An exhaustive habitat assessment combined with appropriate fishery management is the only way to safeguard the future of *S. plumbea* in Pakistan.

1. INTRODUCTION

Despite its widespread occurrence in a variety of coastal habitats in the Indian Ocean, studies on the biology and ecology of the Indian Ocean humpback dolphin, or plumbeous dolphin, *Sousa plumbea* (Cuvier, 1829), are relatively limited. Research has targeted mainly South Africa (Best, 2007; Karczmarski, 1996, 1999, 2000; Ross, 1984, 2002; Ross et al., 1994), Zanzibar (Stensland et al., 2006), Mozambique (Guissamulo and Cockcroft, 2004), Pakistan in the 1970s (Pilleri, 1972, 1973, 1975; Pilleri and Gihr, 1972, 1974; Pilleri and Pilleri, 1979), India (Parsons, 1998; Sutaria and Jefferson, 2004), and the Arabian region (Baldwin et al., 2004).

The taxonomy of the genus *Sousa* experienced great flux over the decades, with recent insights supportive of multiple species, largely in accordance with historical morphology-based conclusions. A view (e.g. Jefferson and Karczmarski, 2001), that lumped all Indo-Pacific humpback dolphins into one widely distributed, highly variable species *Sousa chinensis*, is now considered obsolete (Jefferson and Rosenbaum, 2014; Jefferson and Van Waerebeek, 2004; Parra and Ross, 2009). The western Indian Ocean *S. plumbea* is distributed from South Africa's Cape Province to India and possibly through the Bay of Bengal where it is thought to overlap with

S. chinensis (Jefferson and Van Waerebeek, 2004; Parra and Ross, 2009; Sutaria and Jefferson, 2004).

In Pakistan, exploratory field research implemented by the University of Berne in the 1970s suggested that the Indus Delta may have one of the largest extant populations of *S. plumbea* (Pilleri, 1972, 1973, 1975; Pilleri and Gihr, 1972, 1974). Despite this, limited new field research has been conducted since that time, mainly by the University of Karachi based Cetacean Conservation Pakistan Project (CCP) research team, from 2005 to 2009, providing many of the new insights discussed herein (see also Gore et al., 2012; Kiani, 2014). In addition, a baseline study on the presence of *S. plumbea* in Miani Hor, Sonmiani Bay (25°10′N, 66°30′E), Balochistan, was conducted from July 2011 to June 2012 by the Sonmiani Development Organization (SDO, 2012).

Other than this, Pakistan's humpback dolphins have not been studied in any detail, although they have been reported from several other areas including the Karachi coast, Kund Malir, Ormara, Kalmat Lagoon, Pasni, Gwadar and Jiwani (Gwater Bay) at the Pakistan–Iran border. The current paper reviews the extant knowledge on the biology and status of *S. plumbea* in Pakistani waters with the objective to stimulate studies on this threatened mammal which are especially important and time-sensitive considering the fast deterioration of the environment within its most preferred habitat in Pakistan (the Indus Delta) and along other parts of Pakistan's coast.

2. DISTRIBUTION IN PAKISTAN

2.1 General

Sousa plumbea is known to be present in varying densities along most of Pakistan's 990 km long coastline (Figure 1); however, specific sightings data resulting from dedicated boat surveys and studies of beach cast specimens are scant, except for in the Indus River Delta, Sonmiani Lagoon and Makran coast (Ahmad, 1998; GoP, 2007; Gore et al., 2012; Hasan and Ahmad, 2006; Kiani, 2014; Niazi, 1990; Pilleri and Gihr, 1972, 1974; Roberts, 1977, 1997; Roberts et al., 1983; SDO, 2012; WWF, 2008; Zbinden et al., 1977). Sightings and stranding records in 2005–2009 from the entire coastline demonstrate presence at Karachi coast (24°47′N, 66°59′E), Miani Hor, Sonmiani Bay (25°10′N, 66°30′E), Ormara (25°12′N, 64°38′E), Kalmat Hor (25°23′N, 64°02′E), Pasni (25°15′E, 63°28′E), Gwadar (25°06′N, 62°19′E) and Jiwani (25°02′N, 61°43′E).

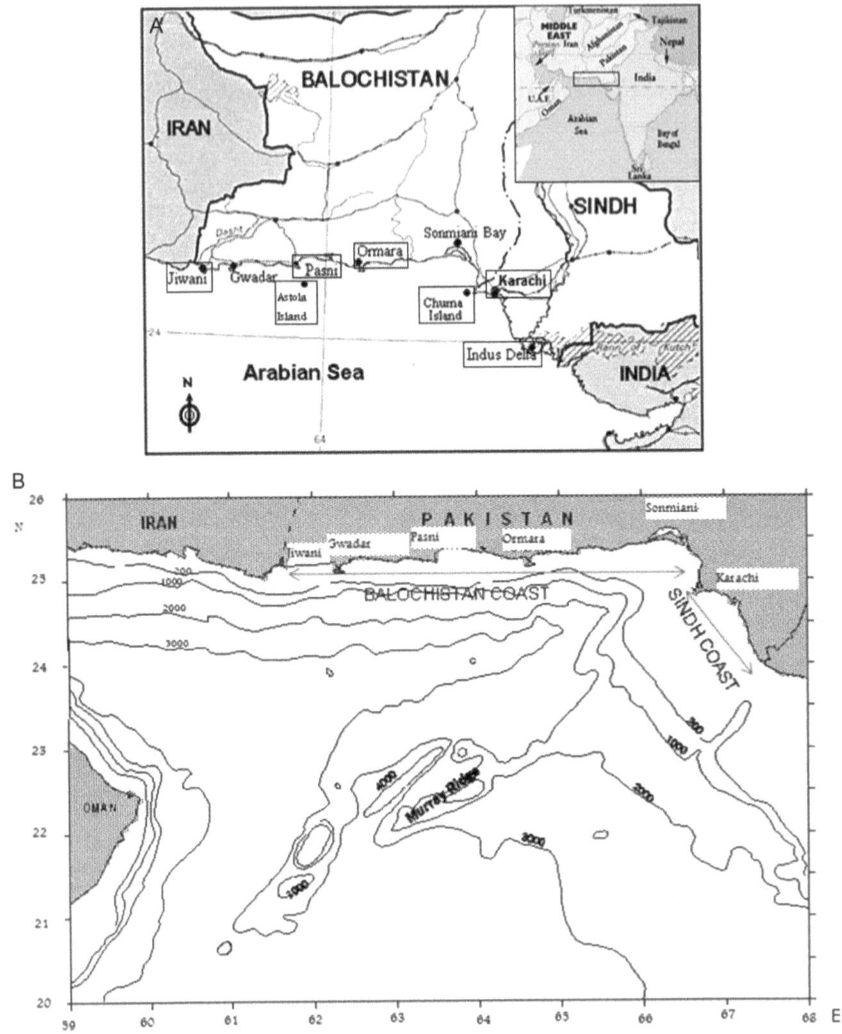

Figure 1 (A) Map showing the Sindh and the Balochistan coastline of Pakistan with all the major coastal areas marked. The Sindh coast in the east extends from the border with India to Churna Island, while the Balochistan coast in the west extends from west of Churna Island and Hub River estuary to Jiwani at Pakistan–Iran border; (B) Hydrography of the Sindh and the Balochistan coastlines of Pakistan.

It is premature to evaluate whether humpback dolphins present in these different areas all represent locally resident populations or whether they are linked by migration, especially for areas in relatively close proximity to each other, such as the Indus Delta and the Karachi coast and several locations along the 670 km long Balochistan coast in western Pakistan.

2.2 Indus River Delta

The Indus River Delta covers an area of about 41,440 km^2 and is among one of the most globally important coastal environments in Pakistan (Ahmad, 1998). It was declared a Ramsar site on 5 November 2002, due to its biological diversity and for being the largest strand of mangrove forest along the Pakistan coast (Ahmad, 1998; Siddiqui et al., 2008). The estuarine creeks and mangrove forests form major nursery grounds for many species of fish and shrimps (Shah et al., 2007). The delta is rich in micro-invertebrates, shrimps, fin-fishes, reptiles, birds and mammals. Recent surveys indicate that humpback dolphins are still common in the Indus Delta (Gore et al., 2012; Kiani, 2014). Between November 2005 and May 2009, a total of 46 small-boat survey day trips were carried out in various major creeks and smaller interconnecting channels of the Indus Delta (Figure 2), during every month of the year except July and October. However, most of the effort (Table 1) was concentrated during the northeast monsoon (November–February) and the spring inter-monsoonal period (March–May), due to favourable sea conditions. Total

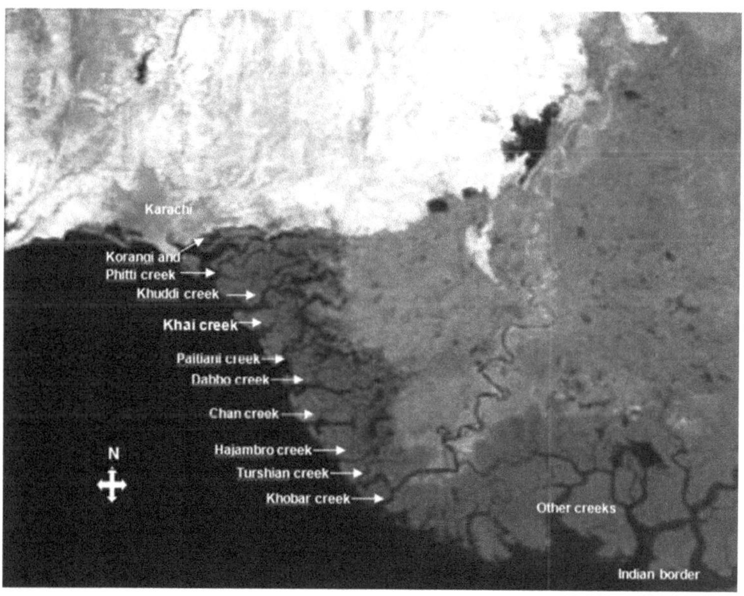

Figure 2 The Indus Delta creek system with names of surveyed creeks where Kiani (2014) confirmed Indian Ocean humpback dolphins, *Sousa plumbea*. The frequent presence of humpback dolphins in "other creeks" near the border with India is supported by anecdotal records from fishermen and other locals.

Table 1 Distribution of Small-Boat Survey Effort for Indian Ocean Humpback Dolphins, *Sousa plumbea*, Across Months and (Monsoon) Seasons, 2005–2009

Month	Jan	Feb	Mar	Apr	May	Jun	Jul	Aug	Sep	Oct	Nov	Dec	
Season	NE	NE	IMS	IMS	IMS	SW	SW	SW	SW	IMA	NE	NE	Total
Number of surveys	5	4	2	3	9	1	0	2	1	0	6	13	46

NE, northeast monsoon; IMS, spring intermonsoonal period; SW, southwest monsoonal period; IMA, autumn intermonsoonal period. See also Gore et al. (2012) and Kiani (2014).

Table 2 Survey Parameters Including Encounter Rates for Indian Ocean Humpback Dolphins, *Sousa plumbea*, During Small-Boat Surveying in the Indus Delta Creek System, 2005–2009

Survey Series	Year	Number of Surveys	Effort Distance (km)	Number of Sightings	Cumulative Number of Dolphins Seen	Number of Sightings (km^{-1})	Number of Dolphins (km^{-1})
1	2005	06	543.4	22	38	0.04	0.07
2	2006	08	342.0	08	51	0.02	0.15
3	2007	12	956.6	43	172	0.04	0.17
4	2008	09	523.9	22	69	0.04	0.13
5	2009	11	656.4	17	109	0.03	0.17
Total/ mean	–	46	3022.1	112	439	0.03	0.14

See Gore et al. (2012) and Kiani (2014).

search effort was 209 h, in addition ca. 47 h were spent following and photographing an estimated 439 individuals (sum of estimated group sizes) during 112 encounters (Table 2).

Humpback dolphins were found in the upper, middle and lower sections of creeks, as well as in nearshore waters contiguous to the creek mouths (Figure 3). Dolphins were always encountered in Turshian Creek whenever it was surveyed ($n=10$). This creek is very close and to the west of Khobar Creek, which is the creek through which the Indus River discharges into the sea (see Figure 2; Gore et al., 2012; Kiani, 2014).

Besides *S. plumbea*, Indo-Pacific bottlenose dolphins *Tursiops aduncus*, and Indo-Pacific finless porpoises *Neophocaena phocaenoides*, are also found in inshore waters, but are never observed inside the Indus Delta creeks (Ahmad, 1998; Shah et al., 2007; Siddiqui et al., 2008). Presently, biodiversity of the Indus Delta, including humpback dolphins, is under a multitude

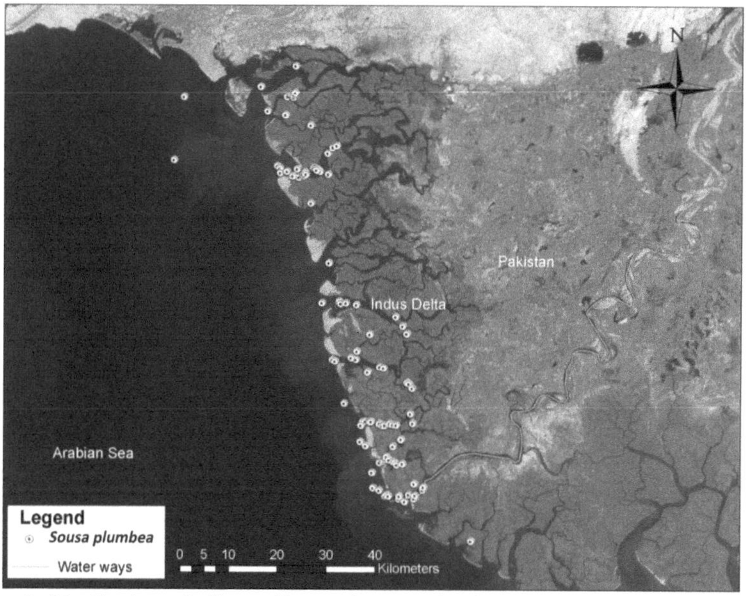

Figure 3 Distribution of all on-effort sightings of Indian Ocean humpback dolphins, *Sousa plumbea*, in the Indus Delta creek system, southeast Pakistan, from November 2005 to May 2009. The Indus River is outlined (long blue line; light grey in the print version). See also Gore et al. (2012) and Kiani (2014).

of stresses due to the reduced influx of freshwater from the Indus River, pollution (domestic, agricultural and industrial sewage), land clearance, over-exploitation of fish resources, sea water intrusion leading to increased salinity, and unsustainable exploitation of mangroves by coastal communities (Ahmad, 1998; Shah et al., 2007).

2.3 Balochistan Coast

A total of 36 humpback dolphin sightings were recorded in waters 3–13 m deep (mean = 7.0 ± 2.79 m) at Miani Hor, Sonmiani Bay (25°10′N, 66° 30′E), Balochistan coast, during nine survey days from July 2011 to June 2012 (Figure 4) as conducted by the SDO (2012). The mean group size was 11.89 ± 13.59 animals (median = 7; n = 36) with the largest group, some 68 animals, encountered in 6 m deep water on 12 November 2010. Dominant behaviour included feeding (77.8%), followed by travelling (25%), surface active behaviour (2.8%), and resting/socializing (8.1%). Threats included by-catch in fisheries (particularly gillnets), degradation of habitat from infrastructure development (e.g. jetty construction),

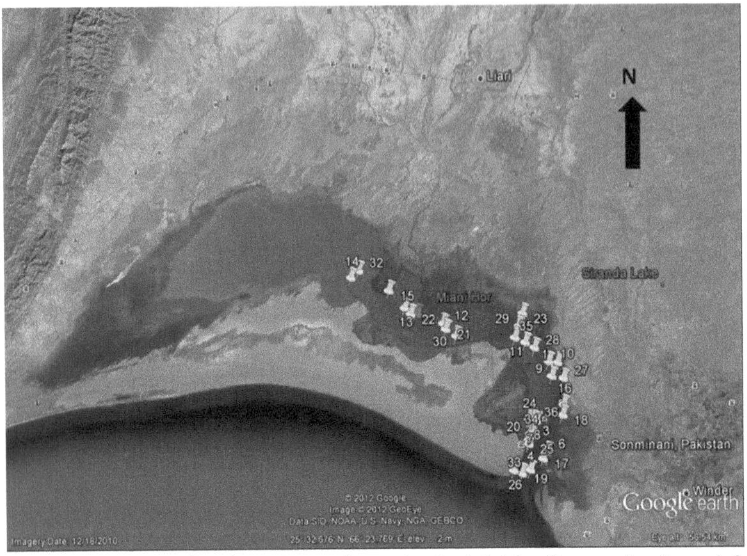

Figure 4 Distribution of Indian Ocean humpback dolphin, *Sousa plumbea*, sightings ($n = 36$) made during nine boat surveys around Miani Hor, Sonmiani Bay, Balochistan, from July 2011 to June 2012 (see also SDO, 2012).

mangrove deforesting, unsustainable fisheries harvest, increasing marine traffic (fishing boats) and contamination (solid waste, domestic sewage). This humpback dolphin population is also targeted by a limited number of dolphin-watching excursions annually by tourists from Karachi, under the SDO umbrella.

3. STOCK STRUCTURE

Zhou et al. (1980) argued early on that morphological differences between populations in the Indian and western Pacific Oceans warranted treating them as different species, *S. plumbea* and *S. chinensis*, respectively (see Jefferson and Rosenbaum, 2014). A multivariate analysis of cranial morphometrics clearly separated *S. plumbea* from both, the Atlantic humpback dolphin, *S. teuszii* and *S. chinensis* (Jefferson and Van Waerebeek, 2004), supported by recent genetics studies (Jefferson and Rosenbaum, 2014; Mendez et al., 2013), which furthermore recognized an Australian species, the Australian humpback dolphin, *S. sahulensis*. Humpback dolphins in Pakistan are manifestly *S. plumbea* with their large distinctive hump and predominantly grey, "plumbeous" colouration. Intraspecific variation of *S. plumbea* remains largely unstudied, but considerable stock structure may exist, including at subspecific level. Based on observations that humpback dolphins in Pakistan, similar to Iran and western India, show mixed phenotypic features, i.e., distinct dorsal humps found in *S. plumbea*, but heavy spotting on different body parts as in *S. chinensis*, Gore et al. (2012) suggested that Pakistan population(s) may represent an intermediate form.

Pilleri and Gihr (1972) first photographically documented the cranial and postcranial skeleton of humpback dolphins from Pakistan and suggested interspecific differences in the thoracic region (12 Th.v., 26% of vertebral column length in *S. chinensis* vs. 11 Th.v., 22–24% in *S. plumbea*) (Pilleri and Gihr, 1974); however, this was based on a small sample size. Statistically, adequate sample sizes of osteological specimens and genetic samples from the entire range will be required to allow for geographic variation analysis.

Observations indicate that neonates and calves are light grey with even paler, often whitish, rostra and with a slightly bigger head in proportion to body size (Figure 5A). Juveniles appear a darker hue of grey than calves, with, compared to adults, few markings on their body and usually no pink spotting on fins, beak or flukes, and no mottling ventrally. Adult animals are dark grey, sometimes brownish grey, often with pink spotting on rostrum, dorsal fin (especially the tip), underside and flukes. Larger, apparently older, animals show markedly increased spotting as compared to juveniles. Adults are usually well marked, apt for photo-ID studies, i.e. with frequent cuts, nicks and serrations on the dorsal fin, and scars of former injuries over much of the body. The dorsal hump is well developed and is most pronounced in the largest, presumably older, individuals, i.e. those dark grey in colour, with robust bodies and strong musculature, specifically in the peduncle area (Figure 5B and C; and see Kiani, 2014).

Figure 5 Examples of individual and ontogenetic variation in external features of Indian Ocean humpback dolphins, *Sousa plumbea*, in Indus Delta, Pakistan. (A) Three animals, one of which is a light grey-coloured calf, in Dabbo Creek, 18 December 2007.
(Continued)

Figure 5—cont'd Note the calf has a proportionally developed hump; (B) Adult with strongly developed dorsal hump, moderate speckling and pinkish (grey in the print version) fin tip, near Phitti Creek, 15 May 2009; (C) Leading individual shows an unusually pronounced dorsal hump and some speckling, near coast of Keti Bunder, 29 May 2008; (D) Mating group in trio formation, Khuddi Creek, 28 January 2009.

4. ABUNDANCE

Encounter rate varied from 0.07 to 0.17 humpback dolphins km^{-1} of small-boat surveying in the Indus River Delta across five study years, 2005–2009 (see Table 1; Gore et al., 2012; Kiani, 2014), without showing any significant trend (Pearson $R=0.2365$; $p=0.701$). Mean encounter rate per distance surveyed was 0.14 humpback dolphins km^{-1} (439 animals over 3022 km survey effort).

No reliable abundance estimates exist for Pakistan's *S. plumbea* populations. Pilleri and Pilleri (1979) explored 30 km of the northern delta and encountered about 100 animals. They estimated that if the delta is 130 km from north to south, the total population from Gizri Creek to the Indian border should be "some 430 or perhaps 500 animals". The guesstimate of 500 individuals was later adapted by Ross et al. (1994) and others. In 2005–2009, a preliminary minimum population estimate obtained by cataloguing uniquely marked (photo-identified) animals of the Indus River Delta population was 87 (see Gore et al., 2012; Kiani, 2014). The number of resightings was extremely low ($n=5$), while the steep discovery curve (Figure 6) indicated that new animals were still being added to the catalogue at a high rate at the end of the study period. Combined, this is strongly

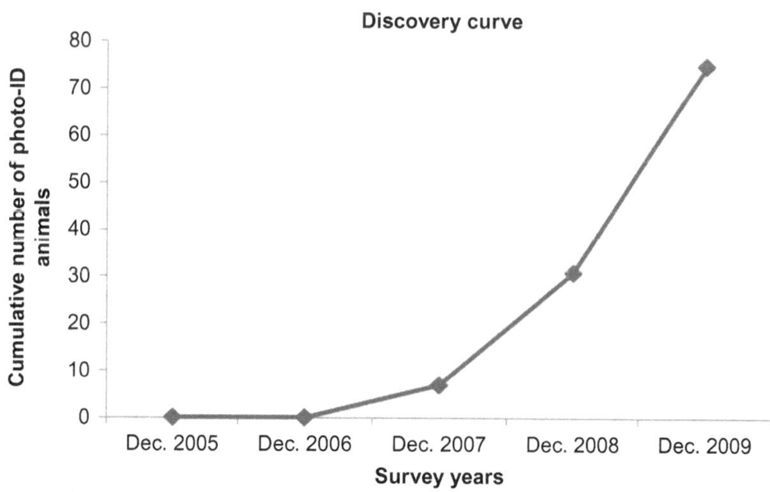

Figure 6 Discovery curve showing cumulative number of uniquely identified humpback dolphins, *Sousa plumbea*, during surveys conducted from 2005–2009 in the Indus Delta (see Gore et al., 2012; Kiani, 2014). No sufficiently high-quality photos were available for reliable individual identification in 2005–2006.

suggestive of a large population in the Indus Delta. In view of incomplete recent spatial and temporal coverage, the guesstimate by Pilleri and Pilleri (1979) of ca. 500 animals, conceivably may still be useful as general reference. If confirmed, the Indus Delta could contain the largest known population of *S. plumbea* in the world, besides Algoa Bay, South Africa, which was estimated as ca. 466 dolphins (Karczmarski, 2000; Karczmarski et al., 1999).

5. HABITAT

The Sindh coast provides shallow water habitat, as preferred by *S. plumbea*, up to a long distance from shore (see Figure 1B). The wide continental shelf ranges from about 77 km at Karachi, up to 160 km near Kori Creek mouth (Gore et al., 2012; see Figure 1B). In the Indus Delta, humpback dolphins were frequently sighted in all the major mangrove creeks, in some smaller interconnecting channels and in nearshore waters (see Gore et al., 2012; Kiani, 2014). Sea surface temperatures at sightings ranged from 12–31 °C with a mean of 23.1 °C $\pm$ 1.2, while salinities ranged from 10–38 ppt. Lowest salinity was recorded in Khobar Creek during SW monsoon after great inflow of rainwater. The dolphins were recorded both in clear and turbid water. Pilleri and Pilleri (1979) noted that *S. plumbea* preferentially remained in muddy, yellow water and "very seldom" was seen in the green water of the Arabian Sea, an environment where the Indo-Pacific bottlenose dolphin, *T. aduncus*, was common. *Sousa plumbea* were observed using smaller interconnecting waterways to move between major creeks in the Indus Delta, but also to take open sea routes between creeks. Prey movement during changing tidal levels could be a driving force behind such activities. In 2005–2009, one humpback dolphin group was seen 20 km from shore, and another distant sighting occurred ca. 9 km from the nearest land, but all other sightings were recorded within a few 100 m from shore, or inside the creeks (see Figure 3; Kiani, 2014).

Mean depth at sightings was 11.24 m (SD = 6.47 m; $n = 105$; range 1.5–30 m). Most dolphins were observed in depths ranging from 1 to 20 m with fewer dolphins encountered in deeper water (Figure 7). A strong negative correlation (Pearson R = 0.822; $n = 6$; $p = 0.045$) was found between water depth and cumulative number of dolphins (Kiani, 2014).

Khobar Creek, which receives fresh water from the Indus River, apparently was heavily used, as three sightings with some of the larger group sizes in Pakistan ($n = 16$, 22, 35) were recorded there (Kiani, 2014). Comparable observations were made in the Sunderbans area, Bangladesh, a delta that also receives fresh water from upstream (Smith et al., 2008).

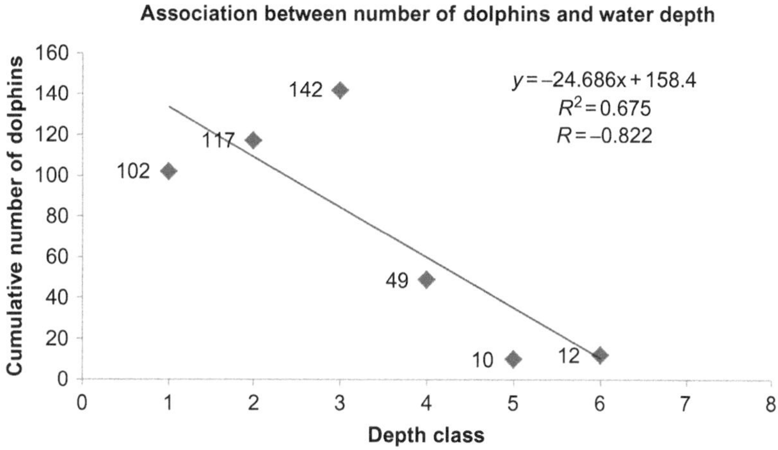

Figure 7 Negative correlation between cumulative number of Indian Ocean humpback dolphins, *Sousa plumbea*, and water depth classes (1 = 0–5 m, 2 = 6–10 m, 3 = 11–15 m, 4 = 16–20 m, 5 = 21–25 m, 6 = 26–30 m) (Kiani, 2014).

Dolphins were sighted in inshore waters contiguous to the Indus Delta, particularly during winter months, suggesting that they possibly use coastal routes to move between creeks and interconnecting channels (Kiani, 2014). The negative correlation between numbers of dolphins and water depth (see Figure 7) is in line with reports from other areas (Ross et al., 1994). However, Afsal et al. (2008) reported humpback dolphins in 50 m deep waters off the Arabian Sea coast of India. All *S. plumbea* observed by Parsons (1998) in Goa, India, were coastal or associated with an estuary. Coastal animals occurred within 3–4 km from shore or within 8 km of the mouth of a river, and all animals were sighted in waters less than 10 m deep (Parsons, 1998).

6. BIOLOGY AND BEHAVIOUR

6.1 General Behaviour

Kiani (2014) observed multiple behaviour patterns, with the most common (% frequency of occurrence): travelling (52%), feeding/foraging (47%) and social or surface active behaviours (33%) (Figure 8). Actions such as breaching, tail slapping and surfacing with full head showing were also commonly observed. On 24 occasions (21.4%), humpback dolphins were sighted, when sea birds (terns, gulls, cormorants) were present near-by. Mean dive time lasted 55.4 s (SD = 55.6 s; median = 29 s; range 5–255 s; $n = 51$).

Dolphins were sighted at all hours throughout the day but were seen moving between creeks primarily during evening hours. Larger group sizes

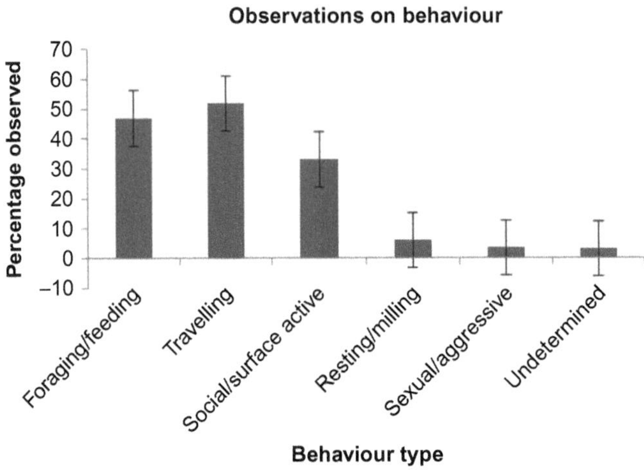

Figure 8 Behaviours recorded in Indian Ocean humpback dolphins, *Sousa plumbea*, in the Indus Delta, as percentage observed in total groups (with standard error bars).

recorded during mornings and midday hours, as compared to evening sessions, may be formed for foraging purposes (co-operative feeding) and for resting and socializing (Kiani, 2014).

Karczmarski (1999) reported seasonal variations in occurrence, abundance and group size in humpback dolphins from South Africa. In Pakistan, most survey effort in 2005–2009 was concentrated during the NE monsoon (November–February) and spring intermonsoonal period (March–May) in view of favourable sea conditions (see Gore et al., 2012; Kiani, 2014; SDO, 2012); therefore, seasonality cannot yet be evaluated.

6.2 Feeding

Feeding behaviour showed very little diversity (Kiani, 2014). In smaller groups, dolphins kept varying distances from each other while chasing prey in different directions. The larger groups were spread over a wide area, and were divided into several subgroups while chasing prey without any specific pattern. No information is available on prey species of *S. plumbea* in Pakistan. Off Natal, South Africa, all prey items were fish, 61% littoral or estuarine species and 25% demersal species primarily associated with reefs (Ross et al., 1994).

6.3 Sexual Behaviour

Courtship and sexual/mating behaviour were recorded during four sightings (3.6%) in the Indus Delta, only during the northeast monsoon period (January and February) (see Kiani, 2014). Median group size for these pods was 9.5

(range, 4–19) and all had a mixed age composition. The individuals that engaged in sexual behaviour tended to remain well separated from the core group (see Figure 5D). During courtship, the pair moved quickly one after another while changing the positions in a helical pattern and was involved in relatively increased aerial and social activity. Animals aligned with each other and made frequent body contacts while spending 10–20 s at the surface. During this, the presumed female at times was seen swimming slowly upside down, exposing flippers and occasionally opening its beak pointed towards the sky. Next, the animals moved some 50–100 m from their previous location and the described sequence was repeated, as evident from multiple observations. The actual mating behaviour following courtship involved slow, subsurface movements in an upside-down posture (possibly by females when approached by males). After body alignment, the presumed male stayed on top of the female, interlocked for about 20–30 s and the pair moved slowly in this position in apparent copulation. The animals separated after this for several minutes. Such copulatory aligning was repeated several times (see Kiani, 2014). Sexual behaviour was estimated to last for more than 2 h. Similar behaviour has been reported from Algoa Bay (Best, 2007).

On three occasions, only two animals were seen engaged in sexual activity, but on one occasion, competition for mating was observed when two presumed males chased one female (see Kiani, 2014). The pursuing individuals, swimming at high speed in a circular pattern, displayed a lot of high-energy activity including splashes, leaps and body contact. It could not be ascertained whether both males managed to mate with the female or whether only one dominant male repeated mating. At least one pair was involved in courting behaviour, as described above and only one pair was generally engaged in mating at a given point in time. Other animals were not observed to interrupt such mating sessions. Sexual behaviour was very similar on all four occasions. Occasionally, an atypical posture was observed, when an animal was lying motionless at the surface on its back exposing throat, belly and flippers and at times briefly opening its beak. When interrupted for respiration the dolphin often resumed that same posture. Sometimes a vertical position was displayed with the tail and peduncle held above the surface for some moments before gently diving.

6.4 Group Size and Composition

In the Indus River Delta, group sizes varied considerably from 1 to 35 animals (mean $= 3.92 \pm 4.60$; median $= 2$; $n = 111$) (see Kiani, 2014). Singletons were the most frequent group size class (27.6%) with gradually declining frequency for larger groups (Figure 9) and typically consisted of large solitary animals, probably mostly adult males, with well-developed humps and musculature, particularly in the peduncle. Overall, groups of 1–10 animals made up 91% of total sightings. Generally in the bigger groups, some subgroups stayed close to each other and merged occasionally while chasing prey and feeding. Group composition was highly variable, i.e., only adults, only juveniles, mixed adults and juveniles, mother and calf pairs and undetermined (Figure 10). The

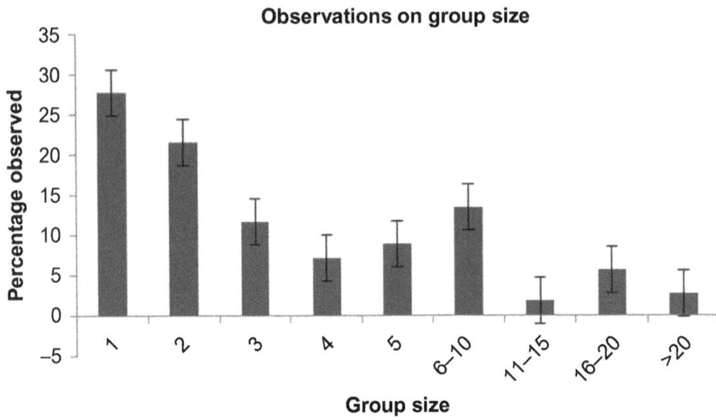

Figure 9 Distribution of group sizes (as percentage of total observations), with standard error bars, in Indian Ocean humpback dolphins, *Sousa plumbea*, observed in the Indus Delta (see Kiani, 2014).

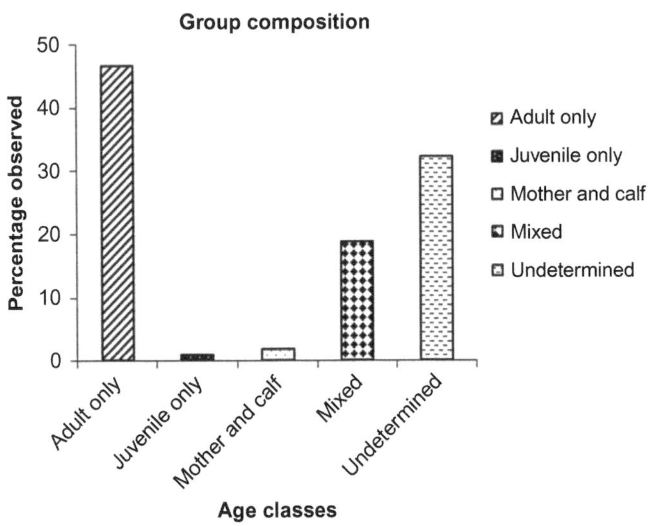

Figure 10 Group composition (age classes) in Indian Ocean humpback dolphins, *Sousa plumbea*, observed in the Indus Delta, 2005–2009. See also Kiani (2014).

presence of neonates/calves in a large proportion of sightings (21.9%) is a likely indication of the relevance of the Indus Delta as calving and nursing area for the species.

Mean group size (3.92) was comparable to those observed along the west coast of India, i.e., 3.9 and 9.5 dolphins, respectively in shallow and deeper waters (Sutaria and Jefferson, 2004) and 2.6 individuals in Goa, western India (Parsons, 1998). In larger groups, animals of all age classes were present, with adults being the predominant class, in accordance with findings from elsewhere (Goodwin, 1997; Jefferson and Karczmarski, 2001).

6.5 Residency

Residency has been documented in most places where *S. plumbea* has been studied, including Algoa Bay and KwaZulu-Natal, South Africa (Karczmarski et al., 1999b), Zanzibar in Tanzania (Stensland et al., 2006) as well as in India (Parsons, 1998). As only five animals were resighted, each only once, no quantification of residence pattern or site fidelity in the Indus Delta was possible.

Humpback dolphins are seen throughout the year in the Indus Delta. Individual dolphins may be resident and use the area opportunistically. Furthermore, humpback dolphins of the Indus Delta may be using some areas of the Indian coastline adjacent to the Pakistan border and vice versa, such as Gulf of Kachchh in Gujarat State, and even beyond. There is also cross-border movement potential for animals observed in Dasht River estuary around the coastal town of Jiwani (Gwater Bay) at the Pakistan–Iran border (see Kiani, 2014). Comparison with photo-ID databases between both neighbouring countries could help improve the understanding of any long-range, cross-border movements.

6.6 Interaction with Boats

Humpback dolphins do not approach small local wooden fishing vessels while moving, but do sometimes approach when they are stationary with the engine switched off. Bow riding was never observed, although occasionally animals may ride the stern wave of a boat.

7. PARASITES AND PATHOLOGY

No information is available on the pathology and helminth or other parasites of *S. plumbea* in Pakistan or the Arabian Sea. During photo-identification work, no striking macroscopical skin conditions were noted.

However, a detailed examination of photographs, especially for dolphins from the heavily contaminated waters near Karachi, might provide insight regarding skin condition and overall health in these dolphins. Several dolphin communities inhabiting polluted coastal waters world-wide have developed multiple cutaneous diseases, some life-threatening (e.g. Van Bressem et al., 2009, 2015). Lane et al. (2014) reported lobomycosis-like disease in a *S. plumbea* from KwaZulu-Natal Coast.

8. ANTHROPOGENIC THREATS AND CONSERVATION

8.1 General

There are likely limits to the adaptability of dolphins to habitat deterioration (Wells, 1993); therefore, long-term monitoring and quantification are essential. As a cetacean with an obligate nearshore habitat, ecologically constrained to a narrow coastal strip, *S. plumbea* is facing severe threats to its conservation in Pakistan, as elsewhere, due to negative impacts from accelerated coastal development and associated human pressure. Known and suspected threats include accidental entanglements in fishing gear, habitat degradation, pollution, probable competition for food due to widespread use of non-selective fishing gears and increasing vessel traffic with associated noise pollution and risk of boat strikes (Iqbal, 2014; Kiani, 2014). Coastal areas in close proximity to industries, port installations (e.g. Port Qasim) and major human settlements (e.g. Rehri and Ibrahim Hyderi villages) are of highest concern.

8.2 Fisheries Interactions

Along the entire coastline of Pakistan, accidental entanglement in fishing nets, mostly gillnets, appears to be the principal threat to humpback dolphins (Figure 11A). The impact is most severe during peak fishing season, i.e., the northeast monsoon from November to February. From 2005 to 2009, humpback dolphins were one of the most commonly stranded cetacean species in Pakistan (24.5%, $n=106$) (Kiani, 2014).

A one year study (2010–2011) carried out along the Gwadar west Bay, located at 25°02′N, 62°19′E on the Balochistan coast, showed that out of 73 dead dolphins recorded stranded on a 20 km stretch of coastline, 22 (30.1%) were Indian Ocean humpback dolphins (A. Shah, IUCN-Pakistan, personal communication, June 2011).

Figure 11 Examples of fisheries interactions with *Sousa plumbea* in Pakistan. (A) Live humpback dolphin entangled in local fishers net, rescued and released in Hajamro Creek near coastal town of Keti Bunder, Indus Delta, 6 June 2008; (B) Illegal fisheries may deplete prey species of creek-dwelling humpback dolphins, such as this bottom trawling inter Phitti and Khuddi Creeks, Indus Delta, 15 April 2009.

Fisheries in Pakistan are generally open access, which has resulted in overcapacity of the fishing fleet. A 2011, boat census in the Indus Delta carried out by Pakistan's Marine Fisheries Department (MFD) and WWF-Pakistan has shown ca. 167% increase in number of boats since the previous census of 1986, from 6488 to 17,325 fishing boats (M. Khan, former Director General MFD, personal communication, 2011). Unsustainable exploitation of coastal fishing resources occurs by legal and illegal fishing, specifically in the Indus Delta (Ahmad, 1998; Hasan and Ahmad, 2006; Siddiqui et al., 2008). Artisanal fisheries in Pakistan mainly operate in coastal and nearshore areas also occupied by and target small pelagic, bentho-pelagic and demersal fish species (Ahmad and Hasan, 1997; FAO, 2005; Hasan, 1993; Van Zalingae et al., 1987; WWF, 2005). Only about 25% of the Pakistan fishing fleet operates in deeper offshore waters (WWF, 2005). Use of gillnets of different types is very common in nearshore as well as offshore fisheries, with high risk for cetacean entanglements.

With low catch per unit effort (CPUE) and establishment of a big market for "trash fish", there is a tendency to use those fishing gears that require less input in terms of manpower, time and costs. As a result, the use of banned fishing gears such as *Bullo* (estuarine set bag net), *Gujjo* (small-mesh size gillnet in various forms) and *Katra* nets (purse seine) targeting small pelagic species is widespread, particularly along the Sindh coast (Abildgaard and Khan, 1986; Gestsson, 1994). In all these nets, indiscriminate bulk catches of juveniles of commercial fish species are reported. In 1998 and 2003, the Government of Pakistan banned *Bullo*, *Gujjo* and *Katra* nets of less than 1.5 cm mesh size and nets over 250 cm in length and 12.2 m in width in the creeks and coastal waters of Sindh, a known habitat for humpback dolphins. *Katra* nets less than 5 cm mesh are prohibited in all creeks and inshore territorial waters. Despite this, their use is extensive and is causing serious damage to coastal habitats. Fishing at the time of spawning is also common along the Sindh coast. Trawlers use bottom dragging, which disturbs the sea bed and causes long-term damage (Figure 11B; M.S. Kiani, personal observations). Fish stocks are depleted or are moving away from the coast to deeper waters, while shrimp stocks are shrinking. Overfishing, well documented in the mangrove areas and the Indus Delta along the Sindh coast (Hasan, 1993; Van Zalingae et al., 1987), resulted in a CPUE decrease, mainly because of massive takes of juvenile fishes and shrimps.

At fisher community interviews, 28% of the fishers identified by-catch as a major threat to cetaceans, particularly coastal species such as humpback dolphin and Indo-Pacific finless porpoise, as their habitats directly overlap with fishing grounds (Kiani, 2014).

Entangled animals are used opportunistically for various purposes. Fishers in the Indus Delta and Balochistan coast use humpback dolphin oil for treatment of rheumatism and also apply it on boats for waterproofing. Some limited evidence of the use of dolphins as shark bait was recorded in Balochistan. However, according to fishers, this practice has reduced considerably due to fast dwindling stocks of sharks, a consequence of overfishing. Indo-Pacific finless porpoise was primarily used as shark bait but also humpback and bottlenose dolphins. Species that closely interact with boats, such as spinner dolphins (*Stenella longirostris*), were killed using hand-held harpoons called *Kaabar* (Kiani, 2014; A. Rahim, Coastal Scientific Society Gwadar Balochistan, personal communication to M.S. Kiani, December 2012; A. Shah, IUCN, personal communication to M.S. Kiani, June 2012). Systematic monitoring will be desirable as these examples suggest that Pakistan's *S. plumbea* population(s) may be severely affected by interactions with coastal artisanal fisheries. Evidence is lacking, at least presently, of human consumption of cetacean meat (marine bushmeat), which was, and may still be, a major concern in Calicut on the SW coast of India (Lal Mohan, 1994) and continues to be a primary threat to *S. teuszii* conservation in western Africa (Ayissi et al., 2014; Van Waerebeek et al., 2004; see Collins, 2015).

8.3 Water Pollution

Industrial and domestic sewage pollution in coastal waters and tidal creeks is a major concern specifically around the populous and rapidly expanding industrial city of Karachi, on the northwestern most part of the delta. In the southeastern part, major pollution sources include pesticides like organochlorines (e.g. DDT), various hexachlorocyclohexanes and domestic water flushed into the Indus River from inland areas (Ahmed, 1977; Akhtar et al., 1997; Beg, 1997; Beg et al., 1984; Haq, 1976; Mashiatullah et al., 2010; Rizvi et al., 1988; Saifullah et al., 2002; Saleem and Kazi, 1995; Sultana et al., 2012, 2014; Tariq et al., 1993; Zaigham, 2004). Domestic sewage from human population centers in the southeastern Indus Delta bordering the Indian coast, such as Keti Bunder and Shah Bunder, is thought to be another major source of water pollution, in humpback dolphin habitat (Kiani, 2014).

Untreated industrial effluents and agricultural run-off are the main sources of coastal and marine pollution in Sindh. Effluents containing lead, mercury and other heavy metals, mainly from tanneries, are extremely harmful to marine life. It is estimated that annually about 37,000 tonnes of industrial waste are being dumped in the coastal environment of Karachi, whereas 20,000 tonnes of oil finds its way to beaches and harbours and the fishing grounds of Karachi annually (Indus Ecoregion Programme).[1]

Thermal pollution is another cause of concern from at least four power plants and a big steel mill on the Karachi coast, which discharge more than 1500 million m^3 of warm water annually, affecting life cycles of many marine

[1] http://foreverindus.org/ie_ecosystem_coastal.php. Accessed 1 March 2015.

species (HDIP, 2008; Khalil, 1999). This may indirectly result in loss of habitat and food resources for coastal cetaceans including humpback dolphins.

The Balochistan coast, relatively speaking, has been less polluted with its small rural towns and villages along the coast and being devoid of major rivers (Ali and Jilani, 1995). However, currently the Balochistan environment is degrading due to poor law enforcement and a general lack of stewardship for marine environment protection (M.S. Kiani, pers. observations, 2005–2014). Ship breaking industry in Gaddani near Miani Hor (Sonmiani Bay) and inauguration of a new port in Gwadar pose potential threats to local humpback dolphins. Domestic sewage and discarded solid waste are predominant in all known humpback dolphin areas along the Balochistan coast, which include Miani Hor, Kund Malir, Ormara, Kalmat, Pasni, Gwadar and Jiwani.

8.4 Other Threats and Legislation

Other threats to humpback dolphins in Pakistan include loss of habitat (mangrove forest) due to land reclamation for development projects, deforestation and camel grazing, which causes damage to new mangrove plants due to trampling. For almost half of the year, no freshwater flows through the delta from the Indus, because its entire flow is diverted for irrigation purposes. This has devastated the ecology of the delta and has almost certainly negatively impacted humpback dolphin feeding ecology. Mangrove deforestation results in the destruction of vital nursery grounds of small fish and perhaps other neritic species that are prey to *S. plumbea* (see Kiani, 2014).

Pakistan National Action Plan for Conservation of Marine Cetaceans (Gore, 2008) states that specific laws for the protection of marine cetaceans in Pakistan are lacking, though Indus river dolphin, *Platanista gangetica minor*, is protected under Sindh Wildlife Act of 1972. It is unclear whether cetaceans are covered by The Balochistan Wild Life Protection Act (No. XIX of 1974) or are covered under the term "game" (The Balochistan Gazette No. 64, 1974), as they are not explicitly listed. Marine cetaceans are not listed in The Sindh Wildlife Protection Ordinance 1972 (with Amendments up to 1 June 2001, printed 2003) (The Gazette of Pakistan, Extra., Dec 6, 1997). The Gazette of Pakistan (Extra., June 30, 1998) is also unclear about protection for marine cetaceans, e.g., whether they are legal "fish" for the purposes of the act, which states under Act No. 35 of 1997, 2nd Paragraph, that "fish means all aquatic animals of inland, marine and brackish water origin ...". This Act regulates quality and promotes export of fish and fishery products from Pakistan (The Gazette of Pakistan, Extra.,

June 30, 1998, Part II Statutory Notifications S.R.O. 739 (I)/98 under section 18 (I) of the Pakistan Fish Inspection and Quality Control Act, 1997). However, Paragraph 5 (Export Restriction Point c) states that dolphins, porpoises and whales are totally forbidden for export and domestic consumption (Environmental Law in Pakistan Part 1 Federal: governing natural resources, processes and institutions affecting them, 2005). Arguably relevant to whale- and dolphin watching, the 15. 6.1.9.4 Pakistan Tourist Guides Act, 1976, requires tourist guides to have a licence and a professional code of conduct. Further, 6.1.9.5 Travel Agencies Act, 1976 covers transport, lodging and conducting guided tours, but does not provide for ecotourism or protection of natural resources from industry. Preparation of a specific national policy and legislation for the conservation of cetaceans after consultation with all relevant stakeholders (both government and non-government) can play a vital role for long-term protection of these animals against various threats.

9. CONCLUSIONS AND RECOMMENDATIONS

In Pakistan, dedicated research effort in all cetacean biology disciplines needs to be either initiated or vastly expanded. Specimens should be collected from by-catches and strandings for natural history studies and biological research. In order to reliably detect population trends abundance or density will need to be determined using distance sampling or capture–recapture techniques based on photo-identification of naturally marked individuals (e.g. Buckland and Turnock, 1992; Gope et al., 2005; Würsig and Jefferson, 1990).

Marine sanctuaries and marine protected areas (MPAs) can play a vital role in promoting conservation by directly protecting the dolphins by implementable legislation and enforcement (Karczmarski et al., 1998; Simmonds and Hutchinson, 1996). However, strengthening of awareness will prove equally essential, as without community support any protection measures will remain inconsequential. The Indus Delta is among one of the proposed sites for the declaration of a first marine protected area in Pakistan, but additional studies should be conducted before this will happen (Siddiqui et al., 2008). Slowly developing, but poorly managed, dolphin-watching tourism in the vicinity of the industrial city of Karachi and in Miani Hor (Sonmiani Bay) along the Balochistan coast needs to be properly monitored and regulated if the purpose of alternate livelihood for coastal communities is to be achieved. Given adequate legislation and properly regulated management, Indian Ocean humpback dolphins may become seen as a valuable aquatic resource exploitable for ecotourism (Karczmarski et al., 1998).

Absolute abundance estimates and trends and stock identification are currently lacking, but circumstantial evidence suggests that direct and indirect interactions from fisheries, pollution, habitat degradation and possibly decreasing food resources due to overfishing (including illegal fishing practices) may be causing species decline. An exhaustive habitat assessment coupled with appropriate fishery management is the only way to safeguard the future of *S. plumbea* and many other neritic species in the coastal waters of Pakistan.

ACKNOWLEDGEMENTS

Critical comments by external reviewers are gratefully acknowledged.

REFERENCES

Abildgaard, N.L., Khan, W., 1986. Production Statistics of the Small Pelagic Fish in Pakistan Waters: Marine fisheries development project Karachi. 7 p.

Afsal, V.V., Yousuf, K.S.S.M., Anoop, B., Anoop, A.K., Kannan, P., Rajagopalan, M., Vivekanandan, E., 2008. A note on cetacean distribution in the Indian EEZ and contiguous seas during 2003–07. J. Cetacean Res. Manage. 10, 209–215.

Ahmad, E., 1998. The Root Causes of Biodiversity Loss in the Indus Delta. Analysis of a Study Conducted by WWF-Pakistan with Financial Support of DANIDA, DGIS, BMZ, GEF and WWF-Sweden.

Ahmad, M.F., Hasan, A., 1997. Effect of Bhoola and Katra fishing nets in the Indus delta Pakistan. Rec. Zool. Surv. Pak. 13, 1–17.

Ahmed, M., 1977. An assessment of the magnitude of coastal pollution in Pakistan through a study of its fauna and fisheries. Thalas. Jugoslav. 13, 395–412.

Akhtar, W., Ali, I., Zaidi, S.S.H., Jilani, S., 1997. The state of pollution levels of Karachi harbour and adjoining coastal water. Water Air Soil Pollut. 94, 99–107.

Ali, I., Jilani, S., 1995. Study of contamination in the coastal waters of Karachi. In: Thompson, M.F., Tirmizi, N.M. (Eds.), The Arabian, Living Marine Resources and the Environment. Vanguard Books (Pvt) Ltd, Lahore, Pakistan, pp. 653–658.

Ayissi, I., Segniagbeto, G.H., Van Waerebeek, K., 2014. Rediscovery of Cameroon dolphin, the Gulf of Guinea population of *Sousa teuszii* (Kükenthal, 1892). ISRN Biodivers. 2014, 1–6. http://dx.doi.org/10.1155/2014/819827.

Baldwin, R.M., Collins, T.J.Q., Van Waerebeek, K., Minton, G., 2004. The Indo-Pacific humpback dolphin of the Arabian region: a status review. Aquat. Mamm. 30, 111–124.

Beg, M.A., 1997. Pollution of the Karachi coastal environment. Wildl. Environ. 5, 20–22.

Beg, M.A., Mehmood, A., Sitwat, S.N., Yousufzai, A.H.K., 1984. Land based pollution and the marine environment of Karachi. Pak. J. Sci. Ind. Res. 27, 199–205.

Best, P.B., 2007. Whales and Dolphins of the Southern African Subregion. Cambridge University Press, Cambridge, 338 pp.

Buckland, S.T., Turnock, B.J., 1992. A robust line transect method. Biometrics 48, 901–909.

Collins, T., 2015. Re-assessment of the conservation status of the Atlantic humpback dolphin, *Sousa teuszii* (Kükenthal,1892), using the IUCN Red list criteria. Adv. Mar. Biol. 72, 47–77.

Cuvier, G., 1829. Le règne animal distribué d'après son organisation, pour servir de base àl'histoire naturelle des animaux et d'introduction à l'anatomie comparée. Chez Deterville, Libraire, Paris, France.
FAO, 2005. Aquaculture production, 2004. Year Book of Fishery Statistics, vol. 96/2 Food and Agricultural Organization of the United Nations, Rome, Italy.
Gestsson, T., 1994. Experimental purse-seining and light fishing in offshore waters of Pakistan. In: Majid, A., Khan, M.Y., Moazzam, M., Ahmed, J. (Eds.), Proceedings of the National Seminar on Fishery Policy and Planning, pp. 438–446.
Goodwin, J.A., 1997. A Molecular Genetic Assessment of the Population Structure and Variation of Two Inshore Dolphin Genera on the East Coast of South Africa. Ph.D. Thesis, Rhodes University, Grahamstown, South Africa.
GoP, 2007. Economic Survey of Pakistan 2006-07. Planning Commission, Government of Pakistan, Islamabad.
Gope, C., Kehtarnavaz, N., Hillman, G., Würsig, B., 2005. An affine invariant curve matching method for photo-identification of marine mammals. Pattern Recognit. 38, 125–132.
Gore, M.A., 2008. Action plan for the conservation of marine Cetaceans of Pakistan. In: Marine Conservation International 5/6 Lang Rigg South Queensferry EH30 9WN UK. 26 pp.
Gore, M.A., Kiani, M.S., Ahmad, E., Hussain, B., Ormond, R.F., Siddiqui, J., Waqas, U., Culloch, R., 2012. Occurrence of whales and dolphins in Pakistan with reference to fishers' knowledge and impacts. J. Cetacean Res. Manag. 12 (2), 235–247.
Guissamulo, A., Cockcroft, V.G., 2004. Ecology and population estimates of Indo-Pacific humpback dolphins (*Sousa chinensis*) in Maputo Bay, Mozambique. Aquat. Mamm. 30 (1), 94–102.
Haq, S.M., 1976. Overview of pollution in the coastal environment of Pakistan and its possible implications for the marine ecosystem. In: Meyers, S.P. (Ed.), Proceedings International Symposium on Marine Pollution Research. Centre for Wetland studies, Louisiana State University, Baton Rouge, pp. 33–53.
Hasan, A., 1993. Fish and Shrimp Harvests and Socio-Economic Survey in Two Mangrove Areas of Sindh Coast. M.Sc. thesis for Agriculture, University of Norway.
Hasan, A., Ahmad, S.I., 2006. Some observations on birds and marine mammals of Karachi coast. Rec. Zool. Surv. Pak. 17, 15–20.
HDIP, 2008. Pakistan Energy Book. Hydrocarbon Development Institute of Pakistan (HDIP), Islamabad.
Iqbal, P., 2014. Distribution and Diversity of Organisms Along Pakistan Coast. PhD thesis, University of Karachi, Karachi, Pakistan.
Jefferson, T.A., Karczmarski, L., 2001. Sousa chinensis. Mamm. Species 655, 1–9.
Jefferson, T.A., Rosenbaum, H.C., 2014. Taxonomic revision of the humpback dolphins (*Sousa* spp.), and description of a new species from Australia. Mar. Mamm. Sci. 30 (4), 1494–1541. http://dx.doi.org/10.1111/mms.12152.
Jefferson, T.A., Van Waerebeek, K., 2004. Geographic variation in skull morphology of humpback dolphins (*Sousa* spp.). Aquat. Mamm. 30 (Special Issue), 3–17.
Karczmarski, L., 1996. Ecological Studies of Humpback Dolphins *Sousa chinensis* in the Algoa Bay Region, Eastern Cape, South Africa. Ph.D. thesis, University of Port Elizabeth.
Karczmarski, L., 1999. Group dynamics of humpback dolphins (*Sousa chinensis*) in the Algoa Bay region, South Africa. J. Zool. Lond. 249, 283–293.
Karczmarski, L., 2000. Conservation and management of humpback dolphins: the South African perspective. Oryx 34, 207–216.
Karczmarski, L., Cockcroft, V.G., McLachlan, A., Winter, P.E.D., 1998. Recommendations for the conservation and management of humpback dolphins *Sousa chinensis* in the Algoa Bay region, South Africa. Koedoe 41, 121–129.

Karczmarski, L., Cockcroft, V.G., McLachlan, A., 1999. Group size and seasonal pattern of occurrence of humpback dolphins *Sousa chinensis* in Algoa Bay, South Africa. S. Afr. J. Mar. Sci. 21, 89–97.

Khalil, S., 1999. Economic valuation of the mangrove ecosystem along the Karachi coastal areas. In: Hecht, J.E. (Ed.), The Economic Value of the Environment: Cases from South Asia. IUCN, Nepal, Katmandu. 73 pp.

Kiani, M.S., 2014. Studies on Marine Cetaceans in Coastal Waters of Pakistan. Ph.D. thesis, University of Karachi, Pakistan. 195 pp.

Lal Mohan, R.S., 1994. Review of gillnet fisheries and cetacean bycatches in the Northeastern Indian Ocean. Rep. Int. Whal. Comm. (Special Issue 15), 329–343.

Lane, E.P., de Wet, M., Thompson, P., Siebert, U., Wohlsein, P., Plön, S., 2014. A systematic health assessment of Indian Ocean bottlenose (*Tursiops aduncus*) and Indo-Pacific humpback (*Sousa plumbea*) dolphins incidentally caught in shark nets off the KwaZulu-Natal Coast, South Africa. PLoS One 9, e107038.

Mashiatullah, A., Javed, T., Chaudhary, M.Z., Fazil, M., Qureshi, R.M., 2010. Stable carbon isotope ration ($\delta^{13}C$) of shallow marine bottom sediments as evidence of pollution in Manora channel, Karachi, Pakistan. Nucleus 47, 233–238.

Mendez, M., Jefferson, T.A., Kolokotronis, S.O., Krützen, M., Parra, G.J., Collins, T., Minton, G., Baldwin, R., Berggren, P., Särnblad, A., Amir, O.A., Peddemors, V.M., Karczmarski, L., Guissamulo, A., Smith, B., Sutaria, D., Amato, G., Rosenbaum, H.C., 2013. Integrating multiple lines of evidence to better understand the evolutionary divergence of humpback dolphins along their entire distribution range: a new dolphin species in Australian waters. Mol. Ecol. 22, 5936–5948. http://dx.doi.org/10.1111/mec.12535.

Niazi, M.S., 1990. A brief on major fisheries of Pakistan and mortality of dolphins in different fishing gear. In: IWC Workshop on Mortality of Cetaceans in Passive Fishing Nets and Traps. La Jolla, CA, pp. 22–250. ct. DOC SC/090/630.

Parra, G.J., Ross, G.J.B., 2009. Humpback dolphins *S. chinensis* and *S. teuszii*. In: Perrin, W.F. (Ed.), Encyclopedia of Marine Mammals, Second ed. Academic Press, London, pp. 576–582.

Parsons, E.C.M., 1998. Observations of Indo-Pacific humpbacked dolphins, *Sousa chinensis*, from Goa, Western India. Mar. Mamm. Sci. 14, 166–170.

Pilleri, G., 1972. Zoologisch-cetologische Expedition nach Westpakistan und Reise nach India im Jahre 1971. Hirnanatomisches Institut, Waldau, Bern. 28 pp.

Pilleri, G., 1973. Cetologische Expedition zum Indus und Persischen Golf und Forschungsreise nach Goa und Thailand im Jahre 1973. Hirnanatomisches Institut, Waldau-Bern. 52 pp.

Pilleri, G., 1975. Zoologisch-cetologische Expedition zum Indus und Brahmaputra im Winter 1973-1974. Hirnanatomisches Institut, Ostermundigen, Schweiz. 55 pp.

Pilleri, G., Gihr, M., 1972. Contribution to the knowledge of cetaceans of Pakistan with particular reference to the genera *Neomeris*, *Sousa*, *Delphinus* and *Tursiops* and description of a new Chinese porpoise (*Neomeris asiaeorientalis*). Invest. Cetacea 4, 107–162.

Pilleri, G., Gihr, M., 1974. Contribution to the knowledge of the cetaceans of Southwest and Monsoon Asia (Persian Gulf, Indus Delta, Malabar, Andaman Sea and Gulf of Siam). Invest. Cetacea 5, 95–149.

Pilleri, G., Pilleri, O., 1979. Observations on the dolphins in the Indus Delta (*Sousa plumbea* and *Neophocaena phocaenoides*) in winter 1978–1979. Invest. Cetacea 10, 129–135.

Rizvi, S.H.N., Saleem, M., Baquer, J., 1988. Steel mill effluents: influence on the Bakran Creek environment. In: Thompson, M.F., Tirmizi, N.M. (Eds.), Proceedings of Marine Science of the Arabian Sea. American Institutes of Biological Sciences, Washington DC, pp. 549–569.

Roberts, T.J., 1977. Cetacean records for Pakistan. Invest. Cetacea 8, 95–99.
Roberts, T.J., 1997. The Mammals of Pakistan. Ernest Benn Limited, London. 361 pp.
Roberts, T.J., Khan, M.K., Knuckey, J., 1983. Mating behaviour of the plumbeous dolphin *Sousa plumbea* (Cuvier, 1829). J. Bombay Nat. Hist. Soc. 80, 210–212.
Ross, G.J.B., 1984. The smaller cetaceans of the south east coast of southern Africa. Ann. Cape Prov. Mus. Nat. Hist. 15, 173–410.
Ross, G.J.B., 2002. Humpback dolphins *Sousa chinensis*, *S. plumbea* and *S. teuszii*. In: Perrin, W.F., Würsig, B., Thewissen, H.G.M. (Eds.), The Encyclopedia of Marine Mammals. Academic Press, San Diego, pp. 585–589.
Ross, G.J.B., Heinsohn, G.E., Cockcroft, V.G., 1994. Humpback dolphins *Sousa chinensis* (Osbeck, 1765), *Sousa plumbea* (G. Cuvier, 1829) and *Sousa teuszii* (Kükenthal, 1892). In: Ridgway, S.H., Harrison, R. (Eds.), Handbook of Marine Mammals. In: The First Book of Dolphins, vol. 5. Academic Press, London, pp. 23–42.
Saifullah, S.M., Khan, S.H., Ismail, S., 2002. Distribution of nickel in a polluted mangrove habitat of the Indus Delta. Mar. Pollut. Bull. 44, 551–576.
Saleem, M., Kazi, G.H., 1995. Distribution of trace metals in the seawater and surficial sediments of the Karachi harbor. In: Thompson, M.F., Tirmizi, N.M. (Eds.), Proceedings of the Arabian Sea, Living Marine Resources and the Environment. Vanguard Books (pvt.) Ltd, Lahore, Pakistan, pp. 659–666.
Shah, A.A., Kawasaki, I., Kamaruzaman, J., 2007. Degradation of Indus Delta mangroves in Pakistan. Int. J. Geol. 1, 27.
Siddiqui, P.J.A., Farooq, S., Shafique, S., Burhan, Z.N., Farooqi, Z., 2008. Conservation and management of biodiversity in Pakistan through the establishment of marine protected areas. Ocean Coast. Manage. 51, 377–382.
Simmonds, M.P., Hutchinson, J.D. (Eds.), 1996. The Conservation of Whales and Dolphins. John Wiley and Sons, Chichester.
Smith, B.D., Ahmed, B., Mowgli, R.M., Strindberg, S., 2008. Species occurrence and distributional ecology of nearshore cetaceans in the Bay of Bengal, Bangladesh, with abundance estimates for Irrawaddy dolphins *Orcaella brevirostris* and finless porpoises *Neophocaena phocaenoides*. J. Cetacean Res. Manage. 10, 45–58.
Sonmiani Development Organization (SDO), 2012. Marine Dolphin Conservation Through Community Education and Capacity Building in Miani Hor, Pakistan: Final Report to Ocean Park Conservation Foundation (OPCF), Hong Kong.
Stensland, E., Carlen, I., Särnblad, A., Bignert, A., Berggren, P., 2006. Population size, distribution and behaviour of Indo-Pacific bottlenose dolphin (*Tursiops aduncus*) and humpback (*Sousa chinensis*) dolphins off the south coast of Zanzibar. Mar. Mamm. Sci. 22, 667–682.
Sultana, R., Ali, W., Ameer, F., Munshi, A.B., Nasir, M., 2012. Accumulation of pesticide residues by shrimp, fish and brine shrimp during pond culture at Ghorabari (District Thatta). J. Chem. Soc. Pak. 34 (3), 541–549.
Sultana, J., Syed, J.H., Mahmood, A., Ali, U., Rehman, M.Y.A., Malik, R.N., Li, J., Zhang, G., 2014. Investigation of organochlorine pesticides from the Indus Basin, Pakistan: sources, air–soil exchange fluxes and risk assessment. Sci. Total Environ. 497–498, 113–122.
Sutaria, D., Jefferson, T.A., 2004. Records of Indo-Pacific humpback dolphin (*Sousa chinensis*, Osbeck, 1765) along the coasts of India and Sri Lanka. An overview. Aquat. Mamm. 30 (Special Issue), 125–136.
Tariq, J., Jaffar, M., Moazzam, M., 1993. Heavy metal concentration in fish, shrimp, seaweed, sediment and water from the Arabian sea, Pakistan. Mar. Pollut. Bull. 26, 644–647.
Van Bressem, M.-F., Raga, J.A., Di Guardo, G., Jepson, P.D., Duignan, P., Siebert, U., Barrett, T., de Oliveira Santos, M.C., Moreno, I.B., Siciliano, S., Aguilar, A. Van Waerebeek, K., 2009. Emerging infectious diseases in cetaceans worldwide and the possible role of environmental stressors. Dis. Aquat. Organ. 86, 143–157.

Van Bressem, M.-F., Simões-Lopes, P.C., Félix, F., Kiszka, J., Daura-Jorge, F.G., Avila, I.C., Secchi, E.R., Flach, F., Fruet, P.F., du Toit, K., Ott, P.H., Elwen, S., Di Giacomo, A.B., Wagner, J., Banks, A., Van Waerebeek, K., 2015. Epidemiology of lobomycosis-like disease in bottlenose dolphins (*Tursiops* spp.) from South America and southern Africa. Dis. Aquat. Organ. http://dx.doi.org/10.3354/dao02932. (In press).

Van Waerebeek, K., Barnett, L., Camara, A., Cham, A., Diallo, M., Djiba, A., Jallow, A.O., Ndiaye, E., Samba Ould Bilal, A.O., Bamy, I.L., 2004. Distribution, status and biology of the Atlantic humpback dolphin *Sousa teuszii* (Kükenthal, 1892). Aquat. Mamm. 30 (1), 56–83. http://dx.doi.org/10.1578/AM.30.1.2004.56.

Van Zalingae, N.P., Khalidudin, M., Khan, W., 1987. Description of Shrimp Fishery Including a Stratified Sampling Scheme for Shrimp Landings and Effort at Karachi Fish Harbour. FAO, Rome. FI:DP/PAK/77/033/7. 73 pp.

Wells, R.S., 1993. The marine mammals of Sarasota Bay. In: Roat, P., Ciciccolella, C., Smith, H., Tomoasko, D. (Eds.), Sarasota Bay: 1992 Framework for Action. The Sarasota National Estuary Program, Tamiami Trail, Sarasota, FL, pp. 9.1–9.23.

Würsig, B., Jefferson, T.A., 1990. Methods of photo-identification for small cetaceans. In: Hammond, P.S., Mizroch, S.A., Donovan, G.P. (Eds.), Individual Recognition of Cetaceans: Use of Photo-Identification and Other Techniques to Estimate Population Parameters. In: Reports of the International Whaling Commission, Special Issue 12, pp. 43–52.

WWF, 2005. Study on Value Added to the Catch of Fisher Folk (Fisheries Development for Sustainable Livelihood): Final report. 35 pp.

WWF, 2008. Detailed Ecological Assessment of Fauna, Including Limnology Studies at Keti Bandar. 188 pp (Unpublished report). Available at http://foreverindus.org/pdf/eaf_Kharochann.pdf.

Zaigham, N.A., 2004. Unauthorized squatter settlements are one of major sources for polluting surface and subsurface waters in Karachi. In: Proceedings of the WSSD Workshop on Human Settlement and Environment (Pakistan's Response to Its Obligations Under the WSSD Plan of Implementation), Islamabad, December 14-15, pp. 100–112.

Zbinden, K., Pilleri, G., Krause, C., Bernath, O., 1977. Observations on the behaviour and underwater sounds of the Plumbeous dolphin (*Sousa plumbea*, Cuvier, 1829) in the Indus Delta. Invest. Cetacea 8, 259–286.

Zhou, K., Li, Y., Qian, W., Yang, G., 1980. Notes on three species of dolphins from the South China Sea and Jiulongjiang River. Oceanol. Limnol. Sinica 11 (4), 306–313.

CHAPTER NINE

Humpback Dolphins (Genus *Sousa*) in India: An Overview of Status and Conservation Issues

Dipani Sutaria*,1, Divya Panicker†, Ketki Jog‡, Mihir Sule‡, Rahul Muralidharan§, Isha Bopardikar‡

*College of Marine and Environmental Sciences, James Cook University, Townsville, Queensland, Australia
†Affiliated to Wildlife Programme at National Centre for Biological Sciences, Bangalore, Karnataka, India
‡Konkan Cetacean Research Team, Mumbai, Maharashtra, India
§Ashoka Trust for Research in Ecology and the Environment, Bangalore, Karnataka, India
[1]Corresponding author: e-mail address: dipani.sutaria@gmail.com

Contents

1. Introduction 230
   1.1 General Background 230
   1.2 Characteristics of the Indian Coast 232
2. Taxonomy 233
   2.1 Taxonomic Notes 233
   2.2 Common Names 234
3. Distribution Records 235
   3.1 West Coast 235
   3.2 East Coast 236
4. Population Status 237
   4.1 West Coast 237
   4.2 East Coast 241
5. Human Interactions and Threats 242
   5.1 Fisheries Interactions 242
   5.2 Coastal Development 245
   5.3 Dolphin-Watching Tourism 249
   5.4 Noise 251
6. Conservation and Management 252
References 253

Abstract

This chapter aims to collate recent work done by different research teams along the Indian coast and presents research plans for the conservation and management of the genus *Sousa* in Indian waters. Humpback dolphins are the most common nearshore cetaceans found along the Indian coast. The taxonomy is confused, but two or more species of humpback dolphins may be present in India. Dedicated research on

Advances in Marine Biology, Volume 72
ISSN 0065-2881
http://dx.doi.org/10.1016/bs.amb.2015.08.006

humpback dolphins and other cetaceans has been initiated only in the past few years and vast gaps in the ecology and conservation of the genus from the region remain. Dedicated and opportunistic research indicates that humpback dolphin presence is continuous along the west coast of India, owing to the contiguous favourable habitat of shallow nearshore waters, while along the east coast humpback dolphins are apparently found in pockets. Humpback dolphins are also the most numerous in incidental catch records from the coast, owing to the large overlap in space use with nearshore fisheries like small gillnets, trawls, shore seines and purse seines. Along many coastal sites, humpback dolphins are known to cause damage and depredation of fish catch of certain fishing gears, making them unpopular. At the same time, many fishers along the west coast have developed local dolphin-watching programmes as an alternate source of livelihood, providing positive impetus for conservation. However, research on the long-term effects of dolphin watching and its management is required. Some recommendations for more effective management of this species are made.

1. INTRODUCTION

1.1 General Background

Humpback dolphins (*Sousa* spp.) are the most common cetaceans found in Indian waters and are highly vulnerable to a number of anthropogenic pressures. Interactions between humans and marine life have greatly increased due to large infrastructural developments and intensive fisheries (De Boer et al., 2003). As coastal infrastructure rapidly develops it is crucial to study the impacts anthropogenic pressures pose on population dynamics, socialising and foraging behaviour of these animals. Research projects that have been initiated in India to study the basic biology and ecology of humpback dolphins are listed in Table 1. West Bengal and Andhra Pradesh are the only two coastal states of east India where no research has been conducted for this species. This focused research has helped address gaps in knowledge of humpback dolphins in the waters of India and has set the stage for continuous monitoring and further detailed studies.

Because there is a paucity of systematic and long-term ecological studies for humpback dolphins in Indian waters, we have found that anecdotal evidence or traditional ecological knowledge (TEK) is useful to bridge an understanding of past conditions with those of the present (e.g. Burbano et al., 2014; Coll et al., 2014). Information collected by surveys of fishing communities can provide insight, including that regarding dolphin behaviour and occurrence, human attitudes toward the animals, and the impacts of habitat degradation. For example, results of fisher surveys have indicated the need to focus attention on destructive fishing gear (trawling, small mesh

Table 1 Dedicated Research on Humpback Dolphins (*Sousa* spp.) in India Since the Year 2000

Study	Region	Methods	References
Cetacean populations in the Gulf of Kachchh and along the coast of Goa	Goa, west India, Gulf of Kachchh, Gujarat 2001–2002	Boat-based visual survey	Sutaria and Jefferson (2004)
Humpback dolphins in Ashtamudi estuary, Kerala	Kerala—Ashtamudi Estuary, southwest India 2011–2012	Interview survey, land-based observation	Bijukumar and Smrithy (2012)
Ecological and anthropogenic factors affecting habitat use and behaviour of humpback dolphins along the southwest coast of India	Kerala—Cochin, Munambam and Ashtamudi Port, southwest India 2012–2013	Interview survey, land-based observation, boat-based visual survey	Panicker and Sutaria (2013)
Distribution and habitat use of humpback dolphins on the south-east coast of India	Tamil Nadu, Andhra Pradesh 2012–ongoing	Boat-based survey, interview survey	Muralidharan (2013)
Population of cetaceans along the coast of Sindhudurg Small cetaceans of Maharashtra: Study to identify threats, habitat and local perception to facilitate conservation efforts	Maharashtra—Vijaydurg to Redi (Sindhudurg district), west India 2012–ongoing	Interview survey, mortality studies, boat-based visual survey opportunistic acoustic sampling	Sule et al. (2015)

purse seine), coastal aquaculture and industries to understand their role in depletion of nearshore fisheries in India (R. Muralidharan, unpublished data). Fisheries surveys combined with shore- and boat-based observation are a powerful tool to reconcile humpback dolphin conservation efforts across the Indian coast.

This chapter reviews past information and also includes preliminary information from current research related to status and conservation of humpback dolphins from India. We review records to provide distribution information for humpback dolphins. For sites where dedicated research has been initiated and long-term data are being collated, we provide details regarding group size, encounter rate and space use. Threats are reviewed, adding preliminary information from our TEK surveys of fishers. We also give a brief overview and recommendations for conservation and policy measures relating to humpback dolphins in the region.

1.2 Characteristics of the Indian Coast

The Indian coastline is approximately 7517 km in length and is made up of nine coastal states with two groups of offshore island archipelagos (Figure 1; Sanilkumar et al., 2006). The peninsular Indian coastline is about 5422 km long (west coast 2877 km, east coast 2545 km; Sanilkumar et al., 2006). The continental shelf on the west coast is very wide, with its widest in southern Gujarat at 374 km to the 100 m isobath, it narrows to about 97 km near southern Maharashtra and is narrowest near central Kerala at about 51 km

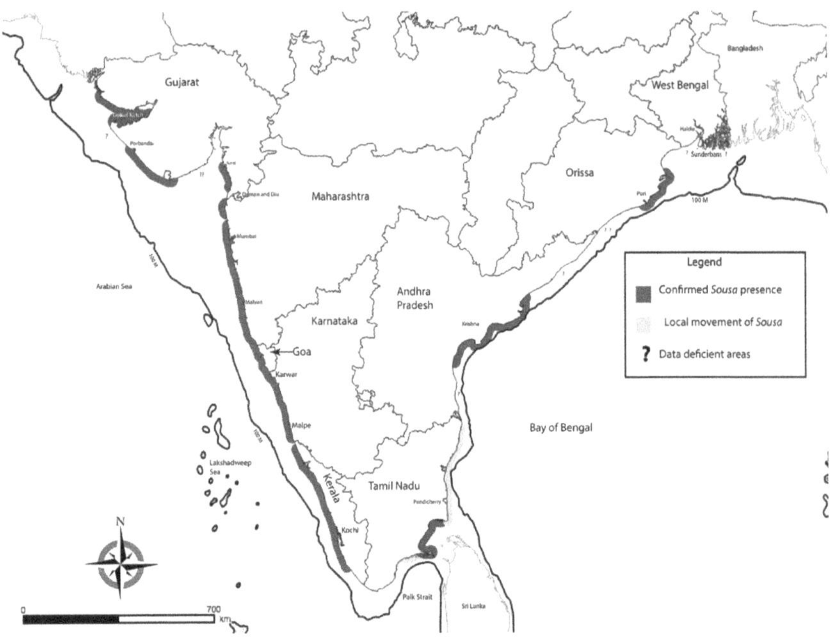

Figure 1 Map showing the probable distribution of humpback dolphins (genus *Sousa*) along the coast of India based upon published records and those recorded in the Marine Mammal Conservation Network of India Database (http://www.marinemammals.in/database/sightings-a-strandings).

(Kumar et al., 2006). Around central Ratnagiri, Maharashtra, the continental shelf (100 m isobath) narrows from about 300 to 150 km. Bottom profile in this area changes as well, from muddy to sandy/rocky (Sanilkumar et al., 2006). The continental shelf on the east coast is narrow, compared to the west coast (Sanilkumar et al., 2006), ranging from 28 to 41 km to the 100 m isobaths, with a few wider continental shelf sections near Point Calimere, Palk Bay, central and north Orissa and West Bengal (Kumar et al., 2006). The west coast of India thus has a much larger nearshore shallow stretch—the preferred humpback dolphin habitat, than the east coast of India.

2. TAXONOMY

2.1 Taxonomic Notes

Until recently, within the genus *Sousa*, the International Union for Conservation of Nature and the International Whaling Commission only recognised two species, *S. chinensis* and *S. teuszii,* with two distinct morphological types in *S. chinensis*; the *chinensis* and *plumbea* types. The scientific community though has historically considered a range of species, from a single, highly variable species (*S. chinensis*) to nine species in total (Jefferson and Rosenbaum, 2014). Frère et al. (2008, 2011), using mtDNA, showed that South African, Chinese and Australian populations of *S. chinensis* were paraphyletic, with *S. chinensis* from China and *S. plumbea* from South Africa more closely related to the Atlantic humpback dolphin *S. teuszii* than to Australian animals (Frère et al., 2008, 2011). Most experts have agreed that the taxonomy needed to be revised (Frère et al., 2008, 2011; Jefferson and Karczmarski, 2001; Jefferson and Van Waerebeek, 2004; Mendez et al., 2013).

Based on a thorough review of genetic, morphological, coloration and biogeographic information, Jefferson and Rosenbaum (2014) split the genus into four species, and the Society for Marine Mammalogy (2014) has recognised this split. The four species are *S. teuszii* (Atlantic humpback dolphin), *S. chinensis* (Indo-Pacific humpback dolphin), *S. plumbea* (Indian Ocean humpback dolphin) and *S. sahulensis* (Australian Humpback dolphin) (Jefferson and Rosenbaum, 2014).

In India, Sutaria and Jefferson (2004) first showed the morphological difference between east and west coast *Sousa*, stating that two distinct types might exist along the Indian coast, *chinensis* type extending from West Bengal to the Tamil Nadu on the East coast and *plumbea* type along the entire west coast, extending to Rameshwaram, Tamil Nadu, along the East coast.

Morphologically, the individuals observed off the coast of Pondicherry, Chennai and Mandapam (Strait between Palk Bay and Gulf of Mannar) and Orissa (Bay of Bengal) along the east coast do not have a prominent hump. However, the animals from Orissa differ from Tamil Nadu in colouration and are far more pink/white (Sutaria and Jefferson, 2004). Animals observed along the west coast of India (Arabian Sea) show a large hump and appear dark grey, thus resembling *S. plumbea* (see Sutaria and Jefferson, 2004). Dolphins that appear to be *S. plumbea* have also been recorded from the northwest coast of Sri Lanka in the Puttalam Lagoon, Bar Reef area, Palk Bay and Talaimannar, adjacent to Adam's Bridge (Bröker and Ilangakoon, 2008; Ilangakoon, 2005, 2006; Nanayakkara et al., 2014).

Mendez et al. (2013) speculated that the Indian population, along with Arabian-Oman and Thailand populations, formed a single cluster, although with distinct haplotypes. However, sample size for specimens from the Indian subcontinent from this study was insufficient ($n=3$) to obtain greater clarity on this matter. Based on genetic analysis of the mtDNA control region, the population of humpback dolphins found off the coast of Bangladesh in the northern Bay of Bengal, constitutes a separate phylogenetic (and management) unit within the genus *Sousa* (Amaral et al., 2015; Mendez et al., 2013). Currently, it is unknown whether this distinct taxon (species or subspecies) is found elsewhere in the Bay of Bengal, including Indian waters (see Amaral et al., 2015; Jefferson and Curry, 2015; Mendez et al., 2013).

Based on their study of a resident group of six individuals of *S. plumbea* in Puttalam Lagoon, northwest Sri Lanka, Nanayakkara et al. (2014) speculated that those dolphins may be of the *lentiginosa*-type (possibly a subspecies) as defined by Jefferson and Rosenbaum (2014). While recent speculation that the east coast of India may have a distinct species/subspecies, possibly *S. lentiginosa* (see Jefferson and Smith, 2016) is intriguing, to date there has been no large-scale study examining the genetic makeup of the *Sousa* population complex across the Indian subcontinent. Clearly, such an examination of this area is wanting.

2.2 Common Names

Other common names used in the area include: *fukariyo* (Gujarat), *gada reda* (Maharashtra), *gad reda* or *gaadha reda* (Karnataka, northern areas), *kadal panni/kadal ongi* (Kerala), *sori vedan/paru vedan/vella ongi* (Tamil Nadu) and the *Thella thoralu/goonu* (Andhra Pradesh).

3. DISTRIBUTION RECORDS

The Marine Mammals of India database provides details on past published and anecdotal evidence of humpback dolphins along coastal stretches of mainland India (http://www.marinemammals.in/database/sightings-a-strandings). There are records of humpback dolphins from all of the coastal states, with the exception of West Bengal. In Figure 1, we display all these records along with records from ongoing studies.

3.1 West Coast

3.1.1 Gujarat

Humpback dolphin presence has been reported from the Gulf of Kachchh, coast of Bhuj district, the coast of Saurashtra and the coast of Surat in Gujarat. Based on reports, sightings are more common near Tapi River mouth in south Gujarat (D. Desai, Prayas Nature Club, Surat, Gujarat personal communication, 2013, 2014, 2015), near Chusna and Pirotan islands in the Gulf of Kachchh (Sutaria and Jefferson, 2004), and near Madhavpur Beach, Porbandar (D. Sutaria, unpublished data). There are no reports of humpback dolphins in the Gulf of Khambhat.

3.1.2 Maharashtra

Humpback dolphin mortality and sighting records are available from the entire Maharashtra coastline. Fisher observations and anecdotal evidence suggest that the frequency of humpback dolphin sightings is low along the northern region (Thane and Mumbai districts), whereas the sighting frequencies seem to get higher as one moves southwards, in Raigad, Ratnagiri and Sindhudurg districts. Even within the southern district of Sindhudurg, ongoing studies show that humpback dolphin densities are much higher in the southern region (from Malvan to Redi) and the northernmost edge (Vijaydurg bay) of Sindhudurg district. In northern Maharashtra (i.e. in Thane and Mumbai districts), very few anecdotal and opportunistic sightings have been recorded.

3.1.3 Goa

Sightings and strandings have been recorded from all along the coast of Goa (see Parsons, 1998). Surveys conducted in 2000–2001 (Sutaria and Jefferson, 2004), showed that densities were higher in northern Goa than in the adjoining region of south Maharashtra. However, there has been a lack of recent research on humpback dolphins in Goa. Opportunistic surveys suggest that distribution and density of humpback dolphins along the coast of Goa have declined since 2004 (D. Sutaria, unpublished data).

3.1.4 *Karnataka*

No dedicated research has been carried out in Karnataka. Humpback dolphins have been observed along the Karnataka coast, with groups seen at Karwar, Gokarna, Honavar, Kundarpura, Malpe and Mangalore (D. Sutaria, unpublished data).

3.1.5 *Kerala*

Humpback dolphin presence has been reported from all across the state of Kerala. This includes from Kasargod and Kozhikode in the northern region, from Chavakkad, Cochin, Munambam and Marari in the central region, and from Azheekal, Ashtamudi, Kovalam, Paravur and Varkala in southern part of the Kerala. Thus, the entire coastline appears to be used to some degree. High densities of humpback dolphins are seen around the Kochi, Ashtamudi and Munambam estuaries. However, this may be a result of concentrated research effort in these areas (Afsal et al., 2008; Bijukumar and Smrithy, 2012; Panicker and Sutaria, submitted).

3.2 East Coast

3.2.1 *Pondicherry*

Opportunistic records since 2012 show that humpback dolphins are often observed within 100 m from shore mostly travelling north in direction along the Pondicherry coast (D. Sutaria, unpublished data). We speculate that humpback dolphins carry out seasonal movements along the Pondicherry–Tamil Nadu coast, as favourable shallow habitat is sparse and prey movement is seasonal in these waters.

3.2.2 *Tamil Nadu*

Our interview surveys (R. Muralidharan, unpublished data; D. Sutaria, unpublished data) show that humpback dolphins are present along the coast, but space use varies seasonally along the coast. In the southern part of Tamil Nadu (along Gulf of Mannar), humpback dolphins are sighted in areas with muddy bottom and around coral reef islands and occur year-round only in the shallow waters off Kodikkarai and Nagapattinam and the islands adjacent to Rameshwaram, in Gulf of Mannar and Palk Bay. In central Tamil Nadu, they are observed in shallow areas (around Point Calimere in Nagapattinam) characterised by turbid waters and mud flats.

3.2.3 Andhra Pradesh

Interviews being carried out by us in Andhra Pradesh from 2014–2015 (R. Muralidharan, unpublished data), suggest that humpback dolphins are found in the shallow estuaries with thick mangrove patches, adjoining the Krishna–Godavari basin.

3.2.4 Orissa

Dedicated boat surveys of the coast of Orissa and four rivers (Rushikulya, Devi, Budhabalanga and Subernekha rivers) of Orissa were carried out in 2004–2005 (Sutaria, 2009). Humpback dolphins were sighted only in the central coast of Orissa, between Devi River mouth and Dhamra, including Bhitarkanika Wildlife Sanctuary and Gahirmatha Marine Protected Area.

3.2.5 West Bengal

Considering their presence in the Bangladesh portion (see Smith et al., 2015), we presume that coastal West Bengal and the Sunderbans of West Bengal might have humpback dolphins, but no research has been carried out in this region.

4. POPULATION STATUS

There are no robust estimates of total abundance for humpback dolphins in Indian waters yet. Photo-identification based mark-recapture analysis is currently being conducted to estimate population size for humpback dolphins in Sindhudurg in Maharashtra and Kerala (Jog et al., submitted; Muralidharan, 2013; Panicker and Sutaria, 2013, submitted; Sutaria and Jefferson, 2004). There are preliminary density estimates and encounter rates for humpback dolphins off Gujarat, Maharashtra, Goa, Kerala, Tamil Nadu, Pondicherry and Orissa (Table 2).

4.1 West Coast

4.1.1 Gujarat

Sutaria and Jefferson (2004) observed a total of 21 groups of humpback dolphins during a search effort of 570 km. A total of 15 groups were sighted along the coast and six groups were sighted around the islands. Fifty-seven percent of the sightings were in water depths of 0–10 m. It was found that

Table 2 Encounter and Group Size Data for Dedicated Humpback Dolphin (*Sousa* spp.) Studies in India Since the Year 2000

State	Location	Number of Encounters	Group Size	Effort	Method of Survey	Month/Year	Source
Gujarat	Gulf of Kachchh	21	Avg: 3.9 Min: 1 Max: 11	570 km	Boat-based visual survey	January 2001	Sutaria and Jefferson (2004)
Maharashtra	Sindhudurg coast	99	Avg: 11.46 Min: 1 Max: 120	781 km	Boat-based visual survey	May 2014 to May 2015	Sule et al. (2015)
Goa	Entire coast	135	Avg: 6.3 Min: 1 Max: 35	573 km	Boat-based visual survey	February 2001	Sutaria and Jefferson (2004)
Karnataka	Encounter and group size data are unavailable for this region						
Kerala	Ashtamudi, Kochi and Munambam estuaries	934	Avg: 2 Min: 1 Max: 8	432 h (2592 10-min scans)	Shore-based scan sampling	April 2012 to January 2013	Panicker and Sutaria (2013, 2015, in review)
	Coastal waters off Kochi and Munambam	34	Avg: 2.78 Min: 1 Max: 15	76.5 km	Boat-based visual survey	March, April 2013	Panicker and Sutaria (2013)
Tamil Nadu	Chennai coast	4	Avg: 20	120 km	Boat-based visual survey	February 2013	Muralidharan (2013)

Pondicherry	Entire coastline	31	Avg: 15 Min: 2 Max: 100 Density: 0.27/ km^2	Opportunistic observations	Shore-based surveys	August 2012 to May 2015	D. Sutaria (unpublished data)
Andhra Pradesh	Encounter and group size data are unavailable for this region						
Orissa	Entire coastline	4	Avg size: 7.28 Min: 3 Max: 70 Density: 0.27/ km^2	810 km	Boat-based visual survey	December 2004 to January 2005	Sutaria (2009)
West Bengal	Encounter and group size data are unavailable for West Bengal. Past records of presence are also unavailable. However, the neighbouring country of Bangladesh, which shares the Ganges–Brahmaputra delta with India, has a sizeable population of *Sousa* (see Jefferson and Smith, 2016).						

in waters ranging from 1 to 20 m, the average group size remained between 1.5 and 3.0 individuals, while in waters 21–30 m deep, average group size was 9.5.

4.1.2 Maharashtra

Along the Sindhudurg coast, humpback dolphin densities seem to be higher around river mouths (Redi, Devbag sangam and Vijaydurg estuary), and densities were relatively lower in the north Sindhudurg region as compared to the southern region (Jog et al., submitted). Group sizes were larger in southern Sindhudurg, with subgroups showing tighter associations. Sightings in north Sindhudurg were closer to shore as compared to the south. Based on current photo-identification surveys, 153 individuals have been identified as of December 2014 in Sindhudurg region (Jog et al., submitted). Humpback dolphins were sighted at a maximum depth of 15.6 m, with 47.05% of sightings within 10 m depth and 52.94% of sightings between 10 and 20 m ($n=85$ sightings). Fourteen humpback dolphin carcasses were reported between February 2012 and April 2015 (11 from Sindhudurg district and 3 from Mumbai).

4.1.3 Goa

In Goa, a total of 135 humpback dolphin groups were sighted by Sutaria and Jefferson (2004). Out of the 135 groups sighted, 100 groups were sighted in the north, closer to the river mouths, and 60% of sightings were in water depths of 6–10 m. One dead humpback dolphin was found during this study in 2001–2002.

4.1.4 Kerala

Recent work indicates that there is seasonal variation in the occurrence of humpback dolphins in estuarine regions of Kerala, namely, Ashtamudi, Munambam and Kochi. Individuals used estuaries to a higher extent in summer and the pre-monsoon season, and a much lower extent in the monsoon season (Panicker and Sutaria, 2013, submitted). The high influx of freshwater leading to a decreased salinity was found to be the determining factor that explains this change of use (Panicker and Sutaria, submitted). Salinity levels are not considered to be a physiological barrier affecting dolphin distribution, but are likely to influence prey distribution and abundance. Although dolphins were observed travelling, milling and socialising in the estuaries, the majority of time was spent foraging in all seasons (Panicker and Sutaria, 2013, submitted). Dolphins feed mainly on fish of the mullet family

(*Liza macrolepis, L. parsia, Mugil cephalus, Valamugil* spp.) in the estuaries (Bijukumar and Smrithy, 2012; Panicker and Sutaria, submitted). This suggests that estuaries are important feeding grounds for the species (Panicker and Sutaria, 2013, submitted). In Ashtamudi, apart from foraging, milling (including resting) activity was seen to a higher degree (than Munambam and Kochi) showing that dolphins use this estuary not just for feeding (Panicker and Sutaria, 2013). A high number of calves/juveniles were observed in Ashtamudi estuary and adjacent Sakthikulangara Bay through the study period, suggesting this site may serve as a nursing ground. One dead humpback dolphin was found during this study off Kochi. Possible cause of death was net entanglement, as fishing line was found around the body.

4.2 East Coast

The distribution pattern of humpback dolphins on the east coast of India is more discontinuous that on the west coast. Dolphins on the east coast appear to form discrete populations. The distinct bathymetric variation patterns make the coast less suitable as humpback dolphin habitat, and this likely affects the distribution. Recent studies (Sutaria, 2009) seem to suggest that the distribution of humpback dolphins along the east coast of India (Orissa to Tamil Nadu) is concentrated in a few small pockets: Gahirmatha to Puri, Orissa; Point Calimere, Chennai, Pondicherry, Rameshwaram, Palk Bay in Tamil Nadu; a few locations in Andhra Pradesh.

4.2.1 Tamil Nadu and Andhra Pradesh

Information regarding sightings and group size from Tamil Nadu coast has been meagre. Occasionally, humpback dolphins have been observed from shore (mostly during morning and evening hours) in larger groups. Humpback dolphins are more frequently observed in Tamil Nadu and Andhra Pradesh between the months of December and April, coinciding with the occurrence of nearshore fish species along the coast (R. Muralidharan, unpublished data).

4.2.2 Pondicherry

Of the 31 groups of humpback dolphins reported from Pondicherry, 76% were milling off the esplanade (D. Sutaria, unpublished data). Most groups were observed travelling or milling, and groups were travelling south to north (see Section 3.2). There seems to be seasonal variation in dolphin occurrence, with maximum sightings between May and September annually (fewer sightings were from December to April).

4.2.3 Orissa

Four groups of humpback dolphins were encountered in the nearshore waters of Orissa in 2004–2005 (Sutaria, 2009). Six dead humpback dolphins were also found between 2004 and 2006 along the Orissa coast. In 2015, the Orissa State Forest Department survey produced a total count of 173 humpback dolphins from Gahirmatha in the coastal waters of central Orissa.

5. HUMAN INTERACTIONS AND THREATS

5.1 Fisheries Interactions

Humpback dolphins are often seen foraging in and around estuarine regions and are particularly impacted by human activities in the coastal zone, which is exploited extensively by artisanal and commercial fisheries (Panicker and Sutaria, 2013; Parra and Jedensjö, 2014; Sutaria and Jefferson 2004). They are generalists, preying on a wide variety of nearshore, estuarine and reef fishes. In Indian waters, they have been observed to primarily feed on mullets, pomfrets, mackerel and sardines (Panicker and Sutaria, submitted; Sule et al., 2015).

5.1.1 Tamil Nadu and Andhra Pradesh Coast

Our observations beginning in 2009 have indicated that humpback dolphins are identifiable by coastal fishers based on colouration patterns, presence around shallow waters and sometimes around fishing nets. The species is referred to as *Kuttiyandi* following a traditional myth. In the story, *Kuttiyandi* (a god or demigod) made a ring out of clay and threw it into the water, asking the dolphins to search for it. It is believed to this day, that the dolphins have been searching for the ring. Interestingly, the ring searching behaviour explained by fishers in their stories coincides with crab-feeding behaviour of humpback dolphins. Humpback dolphins, according to fishers, are known to nudge muddy bottom to forage on crabs forming circular mud plumes on the water surface. In both north Tamil Nadu and Andhra Pradesh, humpback dolphins are known for depredation of small forage fish from fishing gear, thus damaging gear, but there is no animosity in the region toward the species despite the heavy losses incurred by fishers. However, a marked conflict is evident in the southern regions of Tamil Nadu, where a similar behaviour is present.

The greatest number of records of live/dead strandings due to fishing gear entanglement and bycatch is reported for the state of Tamil Nadu.

Older fishers from the region have reported frequent depredation of their fish catch by humpback dolphins, but these encounters seem to be on the decline, probably related to decline in dolphin presence itself (R. Muralidharan, unpublished data). Fishers seem to attribute this decline over time to the collapse of nearshore fisheries, usage of destructive fishing gear and lack of river flow into estuaries (R. Muralidharan, unpublished data).

Artisanal fishers in Tamil Nadu indicate that entanglement of humpback dolphins occurs only in large mesh, multifilament gillnets (R. Muralidharan, unpublished data) but they prefer to forage around small mesh size monofilament gillnets, as this gear type catches a preferred size of fish. We assume that if entangled, humpback dolphins can break free from small mesh monofilament gillnets as also reported by fishers in Kerala (Panicker and Sutaria, 2013, submitted).

5.1.2 Kerala

Humpback dolphins are commonly seen in Kerala waters. Locally they are known as "*kadal panni*" (sea pig). The name seems to have stemmed from two sources, one from the behaviour by which the dolphins dig up the sea floor with their snouts (much like pigs do) and the second one being the similarity of taste between pork and dolphin meat.

Observations of fishers in the region were collected by interview survey during 2012–2013 (Panicker and Sutaria, 2013, submitted; see Table 1). Average number of years of fishing experience was 32.43 years $\pm$ 1.47 (SE) for fishers in this region. Fishers observed that dolphins were found in groups within the estuary throughout the year, with higher usage seen in summer and pre-monsoon months. Fishers told us that dolphins often came close to the nets in the estuary, but mostly did not directly interact with their nets. Those using cast and Chinese dip nets said that the dolphins were a useful ally because they chased fish toward the shore and into their net resulting in higher catch (see also Bijukumar and Smrithy, 2012). It was reported that, very rarely, dolphins removed fish from the nets (bit fish off of the nets) in the estuary. However, fishers said that in coastal waters, stealing behaviour by dolphins was more frequently observed especially by older individuals (ones which were heavily pigmented), often resulted in huge losses. Fishers seemed to think that higher abundance of preferred fish species in the estuary might be one reason for decreased conflict in the estuarine waters. Deterrence of "stealing" behaviour was attempted by introducing dolphin excluder nets, fishers entering the water or by using fireworks illegally to scare the animal away (D. Panicker, D. Sutaria, personal observations). Excluder nets have also been used in Calicut, where it has been popular among ring net fishers (Sivadas and Pradeep Kumar, 2009). The effectiveness of the excluder nets over time is yet to be assessed. Fisher associations have demanded insurance cover for damage done to nets by dolphins and have also suggested that the government should develop fisheries management systems that are informed by scientific research to prevent such human–dolphin conflict (Martin, 2013).

Fishers said that bycatch of dolphins rarely occurs in estuaries, and that dolphins swiftly avoid nets (Panicker and Sutaria, 2013). Fishers also said that in the rare cases that entanglement occurs, dolphins can easily break free from the monofilament gillnets. Bycatch in coastal waters was considered more frequent. Fishers were aware of existing laws that ban consumption or landing of dolphins, and a direct fishery targeting dolphins does not exist in the region. As a result bycatch is usually discarded at sea itself. Although, one-third of fisher respondents told us that dolphins are still consumed at a local scale when accidentally caught in nets.

Although fishers were aware of the law to protect humpback dolphins, their understanding of the law varied. Some fishers perceived an increase in dolphin abundance due to the absence of hunting (and lack of other predators or threats) thereby seeing the law as obsolete. Other fishers attributed decreasing fish stocks, detrimental fishing practices and pollution to a decrease in dolphin abundance.

5.1.3 Maharashtra

In Maharashtra, humpback dolphins are known to forage around small gillnets and shore seines (Jog et al., submitted). Fishers suffer monetary losses because of net damage and loss of catch (Jog et al., submitted). This is a conservation concern and has resulted in conflict between fishers and dolphins, increasing the probability of accidental entanglement and mortalities.

Along the Sindhudurg coast, fishing operations range from shore seines, nearshore gillnets, bottom set gill and cast nets to industrial trawl nets and purse seiners. Trawls and purse seiners mainly operate offshore. Small gillnets and shore seines operate almost exclusively in the narrow coastal zone. Recently, small purse nets have been introduced to the fishery in southern Sindhudurg. Seasonal longline fisheries also operate from gillnet vessels in some villages; small handheld gear, such as hook and line and cast nets are used in the shallows and in estuaries.

Fisher perceptions in Maharashtra towards humpback dolphins are negative, varying significantly across age class and gear type, and appear to be strongly influenced by type of gear used (Jog et al., submitted). Small gillnet users particularly think of humpback dolphins as a nuisance. These nets operate mainly in the shallow coastal waters (less than 20 m in depth), where the humpback dolphins are found. Dolphins often forage near, or from, the nets, damaging them in the process, earning them their local moniker "*Gada Reda*" (*Marathi*), "clumsy bull." This is a serious issue, as small gillnet fishers are generally the poorest class of fishers. Many fishers also say they believe that humpback dolphins are highly intelligent and can tell the difference between various gear and craft and also target specific fish species like pomfret and mullet (Jog et al., submitted). Many fishers have also recounted incidents where dolphins have depressed the top ropes of shore seine nets and vaulted over to escape after foraging from these nets (Jog et al., submitted).

The younger generations of fishers who are actively involved in dolphin-based tourism are more tolerant of these dolphins, while older fishers generally have a negative attitude toward the dolphins. Older fishers have spent more time using non-mechanised boats, which may lead to a greater area overlap with humpback dolphins. The advent of engines allowed the younger fishers to move beyond the range of this species. In addition, the older generation of fishers used hand-woven nets, which were easily torn and were labour intensive and expensive to repair. These negative interactions led to deliberate killing of these dolphins. The carcasses were used for bait, oil (tallow) and fertiliser. Flukes were used to claim a bounty from the government until 1966 (Jog et al., submitted).

Perceptions toward these dolphins were affected by the frequency of interactions and area overlap with fishing (Jog et al., submitted). Fishers in the northern part of Sindhudurg district (Sarjekot to Vijaydurg) are mainly engaged in large gillnet, purse seine and trawl fisheries. This gear is generally used outside the dolphins' range (<20 m depth) and the fishers mostly encounter humpback dolphins only while exiting or entering harbours. Fisher perceptions in this area are therefore influenced largely by depth-based resource partitioning, thereby minimising negative interaction (Jog et al., submitted).

5.2 Coastal Development

People use coastal ecosystems for various purposes such as food production, economic benefits, tourism and defence. Nearly 50% of the world's population lives in coastal areas, this number is steadily increasing, especially in developing countries such as India (Creel, 2003). India has 0.25% of the world's coastline and supports 11% of the global population, housing some of its largest cities such as Kolkata, Mumbai, Chennai, Kochi and Vishakhapatnam (around 250 million people live within 50 km of the coastline; http://www.moef.nic.in/downloads/public-information/Notice_ICZM.pdf). The unplanned and unprecedented rise in large-scale development activities along the Indian coastline has prompted an increase in industry, especially near major ports (Rodriguez and Sridhar, 2010). Physical barriers such as seawalls and breakwaters have increased coastal erosion, especially in states like Kerala. Ports, along with associated manufacturing and processing, affect about 45% of the Indian coastline (Lakshmi et al., 2012). Twelve major ports and 200 non-major ports dot the coastline of India with more proposals being made on developing new or expanding existing ports (www.shipping.nic.in). Often the areas selected for port building and other large-scale developmental activities are medium- or large-sized estuaries due to their sheltered nature and better accessibility (Rodriguez and Sridhar, 2010). The increase in habitat modification accompanying development, and the preference of humpback dolphins for sheltered bays and estuaries, makes it critical to understand the impact of development in the region. Jefferson et al. (2009) have studied the impacts of similar large-scale development on *S. chinensis* in Hong Kong. In India, such studies are wanting and the impacts of coastal development on humpback dolphins are poorly understood.

Panicker and Sutaria (2013, submitted) compared ports of differing sizes, Neendakara port (Ashtamudi estuary), Munambam port and Kochi port, and explored habitat use and behaviour at a macro-scale. Time activity budgets of humpback dolphins in the estuarine ports showed that dolphins spent the majority of their time in foraging and related activities in all three estuaries, highlighting the importance of the estuaries as feeding grounds. No time was spent milling (or resting) in the major port of Kochi (greater human activity), whereas milling was observed in both of the other ports included in the study (Figure 2; Panicker and Sutaria, 2013, submitted). A higher proportion of travelling was also seen in the major port, suggesting increased movement within the estuary. Although immediate behavioural responses were not quantified for this study, temporary displacement due to heavy boat traffic was observed several times (Panicker and Sutaria, 2013, submitted).

Further examination on the effect of boat traffic on humpback dolphins within the major port indicated that there was no spatial avoidance due to the presence of boats (Figure 3; and see Panicker and Sutaria, submitted).

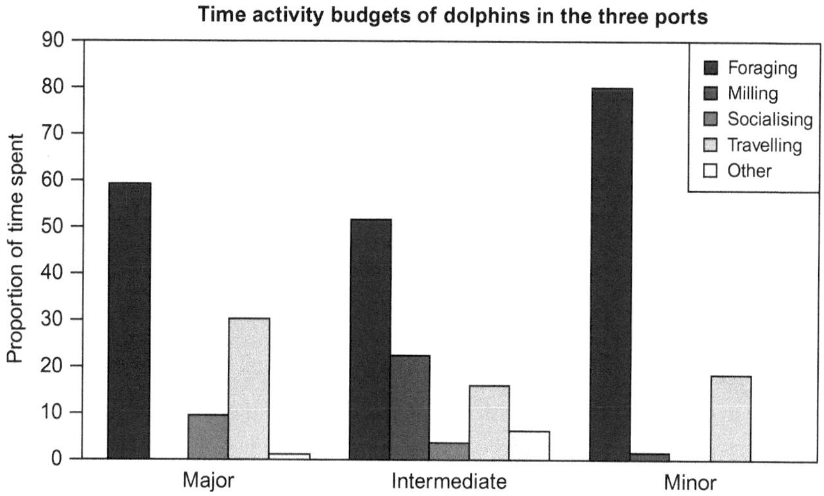

Figure 2 Time activity budget of humpback dolphins (genus *Sousa*) across the three ports (Munambam, Ashtamudi and Kochi estuaries) in Kerala. *From D. Panicker (unpublished data).*

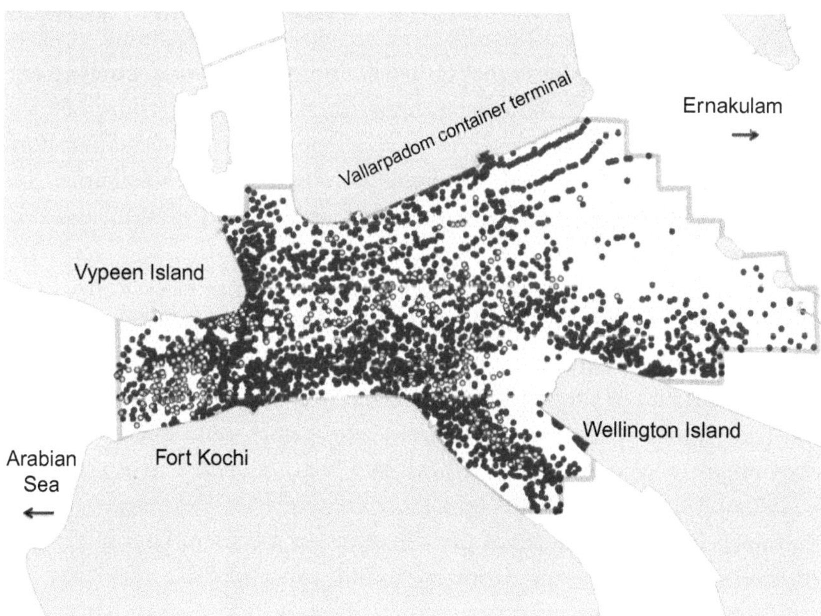

Figure 3 Map of vessels and humpback dolphins (genus *Sousa*) in the Cochin estuary in 2012–2013. *From D. Panicker (unpublished data).*

However, immediate behavioural responses that were not quantified here, would better explain some of the habitat use patterns exhibited by humpback dolphins. Dolphins may be adapting behaviourally to increased vessel traffic through strategies such as increased dive intervals and temporary displacement, which may in turn reduce foraging effectiveness and have long-term impacts on fitness and survival.

With the lack of detailed studies on the impact of habitat degradation on humpback dolphins, it is important to adopt a precautionary approach for conservation efforts. The known habitat requirements and conservation of the species must be included in the policy-making process and relevant legal frameworks. For example, environmental impact assessments (EIAs) of coastal projects must account for humpback dolphin presence/absence. From existing studies and anecdotal reports, it is clear that this species is found throughout the east and west coasts. In areas where local distribution data are lacking, studies should be commissioned to ascertain the same. For example, the EIA of the proposed Vizhinjam port in southern Kerala does not mention humpback dolphins (http://www.vizhinjamport.in/eia-30-5-13.php), despite the fact that the port will be situated close to estuaries and that humpback dolphins have been recorded in the region (Sutaria and Jefferson, 2004).

The Tamil Nadu coast has experienced coastal erosion at several locations, mainly from coastal development activities, like the building of sea walls and harbours (Lakshmi et al., 2012). A number of thermal power plants and shipping harbours have been proposed for the coast (Rodriguez and Sridhar, 2010). Reduction in river flow in the numerous estuaries along the coast, along with constant dredging activities, has led to siltation and to the degradation of mangrove areas (Blasco and Aizpuru, 2002) and suitable humpback dolphins habitat along the coast.

Interviews with small scale, artisanal fishing communities along Tamil Nadu coast have shown that people have observed a massive decline in fish catch, which fishers attribute to nearshore trawling and small mesh size purse seining operations in the region. Interviews along the Andhra Pradesh coast in Godavari and Krishna river estuaries show stark differences in coastal development activities between these two areas. For instance, the depth of Gaderu river estuary in Kakinada, East Godavari district, was about 25 m deep before an industrial complex was set up there almost 15 years ago; however, the present depth is around 3 m and fishers say that one can walk across Gaderu river estuary during low tide. Fishers say that in the past humpback dolphins were common in the area, but have not been observed entering into the Gaderu estuary for the past 10 years. In contrast, fishers in Krishna estuary have observed an increase in humpback dolphin numbers over recent years. There are no major industries, other than vast aquaculture farms around the Krishna estuary.

5.3 Dolphin-Watching Tourism

Nature-based tourism is an increasing global phenomenon, growing rapidly in developing countries as a form of alternate livelihood. Forty-four communities from 19 countries benefit from whale watching in south and south-east Asia. Worldwide, 106 countries and 495 communities are known to earn income from the whale-watching industry (Hoyt, 2001, 2005).

In India, only two species of marine cetaceans are currently the focus of the dolphin-watching industry. These are humpback dolphins and Irrawaddy dolphins, *Orcaella brevirostris*, the latter limited to Chilika Lake, Orissa. The dolphin-watching industry is a locally developed cottage industry creating an opportunistic source of income. Dolphin watching on the west coast is common along Goa and Karnataka, Sindhudurg in Maharashtra and in Cochin and Varkala, Kerala. The proximity to shore, accessibility and prior knowledge of fishers regarding locations where dolphins are commonly sighted, offers an advantage to the growth of humpback dolphin watching in India. However, there is very little information assessing the sustainability of the industry. Research must be conducted to better inform policy in India and to show that alternate livelihoods with a community-driven conservation strategy in place can be sustainable.

To assess the importance of dolphins and tourism, data were collected from two groups of stakeholders: tourist operators who are also fishers and tourists (Sule et al., 2015). Managers and members of locally run dolphin associations were interviewed to understand how they operate and how the industry has grown over recent years. A content analysis of logbook data and interviews with tourists was done to better understand the value of the industry. Questionnaire surveys with fishers involved in tourism were completed to document fisher perceptions towards tourism and dolphin conservation (Sule et al., 2015). Boat surveys are also underway to assess the effect of dolphin watching on the behaviour and health of the dolphin populations being targeted (Sule et al., 2015).

5.3.1 Kerala

Kerala is an important tourist destination. Dolphin watching has been advertised in various tourist areas such as Varkala, Kochi and Ashtamudi. Medium- and small-sized motorboats and speedboats are used for dolphin

Figure 4 Two adult humpback dolphins (genus *Sousa*) and a calf socialising in Kochi Harbour. *Photograph by Divya Panicker.*

watching (Figure 4). Tour operators promise a high likelihood of dolphin sightings with some operators currently giving cash-back (50%) or repeat trips in case of no sightings. The tourism industry in Kerala has not been studied. It is critical to do so before an exclusive dolphin-watching industry is formed to ensure necessary regulations are in place.

5.3.2 Sindhudurg

Tourism as an industry alternate to fisheries has flourished in Sindhudurg, Maharashtra since 2005 (Chakravarty, 2003; Chakravarty et al., 2008). Many home stays and restaurants have turned into a full-fledged industry based on dolphin watching. Tourism activities here are still localised with a few major fishing villages such as Devbag and Tarkarli dominating the dolphin-watching industry. A few tours operate from Medha Nivati, Sagareshwar, Sagatirth, Shiroda Velagar and Aravli Velagar. Talashil has a budding dolphin-watching industry, but the frequency of tourists coming here or of dolphins being seen seems to be very low (Jog et al., submitted). There are a total of 26 tourism units, from Talashil in the north to Shiroda in southern Sindhudurg, where dolphin tours operate. Tarkarli and Devbag see maximum tourist traffic out of all other areas. There are about 15 tourism units in this region with about 300–350 boats operating (for tours) (Jog et al., submitted). November to January and March to May are peak seasons. Based on projections from one tourism unit in Devbag (an association with 52 boats registered under it), each boat makes about

US$2000 per year (Jog et al., submitted). Other places south and north of Malvan have seven small-scale tourist outlets, most of them developed in the past 3–5 years (Jog et al., submitted).

The advent of dolphin-based tourism has led to increased awareness about the presence of cetaceans in this area among the locals as well as tourists, but most of the tourism operators are misinformed about these animals (Jog et al., submitted). In light of the diminishing fishing industry, humpback dolphins have emerged as the new cash source of income for the fishers. This change in attitude of the younger generation and dolphin tourism operators seem to be a good omen for cetaceans in this area. This potential amity of the younger generation towards cetaceans can be tapped for engaging them in community led conservation endeavours. However, this new industry may also introduce unforeseen threats into dolphins' habitat.

Recently, water sports have gained widespread popularity in this region. Most tourist sites offer sports activities like jet ski rides, "banana boat" rides and parasailing in addition to dolphin-watching tours. These high-speed vessels operating in the shallows pose a great risk of boat strikes for humpback dolphins, and the engine noise generated may pose a new set of threats for the dolphins.

5.4 Noise

Vocal repertoire and communication behaviour of humpback dolphins in Indian waters has not been well studied. Because these dolphins are the focus of developing dolphin-watching industries in a few areas and are impacted by habitat degradation, it is essential to gauge how increased ambient noise in the marine environment will affect these animals. Studies on humpback dolphins in Australia and Hong Kong have shown that communication is affected by vessel traffic and boat noise (Sims et al., 2012b; Van Parijs and Corkeron, 2001a,b; Wang et al., 2013).

Recent observations of humpback dolphins along the Sindhudurg coast as well as Cochin harbour have indicated that dolphins forage and socialise within areas of high disturbance and stress, instead of abandoning these areas (Panicker and Sutaria, submitted) If vessel noise leads to auditory masking, individuals may alter vocal behaviour to compensate (e.g. Sims et al., 2012a).

Developing coastal infrastructure and increasing vessel traffic have a heavy overlap with the habitats of these dolphins. Hence there is a need to study the effect of noise as a source of habitat degradation and its effect on the acoustic behaviour of these animals. In May and October of 2014, opportunistic passive acoustic monitoring (PAM) data were collected using an SQ26-MT hydrophone (Cetacean Research Technology), during two sets of visual surveys along the Sindhudurg coast (Sule et al., 2015). Both clicks and whistles were recorded. Whistles were recorded during socialising, as well as foraging states, as observed in other humpback dolphin populations (Van Parijs and Corkeron, 2001b). Further analysis of these and other data are required to better understand the impacts of various sources of anthropogenic noise on humpback dolphins.

6. CONSERVATION AND MANAGEMENT

Humpback dolphins are common residents of the Indian coastline. The paucity of research in the past has also been reflected in a lack of specific conservation or policy measures designed towards protecting humpback dolphins or their habitats in Indian waters. Although the Wildlife Protection Act (WPA 1972) gives at least nominal protection to all cetaceans in Indian waters, there is no provision for detailed protection on any particular species.

Furthermore, the enforcement of laws is poorly developed. The forest department, marine police and coast guard are in charge of monitoring the bycatch or disturbance to cetacean species. The training, if any is given, to these departments to carry out such tasks are unclear. A complete ban on hunting cetaceans is in effect across all the major harbours in India (as per WPA 1972). Fishers all along the coast are well aware of the law and mostly refrain from bringing any dead cetaceans to the harbours, fearing conviction. However, landings in remote areas along the coast (especially by artisanal fishers) are unmonitored and regulation without community support would be difficult. At present there are no targeted conservation efforts for this species by the government or the local community.

Dedicated conservation-oriented studies on humpback dolphins have only just begun in India. Effectively monitoring habitat used for foraging and other important life functions can help facilitate appropriate conservation efforts (Bailey and Thompson, 2009; Markowitz et al., 2004; Pirotta et al., 2015). Recently initiated work includes programmes to estimate population size and calculate trends using mark-recapture analysis in Maharashtra (by the State Forest Department in conjunction with the United Nations Development Program), to monitor the intensity of operational interactions with fisheries and the influence of fisheries on foraging strategies and group dynamics in Maharashtra and Kerala, a PAM study in Maharashtra, a study on interactions between fishing gears and dolphins in Kerala, and a study of fisher perceptions and boat surveys to study movements and habitat use in relation to coastal development in Tamil Nadu. Novel methods, such as those involving PAM, can provide important data on ambient noise, spatial and temporal data on foraging and other activities and on communication without causing disturbance (Richardson et al., 1995).

Finally, given the overlap of humpback dolphin habitat with intense human activities in the coastal waters of India, it is vital to have a specific legal framework pertaining to humpback dolphins, particularly with regulations regarding the major threats as we currently understand them: dolphin-watching tourism, large coastal development projects and fisheries interactions.

REFERENCES

Afsal, V.V., Yousuf, K.S.S.M., Anoop, B., Anoop, A.K., Kannan, P., Rajagopalan, M., Vivekanandan, E., 2008. A note on cetacean distribution in the Indian EEZ and contiguous seas during 2003–07. J. Cetac. Res. Manage. 10, 209–215.

Amaral, A.R., Smith, B., Mansur, R., Brownell Jr., R.L., Rosenbaum, H.C., 2015. Genetic identity of Indo-Pacific humpback dolphins (*Sousa* spp.) in the northern Bay of Bengal, Bangladesh: International Whaling Commission Scientific Committee Report. 13 pp. (unpublished). SC/66a/SM/24.

Bailey, H., Thompson, P.M., 2009. Using marine mammal habitat modelling to identify priority conservation zones within a marine protected area. Mar. Ecol. Prog. Ser. 378, 279–287.

Bijukumar, A., Smrithy, R., 2012. Behaviour of Indo-Pacific humpback dolphin, *Sousa chinensis* (Osbeck) in the Ashtamudi estuary, southwest coast of India. J. Mar. Biol. Assoc. India 54, 5–10.

Blasco, F., Aizpuru, M., 2002. Mangroves along the coastal stretch of the Bay of Bengal: present status. Indian J. Mar. Sci. 31, 9–20.

Bröker, K.C.A., Ilangakoon, A., 2008. Occurrence and conservation needs of cetaceans in and around the Bar Reef Marine Sanctuary, Sri Lanka. Oryx 42, 286–291.

Burbano, D.V., Mena, C.F., Guarderas, P., Vinueza, L., Reck, G., 2014. Shifting baselines in the Galapagos white fin fishery, using Fisher's anecdotes to reassess fisheries management: the case of the Galapagos grouper. In: Denkinger, J., Vinueza, L. (Eds.), The Galapagos Marine Reserve—A Dynamic Social-Ecological System. Springer International Publishing, Switzerland, pp. 227–246

Chakravarty, I., 2003. Marine ecotourism and regional development: a case study of the proposed marine park at Malvan, Maharashtra, India. In: Garrod, B., Wilson, J. (Eds.), Marine Ecotourism: Issues and Experiences. Multilingual Matters, Bristol, UK, pp. 177–197.

Chakravarty, I., Prideaux, B., Timothy, D.J., Chon, K., 2008. Heritage tourism and community participation: a case study of the Sindhudurg Fort, India. In: Prideaux, B., Timothy, D.J., Chon, K. (Eds.), Cultural and Heritage Tourism in Asia and the Pacific. Routledge, London, pp. 189–202.

Coll, M., Carreras, M., Ciércoles, C., Cornax, M.-J., Gorelli, G., Morote, E., Sáez, R., 2014. Assessing fishing and marine biodiversity changes using fisher's perceptions: the Spanish Mediterranean and Gulf of Cadiz Study. PLoS One 9 (1), e85670.

Creel, L., 2003. Ripple Effects: Population and Coastal Regions. Making the Link, Population Reference Bureau, Washington, DC. Retrieved from, http://www.prb.org/Publications/Reports/2003/RippleEffectsPopulationandCoastalRegions.asp.

De Boer, M.N., Baldwin, R., Burton, C.L.K., Eyre, L., Jenner, K.C.S., Jenner, M.N.M., Keith, S.G., McCabe, K.A., Parsons, E.C.M., Peddemors, V.M., Rosenbaum, H.C., Rudolph, P., Thiele, D., Simmonds, M., 2003. Cetaceans in the Indian Ocean Sanctuary: A Review. Whale and Dolphin Conservation Society, UK. 52 pp.

Frère, C.H., Hale, P.T., Porter, L., Cockcroft, V.G., Dalebout, M.L., 2008. Phylogenetic analysis of mtDNA sequences suggests revision of humpback dolphin (*Sousa spp.*) taxonomy is needed. Mar. Freshw. Res. 59, 259.

Frère, C.H., Seddon, J., Palmer, C., Porter, L., Parra, G.J., 2011. Multiple lines of evidence for an Australasian geographic boundary in the Indo-Pacific humpback dolphin (*Sousa chinensis*): population or species divergence? Conserv. Genet. 12, 1633–1638.

Hoyt, E., 2001. Whale Watching 2001: Worldwide Tourism Numbers, Expenditures and Expanding Socio-Economic Benefits. International Fund for Animal Welfare, Yarmouth Port, MA. Pages i–vi; 1–158.

Hoyt, E., 2005. Marine Protected Areas for Whales, Dolphins and Porpoises: A World Handbook for Cetacean Habitat Conservation. Earthscan, London.

Ilangakoon, A.D., 2005. Research and conservation of marine mammals in relation to the Bar Reef Marine Sanctuary North-Western Sri Lanka: Project Completion Report submitted to the Coast Conservation Department, Sri Lanka.

Ilangakoon, A.D., 2006. Cetacean occurrence and distribution around the Bar Reef Marine Sanctuary, north-west Sri Lanka. J. Nat. Sci. Found. Sri Lanka 34, 149–154.

Jefferson, T.A., Curry, B.E., 2015. Humpback dolphins: a brief introduction to the genus *Sousa*. Adv. Mar. Biol. 72, 1–16.

Jefferson, T.A., Karczmarski, L., 2001. Sousa chinensis. Mamm. Species 655, 1–9.

Jefferson, T.A., Rosenbaum, H.C., 2014. Taxonomic revision of the humpback dolphins (*Sousa* spp.), and description of a new species from Australia. Mar. Mamm. Sci. 30, 1494–1541.

Jefferson, T.A., Smith, B.D., 2016. Re-assessment of the conservation status of the Indo-Pacific humpback dolphin (*Sousa chinensis*) using the IUCN Red List criteria. Adv. Mar. Biol. 73, (in press).

Jefferson, T.A., Waerebeek, K.V., 2004. Geographic variation in skull morphology of humpback dolphins (*Sousa* spp.). Aquat. Mamm. 30, 3–17.

Jefferson, T.A., Hung, S.K., Würsig, B., 2009. Protecting small cetaceans from coastal development: Impact assessment and mitigation experience in Hong Kong. Mar. Policy 33, 305–311.

Jog, K., Sule, M., Patankar, V., Bopardikar, I., Sutaria, D., submitted. Living with dolphins—fisher perceptions and conservation implications along the Konkan coast, Maharashtra. Mar. Mamm. Sci.

Kumar, S., Pathak, K.C., Pedneker, P., Raju, N.S.N., Gowthaman, R., 2006. Coastal processes along the Indian coastline. Curr. Sci. 91, 530–536.

Lakshmi, A., Schiavina, A., Banerjee, P., Reddy, A., Mandeen, S., Rodriquez, S., Apte, D., 2012. The challenged coast of India: a report by PondyCAN in collaboration with BNHS and TISS. Retrieved on 29 March 2014 from, http://deepakapte.com/attachments/article/20/Challenged%20Coast%20of%20India_Lowres.pdf.

Markowitz, T.M., Harlin, A.D., Würsig, B., McFadden, C.J., 2004. Dusky dolphin foraging habitat: overlap with aquaculture in New Zealand. Aquat. Conserv. 14, 133–149.

Martin, K.A., 2013. Dolphins Bite into Fishermen's Income. The Hindu. Retrieved on 22 August 2015 from, www.thehindu.com/news/cities/Kochi/dolphins-bite-into-fishermens-income/article5039324.ece.

Mendez, M., Jefferson, T.A., Kolokotronis, S.O., Krützen, M., Parra, G.J., Collins, T., Minton, G., Baldwin, R., Berggren, P., Sarnblad, A., Amir, O.A., Peddemors, V.M., Karczmarski, L., Guissamula, A., Smith, B., Sutaria, D., Amato, G., Rosenbaum, H.C., 2013. Integrating multiple lines of evidence to better understand the evolutionary divergence of humpback dolphins along their entire distribution range: a new dolphin species in Australian waters? Mol. Ecol. 22, 5936–5948.

Muralidharan, R., 2013. Sightings and behavioral observations of Indo-Pacific humpback dolphins *Sousa chinensis* (Osbeck, 1765) along Chennai coast, Bay of Bengal. J. Threat. Taxa 5, 5002–5006.

Nanayakkara, R.P., Kusuminda, T., Jefferson, T.A., 2014. Can the Indian Ocean humpback dolphin (*Sousa plumbea*) survive in Sri Lanka? Occurrence of a relict population in Puttalam Lagoon. Aquat. Mamm. 40, 398–406.

Panicker, D., Sutaria, D., 2013. Developing with dolphins: ecological and anthropogenic factors affecting estuary use and behaviour of humpback dolphins in the southwest coast of India: Report submitted to Rufford Foundation, UK.

Panicker, D., Sutaria, D., submitted. Indo-Pacific humpback dolphin habitat use in Kochi estuary, India. Mar. Mamm. Sci.

Parra, G.J., Jedensjö, M., 2014. Stomach contents of Australian snubfin (*Orcaella heinsohni*) and Indo-Pacific humpback dolphins (*Sousa chinensis*). Mar. Mamm. Sci. 30, 1184–1198.

Parsons, E.C.M., 1998. Observations of Indo-Pacific humpbacked dolphins, *Sousa chinensis*, from Goa, western India. Mar. Mamm. Sci. 14, 166–170.

Pirotta, E., Merchant, N.D., Thompson, P.M., Barton, T.R., Lusseau, D., 2015. Quantifying the effect of boat disturbance on bottlenose dolphin foraging activity. Biol. Conserv. 181, 82–89.

Richardson, W.J., Greene Jr., C.R., Malme, C.I., Thomson, D.H., 1995. Marine Mammals and Noise. Academic Press, San Diego, California.

Rodriguez, S., Sridhar, A., 2010. Harbouring Trouble: The Social and Environmental Upshot of Port Growth in India. Dakshin Foundation, Bangalore. 62 pp.

Sanilkumar, V., Pathak, K.C., Pednekar, P., Raju, N.S.N., Gowthaman, R., 2006. Coastal processes along the Indian coastline. Curr. Sci. 91, 530–536.

Sims, P.Q., Vaughn, R., Hung, S.K., Würsig, B., 2012a. Sounds of Indo-Pacific humpback dolphins (*Sousa chinensis*) in west Hong Kong: a preliminary description. J. Acoust. Soc. Am. 131, EL48–EL53.

Sims, P.Q., Hung, S.K., Würsig, B., 2012b. High-speed vessel noises in west Hong Kong waters and their contributions relative to Indo-Pacific humpback dolphins (Sousa chinensis). J. Mar. Biol. 2012. http://dx.doi.org/10.1155/2012/169103. (Article ID 169103, 11 pp).

Sivadas, M., Pradeep Kumar, K.C., 2009. Dolphin excluder net—an indigenous method to ward off the damage by dolphins in ring net. Mar. Fish. Inf. Serv., Tech. Ext. Ser. 199, 12–13.

Smith, B.D., Mansur, R.M., Strindberg, S., Redfern, J., Moore, T., 2015. Population demographics, habitat selection, and a spatial and photographic analysis of bycatch risk of Indo-Pacific humpback dolphins *Sousa chinensis* and bottlenose dolphins *Tursiops aduncus* in the northern Bay of Bengal, Bangladesh: International Whaling Commission Scientific Committee Report. 30 pp. (unpublished). SC/66a/SM/19.

Society for Marine Mammalogy (Committee on Taxonomy). 2014. List of marine mammal species and subspecies. Available at: http://www.marinemammalscience.org (consulted on 22 March 2014).

Sule, M., Jog, K., Bopardikar, I., Patankar, V., Sutaria, D., 2015. Cetaceans of the Sindhudurg coast: Report submitted to the Government of India, GoI-GEF-UNDP Sindhudurg Project.

Sutaria, D., 2009. Understanding Species Conservation in Complex Socio-Ecological Systems: Case of Irrawaddy Dolphins in Chilika Lagoon, India. PhD thesis, James Cook University, Townsville, Australia.

Sutaria, D., Jefferson, T.A., 2004. Records of Indo-Pacific humpback dolphins (*Sousa chinensis*, Osbeck, 1765) along the coasts of India and Sri Lanka: an overview. Aquat. Mamm. 30, 125–136.

Van Parijs, S.M., Corkeron, P.J., 2001a. Boat traffic affects the acoustic behaviour of Pacific humpback dolphins, *Sousa chinensis*. J. Mar. Biol. Assoc. UK 81, 533–538.

Van Parijs, S.M., Corkeron, P.J., 2001b. Vocalizations and behaviour of Pacific humpback dolphins *Sousa chinensis*. Ethology 107, 701–716.

Wang, Z., Fang, L., Shi, W., Wang, K., Wang, D., 2013. Whistle characteristics of free-ranging Indo-Pacific humpback dolphins (*Sousa chinensis*) in Sanniang Bay, China. J. Acoust. Soc. Am. 133, 2479–2489.

SUBJECT INDEX

Note: Page numbers followed by "*f*" indicate figures and "*t*" indicate tables.

A

Africa, *Sousa teuszii* in. *See* Atlantic humpback dolphin
Andhra Pradesh, humpback dolphins in
  distribution records, 237
  encounter rates, 238*t*
  fisheries interactions, 242–243
  population status, 241
Animal behaviour, humpback dolphin
  behavioural habituation, 21
  defensive tactics, 19
  selective tolerance, 22
  sensitization, 21
  stimuli response, 19–20
Anthropogenic activities
  humpback dolphins
    cetacean–fishery interactions, 29–30
    cumulative effects, 33
    dolphin-based tourism, 28–29
    habitat loss and degradation, 32–33
    marine vessel traffic, 26–27
    noise pollution, 30–32
  marine mammal behaviour
    cetacean–fishery interactions, 24
    cumulative effects, 26
    dolphin-based tourism, 23–24
    habitat loss and degradation, 25
    marine vessel traffic, 23
    noise pollution, 24–25
  noise, Atlantic humpback dolphin, 67–68
  threats and conservation
    fisheries interactions, 218–221
    legislation, 222–223
    mangrove deforestation, 222
    Pakistan National Action Plan for Conservation of Marine Cetaceans, 222–223
    threats, 222–223
    water pollution, 221–222
Atlantic humpback dolphin, 6
  abundance and population sizes, 56*t*
  accuracy code, 81, 82*t*
  at-sea records, 82–85, 87*f*
  conservation actions, 68–69
  countrywise distribution
    Angola, 97
    Benin, 94
    Cameroon, 94–96
    Congo Republic, 97
    Gabon, 97
    Gambia, 92
    Guinea, 93
    Guinea-Bissau, 92–93
    Mauritania, 88–90
    Senegal, 91–92
    Togo, 93
    Western Sahara, 86–87
  description, 80
  distribution gaps, 98
  dorsal hump, 62*f*
  dorsal profile, 81
  ecology and biology, 49
  ecoregions, 82–83, 83*t*, 97
  fine-scale distribution, 89*f*
  food and feeding, 58–59
  geographic range
    countries, 53
    description, 51–53
  global distribution, 86
  group sizes, 98–101, 99*t*, 101*f*
  habitat, 57–58
    distance from shore, 99*t*, 102–104
    sea-surface temperature, 99*t*, 104
    water depth, 99*t*, 102
  holotype, 48–49
  IUCN Red List justification, 69–71
  major threats
    anthropogenic noise, 67–68
    by-catches, 64–65, 65*f*
    coastal developments, 67, 67*f*
    entanglement, 65–66
    non-target-deliberate acquisition, 65–66
    trawlers fishing, 66

Atlantic humpback dolphin (*Continued*)
- marine bushmeat, 48
- migrations and movements, 62–64
- minor threats, 68
- other records, 82, 88*f*
- population
  - abundance, 53–55
  - trend, 55–57
- population size, 7
- prey items, 60*t*
- Quantum Geographic Information System, 82–83
- reproduction, 59–62
- Saloum Delta *vs.* Banc d'Arguin, occurrence, 104–105
- species validation, 81
- taxonomy, 50–51
- temporal trend in records, 85–86
- use and trade, 64

Australian humpback dolphins, 6–8

B

Balochistan coast, Indian Ocean humpback dolphin, 207–208

Beach-based observations, 63

Behavioural response. *See* Anthropogenic activities

Boat-based tourism, 28

By-catch
- Atlantic humpback dolphin, threats, 64–65, 65*f*
- Indian Ocean humpback dolphin
  - consumption and local sale, 186–187
  - data, 194–195
  - encounter rate *vs.* group size, 190*f*
  - fishing gear/technique, 186*f*
  - geographic spread of reports, 192
  - habitat degradation, 195
  - interview protocols, 179–180
  - locations, 181*f*
  - in Northwest, 187*f*, 190–191
  - Nosy Be region, 191–192
  - prey depletion, 195
  - reports, 184
  - socio-ecological interview surveys, 179
  - in Southwest, 187*f*
  - suspected range, 189*f*
  - *Tursiops* spp., 184, 185*t*
  - Vezo, 193–194
  - villages, interviewees in, 180
  - West Coast locations, 182*t*

C

Cetacean–fishery interactions
- humpback dolphins, 29–30
- marine mammal behaviour, 24

Coastal developments threats, 67, 67*f*

Conservation
- Atlantic humpback dolphin, 68–69
- behaviour, 18–19, 32
- Pakistan, Indian Ocean humpback dolphin in
  - fisheries interactions, 218–221
  - legislation, 222–223
  - mangrove deforestation, 222
  - Pakistan National Action Plan for Conservation of Marine Cetaceans, 222–223
  - threats, 222–223
  - water pollution, 221–222

D

Dolphin-based tourism, 23–24, 28–29

Dolphin-watching tourism
- Kerala, 249–250
- Sindhudurg, 250–251

E

Entanglement threats, Atlantic humpback dolphin, 65–66

Environmental contamination, Atlantic humpback dolphin, 68

Environmental impact assessments (EIAs), 248

F

Fisheries interactions
- by-catch, 220
- catch per unit effort (CPUE), 220
- entangled animals, 220–221
- *Sousa plumbea*, Pakistan coastline, 219*f*

G

Goa, humpback dolphins in
- distribution records, 235–236
- encounter rates, 238*t*

population status, 240
Group size
Atlantic humpback dolphin, 98–101, 99*t*, 101*f*
humpback dolphins, 9
Indian Ocean humpback dolphin, 190*f*
South African waters, Indian Ocean humpback dolphins in, 150–151
S. teuszii, 54, 59
Gujarat, humpback dolphins in
distribution records, 235
encounter rates, 238*t*
population status, 237–240

H

Habitat alteration, 25, 32
Habitat loss and degradation
humpback dolphins, 32–33
marine mammal behaviour, 25
Human activities
humpback dolphins
cetacean–fishery interactions, 29–30
cumulative effects, 33
dolphin-based tourism, 28–29
habitat loss and degradation, 32–33
marine vessel traffic, 26–27
noise pollution, 30–32
marine mammal behaviour
cetacean–fishery interactions, 24
cumulative effects, 26
dolphin-based tourism, 23–24
habitat loss and degradation, 25
marine vessel traffic, 23
noise pollution, 24–25
Humpback dolphin
abundance/population size, 7–8
animal behaviour
behavioural habituation, 21
defensive tactics, 19
selective tolerance, 22
sensitization, 21
stimuli response, 19–20
anthropogenic activities, marine mammal behaviour
cetacean–fishery interactions, 24
cumulative effects, 26
dolphin-based tourism, 23–24
habitat loss and degradation, 25
marine vessel traffic, 23
noise pollution, 24–25
Atlantic, 6
Australian, 6–7
calving, 9–10
conservation actions and status, 11–13
distribution, 232–233, 232*f*
feeding habits, 10
geographic range, 6–7
gestation, 9–10
group size and structure, 9
habitat and ecology, 8–10
in India (*see* India, humpback dolphins in)
Indian Ocean, 6
Indo-Pacific, 6–7
IUCN Red List conservation status, 2–3, 3*t*
marine mammal behaviour, human impacts on
odontocetes, 22
natural predators, 10
nearshore habitat, 2
taxonomy, 3–6
threats, 10–11
Hunting, Indian Ocean humpback dolphin
consumption and local sale, 186–187
data, 194–195
encounter rate *vs*. group size, 190*f*
fishing gear/technique, 186*f*
geographic spread of reports, 192
habitat degradation, 195
interview protocols, 179–180
locations, 181*f*
in Northwest, 187*f*, 190–191
Nosy Be region, 191–192
prey depletion, 195
reports, 184
socio-ecological interview surveys, 179
in Southwest, 187*f*
suspected range, 189*f*
Tursiops spp., 184, 185*t*
Vezo, 193–194
villages, interviewees in, 180
West Coast locations, 182*t*

I

India, humpback dolphins in
Andhra Pradesh
distribution records, 237
encounter rates, 238*t*
fisheries interactions, 242–243
population status, 241
common names, 234
conservation and management, 252–253
East Coast
distribution records, 236–237
population status, 241–242
Goa
distribution records, 235–236
encounter rates, 238*t*
population status, 240
Gujarat
distribution records, 235
encounter rates, 238*t*
population status, 237–240
Indian coastline
characteristics, 232–233
development, 245–249
Karnataka
distribution records, 236
encounter rates, 238*t*
Kerala
distribution records, 236
dolphin-watching tourism, 249–250
encounter rates, 238*t*
fisheries interactions, 243–244
population status, 240–241
Maharashtra
distribution records, 235
dolphin-watching tourism, 250–251
encounter rates, 238*t*
fisheries interactions, 244–245
noise impacts, 251–252
population status, 240
Orissa
distribution records, 237
encounter rates, 238*t*
population status, 242
photo-identification based mark-recapture analysis, 237
Pondicherry
distribution records, 236
encounter rates, 238*t*
population status, 241
research projects, 230
Tamil Nadu
distribution records, 236–237
encounter rates, 238*t*
fisheries interactions, 242–243
population status, 241
taxonomy, 233–234
time activity budget, 246, 247*f*
traditional ecological knowledge, 230–231
West Bengal
distribution records, 237
encounter rates, 238*t*
West Coast
distribution records, 235–236
population status, 237–241
Indian Ocean humpback dolphin, 6
abundance, 8
abundance estimates, 125–127
in Algoa Bay, 127
in Arabian Gulf, 127
blast fishing, 135
by-catch and hunting, 166
by-catch and hunting, interview surveys
consumption and local sale, 186–187
data, 194–195
encounter rate *vs.* group size, 190*f*
fishing gear/technique, 186*f*
geographic spread of reports, 192
habitat degradation, 195
interview protocols, 179–180
locations, 181*f*
in Northwest, 187*f*, 190–191
Nosy Be region, 191–192
prey depletion, 195
reports, 184
socio-ecological interview surveys, 179
in Southwest, 187*f*
suspected range, 189*f*
Tursiops spp., 184, 185*t*
Vezo, 193–194
villages, interviewees in, 180
West Coast locations, 182*t*
confidence intervals, 125–126, 126*t*
conservation actions, 135–136
description, 120
destruction of inshore habitats, 133

environmental contaminants, 133–134
food and feeding, 129
genetic analysis, 122
geographic range description, 122–125
habitat, 128–129
hunting, 134–135
incidental fishing mortality, 132–133
inshore boat traffic disturbance, 135
IUCN Red List justification, 136–137
KwaZulu-Natal coast, 127
in Madagascar, 128, 164–167
migration and movements, 130–131
mitochondrial DNA, 121
non-consumptive uses, 131–132
nuclear DNA markers, 121
occurrence in countries, 125
in Pakistan
abundance, 211–212
anthropogenic threats and conservation, 218–223
Balochistan coast, 207–208
behaviour, 213–214
biology, 213–217
distribution, 203–208
feeding, 214
group size and composition, 215–217
habitat, 212–213
Indus River Delta, 205–207
interaction with boats, 217
parasites and pathology, 217–218
residency, 217
sexual behaviour, 214–215
stock structure, 208–210
reproduction, 129–130
reproductive rate, 127
scientific and common names, 121
sightings and encounter rates
Anakao region, 170*f*, 171–176
Nosy Be region, 171*f*, 177–178, 178*t*
Nosy Iranja region, 178–179
Nosy Mitsio region, 178
skull, 121
small boat-based field surveys, 167
in South African waters
age distribution, 149, 149*f*
age structure, 147–149
bycatch threats, 153–154
calves colour, 145–146
chemical pollution, 155–156
coastal development/disturbance, in coastal zone, 154–155
diet, 151–152
gestation, 146
habitat utilisation, 152–153
health, 155–156
longevity, 146
male and female growth parameters, 147, 147*t*
mating system, indicators of, 150–151
morphometrics, 145–146
natural history parameters, 147, 148*t*
neonatal lengths, 145–146
noise pollution, 154
ovulation rate, 146
pathology, 156
persistent organic pollutants, 155–156
sexual dimorphism, 150
smaller organ weights, 145–146
small group size, 150–151
strandings, 166–167
surveys, West Coast of Madagascar, 167–179
effort, 169
methodology, 168
sightings and encounter rates, 169–179
tooth counts, 121
trade and uses, 131–132
Indo-Pacific humpback dolphins, 6–8
Indus River Delta, Indian Ocean humpback dolphin, 205–207
Industrial and domestic sewage pollution, 221
IUCN Red List
2008 criteria, 55
justification, 69–71

K

Karnataka, humpback dolphins in
distribution records, 236
encounter rates, 238*t*
Kerala, humpback dolphins in
distribution records, 236
dolphin-watching tourism, 249–250
encounter rates, 238*t*
fisheries interactions, 243–244
population status, 240–241

M

Madagascar
  S. plumbea in
    bycatch and hunting, 166
    Information, 165–166
    population structure, 164–165
  West Coast surveys, 167–179
    effort, 169
    methodology, 168
    sightings and encounter rates, 169–179
Maharashtra, humpback dolphins in
  distribution records, 235
  dolphin-watching tourism, 250–251
  encounter rates, 238*t*
  fisheries interactions, 244–245
  noise impacts, 251–252
  population status, 240
Marine bushmeat, Atlantic humpback dolphin, 48
Marine vessel traffic
  humpback dolphins, 26–27
  marine mammal behaviour, 23

N

Nature-based tourism, 249
Nearshore habitats
  S. teuszii, 51, 64–65
  zone, oceanographic definition of, 103–104
Noise pollution
  humpback dolphins, 30–32
  marine mammal behaviour, 24–25
Non-target-deliberate acquisition, 65–66

O

Oil exploration, Atlantic humpback dolphin, 68
Orissa, humpback dolphins in
  distribution records, 237
  encounter rates, 238*t*
  population status, 242

P

Pakistan, Indian Ocean humpback dolphin in
  abundance, 211–212
  anthropogenic threats and conservation
    fisheries interactions, 218–221
    legislation, 222–223
    mangrove deforestation, 222
    Pakistan National Action Plan for Conservation of Marine Cetaceans, 222–223
    threats, 222–223
    water pollution, 221–222
  behaviour, 213–214
  biology, 213–217
  Cetacean Conservation Pakistan Project (CCP), 203
  distribution
    Balochistan coast, 207–208
    Indus River Delta, 205–207
    Sindh coastline, 204*f*
  feeding, 214
  group size and composition, 215–217
  habitat, 212–213
  interaction with boats, 217
  parasites and pathology, 217–218
  residency, 217
  sexual behaviour, 214–215
  stock structure
    cranial morphometrics, multivariate analysis, 208–209
    encounter rates, 206*t*
    thoracic region, interspecific differences, 209
    variations, external features, 209*f*
Phosphorite mining, Atlantic humpback dolphin, 68
Photo-identification based mark-recapture analysis, 237
Pollution
  chemical, 155–156
  industrial and domestic sewage, 221
  noise, 24–25, 30–32, 154
  spills and chronic, 68
  thermal, 221–222
  water, 221–222
Pondicherry, humpback dolphins in
  distribution records, 236
  encounter rates, 238*t*
  population status, 241

Q

Quantum Geographic Information System (QGIS), 82–83

S

Sindh coastline, Indian Ocean humpback dolphin, 204*f*
Socio-ecological interview survey, Indian Ocean humpback dolphin
  consumption and local sale, 186–187
  data, 194–195
  encounter rate *vs.* group size, 190*f*
  fishing gear/technique, 186*f*
  geographic spread of reports, 192
  habitat degradation, 195
  interview protocols, 179–180
  locations, 181*f*
  in Northwest, 187*f*, 190–191
  Nosy Be region, 191–192
  prey depletion, 195
  reports, 184
  socio-ecological interview surveys, 179
  in Southwest, 187*f*
  suspected range, 189*f*
  Tursiops spp., 184, 185*t*
  Vezo, 193–194
  villages, interviewees in, 180
  West Coast locations, 182*t*
Sousa plumbea. See Indian Ocean humpback Dolphin; Indian Ocean humpback dolphin
Sousa teuszii. See Atlantic humpback dolphin
South African waters, Indian Ocean humpback dolphins in
  age distribution, 149, 149*f*
  age structure, 147–149
  bycatch threats, 153–154
  calves colour, 145–146
  chemical pollution, 155–156
  coastal development/disturbance, in coastal zone, 154–155
  diet, 151–152
  gestation, 146
  habitat utilisation, 152–153
  health, 155–156
  longevity, 146
  male and female growth parameters, 147, 147*t*
  mating system, indicators of, 150–151
  morphometrics, 145–146
  natural history parameters, 147, 148*t*
  neonatal lengths, 145–146
  noise pollution, 154
  ovulation rate, 146
  pathology, 156
  persistent organic pollutants, 155–156
  sexual dimorphism, 150
  smaller organ weights, 145–146
  small group size, 150–151
Spills and chronic pollution, 68

T

Tamil Nadu, humpback dolphins in
  distribution records, 236–237
  encounter rates, 238*t*
  fisheries interactions, 242–243
  population status, 241
Thermal pollution, 221–222
Threats, Atlantic humpback dolphin
  anthropogenic noise, 67–68
  by-catches, 64–65, 65*f*
  coastal developments, 67, 67*f*
  entanglement, 65–66
  non-target-deliberate acquisition, 65–66
  trawlers fishing, 66
Trawlers fishing, 66

V

Vessel traffic, marine
  humpback dolphins, 26–27
  marine mammal behaviour, 23

W

Water pollution, 221–222
West Bengal, humpback dolphins in
  distribution records, 237
  encounter rates, 238*t*

TAXONOMIC INDEX

Note: Page numbers followed by "*f*" indicate figures and "*t*" indicate tables.

A

Arenaria interpres, 20

B

Bothidae sp., 60*t*

C

Chaetodipterus lippei, 60*t*
Clupeidae, 60*t*

D

Delphinus, 3–4
  D. capensis, 145–146, 151
  D. delphis, 26

E

Ephippidae, 60*t*
Ethmalosa fimbriata, 60*t*

H

Haemulidae, 60*t*

J

Johnius amblycephalus, 151

L

Lagenorhynchus, 5–6
  L. obscurus, 19
Lethrinidae, 60*t*
Lethrinus atlanticus, 60*t*
Lipotes vexillifer, 25
Liza
  L. aurata, 60*t*
  L. macrolepis, 240–241
  L. parsia, 240–241
  L. richardsonii, 60*t*
Lutjanidae, 60*t*
Lutjanus goreensis, 60*t*

M

Megaptera novaeangliae, 156–157
Mesoplodon densirostris, 24–25
Mugil cephalus, 60*t*, 240–241
Mugilidae, 60*t*
Mugil sp., 60*t*

N

Neophocaena phocaenoides, 206–207, 220

O

Orcaella brevirostris, 249
Orcinus orca, 10, 19
Otolithes ruber, 151

P

Paralichthyidae, 60*t*
Pentanemus quinquarius, 60*t*
Peponocephala electra, 165–166
Phocarctos hookeri, 22
Platanista gangetica, 222–223
Polynemidae, 60*t*
Pomadasys
  P. jubelini, 60*t*
  P. olivaceum, 151
Pomadasys sp., 60*t*
Pseudorhombus sp., 60*t*
Pseudotolithus
  P. senegalensis, 60*t*
  P. typus, 60*t*

S

Sarda sarda, 60*t*
Sardinella sp., 60*t*
Scianidae, 60*t*
Scombridae, 60*t*
Sotalia, 3–4
  S. teuszii, 50
Sousa, 2–13, 18–36, 80, 144, 202–203, 230–253
  S. chinensis, 2–6, 3*t*, 5*f*, 7*f*, 8, 12*t*, 26–33, 67–68, 70, 121, 125, 130, 147–150, 154, 164, 202–203, 208–209, 233, 245–246
  S. c. taiwanensis, 2–3, 30, 33, 35
  S. lentiginosa, 4–6, 234

Sousa (*Continued*)
S. plumbea, 2, 3*t*, 4–6, 5*f*, 7*f*, 12*t*, 26–32, 50–51, 59–62, 70, 98, 120–137, 144–157, 164–195, 175*t*, 202–224, 233–234
S. sahulensis, 2, 3*t*, 4, 5*f*, 7*f*, 12*t*, 29–31, 33, 67–68, 208–209, 233
S. teuszii, 2–4, 3*t*, 5*f*, 7*f*, 12*t*, 31–33, 48–71, 80–106, 124, 144–145, 220–221, 233
Sousa sp., 102, 121, 145–146
Sousa spp., 2, 26–27, 36, 154, 230, 231*t*
Spheniscus humboldti, 21
Squilla mantis, 60*t*
Squillidae, 60*t*
Stenella, 5–6, 166
S. coeruleoalba, 151
S. longirostris, 166, 220–221
Stolephorus indicus, 166

T

Thryssa vitrirostris, 151
Trichechus senegalensis, 82
Trichiurus lepturus, 151
Tursiops, 166, 169, 177, 180, 183*f*, 183*t*, 184–186, 185*t*, 187*f*
T. aduncus, 28, 127, 131–132, 134–135, 145–146, 151, 156, 166–167, 169, 171–172, 175–176*t*, 176–180, 178*t*, 181*f*, 190–191, 206–207, 212
T. cf. *aduncus*, 23–24
T. truncatus, 19, 81, 180
Tursiops sp., 169, 175*t*, 178–179, 184–186
Tursiops spp., 2, 177, 180–187, 192

V

Valamugil spp., 240–241

Kolibri 360 GmbH
Ferdinand-Jühlke-Straße 7
99095 Erfurt, Deutschland
produktsicherheit@kolibri360.de